# Integrated Assessment of Ecosystem Health

# Integrated Assessment OF Ecosystem Health

### EDITED BY

**Kate M. Scow**, Ph.D.
*Professor, University of California, Davis*

**Graham E. Fogg**, Ph.D.
*Professor, University of California, Davis*

**David E. Hinton**, Ph.D.
*Professor, University of California, Davis*

**Michael L. Johnson**, Ph.D.
*Professor, University of California, Davis*

LEWIS PUBLISHERS
Boca Raton  London  New York  Washington, D.C.

**Library of Congress Cataloging-in-Publication Data**

Intergrated assessment of ecosystem health / edited by Kate Scow
    p.   cm.
Includes bibliographical references.
ISBN 1-56670-453-7
    1. Ecology. 2. Ecological assessment (Biology) 3. Environmental health. I. Scow, Kate.
QH541.I556 1999
577—dc21                                                                                                              99-29645
                                                                                                                                                             CIP

    This book contains information obtained from authentic and highly regarded sources. Reprinted material is quoted with permission, and sources are indicated. A wide variety of references are listed. Reasonable efforts have been made to publish reliable data and information, but the author and the publisher cannot assume responsibility for the validity of all materials or for the consequences of their use.

    Neither this book nor any part may be reproduced or transmitted in any form or by any means, electronic or mechanical, including photocopying, microfilming, and recording, or by any information storage or retrieval system, without prior permission in writing from the publisher.

    All rights reserved. Authorization to photocopy items for internal or personal use, or the personal or internal use of specific clients, may be granted by CRC Press LLC, provided that $.50 per page photocopied is paid directly to Copyright Clearance Center, 222 Rosewood Drive, Danvers, MA 01923 USA. The fee code for users of the Transactional Reporting Service is ISBN 1-56670-453-7/00/$0.00+$.50. The fee is subject to change without notice. For organizations that have been granted a photocopy license by the CCC, a separate system of payment has been arranged.

    The consent of CRC Press LLC does not extend to copying for general distribution, for promotion, for creating new works, or for resale. Specific permission must be obtained in writing from CRC Press LLC for such copying.

    Direct all inquiries to CRC Press LLC, 2000 Corporate Blvd., N.W., Boca Raton, Florida 33431.

    **Trademark Notice:** Product or corporate names may be trademarks or registered trademarks, and are used only for identification and explanation, without intent to infringe.

© 2000 by CRC Press LLC
Lewis Publishers is an imprint of CRC Press LLC

No claim to original U.S. Government works
International Standard Book Number 1-56670-453-7
Library of Congress Card Number 99-29645
Printed in the United States of America  1  2  3  4  5  6  7  8  9  0
Printed on acid-free paper

# Preface

Virtually all of the earth's ecosystems have been altered by human activities and these changes, in turn, strongly interact with naturally occurring stresses. The concept of "ecosystem health" was developed to reflect the importance of considering an ecosystem truly as a system, rather than as an assemblage of individual species. The well-being of an ecosystem depends on the functioning of its parts, but additionally on higher-order phenomena such as population dynamics, community structure, and landscape processes. Also, the physical component of ecosystems, though often overlooked, can be critical in regulating the severity of stresses and ecosystem recovery. A systems perspective is essential in many environmental disciplines such as ecotoxicology, restoration ecology, and natural resource economics and policy.

Analysis of ecosystem health requires the development and implementation of complex methodologies that go beyond the level of the individual which is the common emphasis in toxicology. Ecosystem health assessments require baseline information about the ecosystem of concern as well as (1) an understanding of the movement and transformation of pollutants and other stressors within the environment and into exposure media; (2) information on concentrations and conditions leading to negative impacts on biotic and abiotic systems; and (3) quantification of risk for use in policy setting and economic analysis. Multidisciplinary research and communication among historically disconnected disciplines are critical to the success of such assessments.

To address the issues described above, a conference, entitled "From Cumulative Impacts Toward Sustainable Solutions: Critical Methodologies for the Study of Ecosystem Health," was held at the University of California, Davis, from September 8–10, 1996. Most of the chapters in this book are the peer-reviewed proceedings resulting from this conference. The objectives of the conference were to bring together scientists and regulators working in various aspects of ecosystem science and ecotoxicology to present new ideas, exchange information, and develop new directions for future research. The conference aimed to provide both an overview of the state of the art and to illuminate new directions for strengthening ecosystem health assessment in theory and in practice.

# The Editors

**Kate M. Scow** is Professor of Soil Science and Microbial Ecology in the Department of Land, Air and Water Resources, at the University of California, Davis. She earned degrees in biology from Antioch College, Ohio (B.S.) and in soil science from Cornell University (M.S., Ph.D.). She also worked for 5 years at Arthur D. Little, Cambridge, as an environmental biologist. Since joining the faculty at UC Davis in 1989, her research program has addressed the biodegradation of organic pollutants in soil and groundwater and the effects of toxicants on soil microbial communities. She is on the editorial board of the *Journal of Microbial Ecology* and *Journal of Soil Biology and Biochemistry*.

**Graham E. Fogg** is Professor of Hydrogeology in the Department of Land, Air and Water Resources, at the University of California, Davis. He earned degrees in hydrology from the University of New Hampshire (B.S.) and the University of Arizona (M.S.) and in geology from the University of Texas at Austin (Ph.D.). After 11 years researching subsurface complexity and groundwater contaminant transport at the University of Texas at Austin, he joined the UC Davis faculty in 1989 and helped start the Hydrologic Sciences Graduate Group. With his students he has been actively researching subsurface hydrology of wetlands, fate, and transport of groundwater contaminants, modeling of subsurface heterogeneity and transport, groundwater vulnerability assessment, and groundwater quality sustainability. Recently, he and his colleagues have been developing methods of estimating toxic chemical exposure levels via groundwater pathways and investigating the transport processes responsible for hindering remediation of contaminated groundwater.

**David E. Hinton** is Professor of Aquatic Toxicology in the Department of Anatomy, Physiology and Cell Biology, School of Veterinary Medicine, University of California, Davis, and is Director, Lead Campus Program in Ecotoxicology, a graduate program of the University of California, Toxic Substances Research and Teaching Program. His research is focused on cellular biological approaches to understanding effects of toxicants on aquatic organisms with emphasis on normal and abnormal development and on carcinogenesis. The Editor-in-Chief of *Aquatic Toxicology*, his laboratory is involved in studies within the Sacramento River, its Delta, and the upper San Francisco Bay.

**Michael L. Johnson** is a Research Scientist in the John Muir Institute for the Environment at the University of California, Davis. He earned degrees from the University of Colorado (B.A. and M.A.) and the University of Kansas (Ph.D.). After working at the Kansas Biological Survey on the effects of agriculture on the environment of Kansas, he moved to UC Davis in 1992. His research interests include risk assessment, ecotoxicology, and conservation biology. Recent projects include evaluating the effects of stormwater runoff on aquatic systems, and the interaction of immune function, disease status, exposure to contaminants, and demography of wildlife populations.

# Contributors

**Michael J. Arbaugh**
USDA Forest Service
U.S. Department of Agriculture
Riverside, California

**Jesse C. Becker**
U.C. Davis Clear Lake Environmental
 Research Center
Lakeport, California
Department of Environmental Science
 and Policy
University of California
Davis, California

**Larry R. Brown**
Water Resources Division
U.S. Geological Survey
Sacramento, California

**Mary Ann Bruns**
Department of Land, Air and Water
 Resources
University of California
Davis, California

**Timothy A. Burton**
Boise National Forest
Boise, Idaho

**Andrzej Bytnerowicz**
Pacific Southwest Research Station
USDA Forest Service
U.S. Department of Agriculture
Riverside, California

**John Cairns, Jr.**
Department of Biology
Virginia Polytechnic Institute and State
 University
Blacksburg, Virginia

**John J. Carroll**
Department of Land, Air and Water
 Resources
University of California
Davis, California

**Kevin J. Daye**
Environmental Projects Laboratories
Merck & Company, Inc.
Elkton, Virginia

**John F. DeGeorge**
Resource Management Associates, Inc.
Suisun, California

**Deirdre M. Dether**
Boise National Forest
Boise, Idaho

**Neil M. Dubrovsky**
Water Resources Division
U.S. Geological Survey
Sacramento, California

**John R. Erickson**
Boise National Forest
Boise, Idaho

**Teresa W.-M. Fan**
Department of Land, Air and Water
 Resources
University of California
Davis, California

**Andrea Farwell**
Department of Environmental Biology
University of Guelph
Guelph, Ontario, Canada

**Mark E. Fenn**
Pacific Southwest Research Station
USDA Forest Service
U.S. Department of Agriculture
Riverside, California

**Joseph P. Frost**
Boise National Forest
Boise, Idaho

**Wade N. Gibbons**
Department of Biology
University of Waterloo
Waterloo, Ontario, Canada

**Raleigh Guthrey**
Pacific Southwest Research Station
USDA Forest Service
U.S. Department of Agriculture
Riverside, California

**Jessica R. Hanson**
Department of Land, Air, and Water
  Resources
University of California
Davis, California

**E. James Harner**
Department of Statistics
West Virginia University
Morgantown, West Virginia

**William. S. Herms**
Department of Political Science
University of California
Santa Barbara, California

**Alan C. Heyvaert**
Department of Environmental Science
  and Policy
University of California
Davis, California

**Richard M. Higashi**
Crocker Nuclear Laboratory
University of California
Davis, California

**Scott W. Hooper.**
Environmental Projects Laboratories
Merck & Company, Inc.
Elkton, Virginia

**Alan D. Jassby**
Department of Environmental Science
  and Policy
University of California
Davis, California

**Jae G. Kim**
Department of Environmental Science
  and Policy
University of California
Davis, California

**Ian P. King**
Professor Emeritus
Department of Civil and Environmental
  Engineering
University of California
Davis, California

**Charles R. Kratzer**
Water Resources Division
U.S. Geological Survey
Sacramento, California

**Bradley A. Lamphere**
Department of Environmental Science
  and Policy
University of California
Davis, California

**Allen S. Lefohn**
A.S.L. & Associates
Helena, Montana

**Xiaoping Li**
Department of Environmental Science
 and Policy
University of California
Davis, California

**Michael V. McGinnis**
Associate Director, Ocean and Coastal
 Policy Center
Senior Research Social Scientist
Marine Science Institute
University of California
Santa Barbara, California

**Mark E. McMaster**
AECB
National Water Research Institute
Environment Canada
Burlington, Ontario, Canada

**Laurent M. Meillier**
U.C. Davis Clear Lake Environmental
 Research Center
Lakeport, California
Department of Wildlife, Fish and
 Conservation Biology
University of California
Davis, California

**Paul R. Miller**
Pacific Southwest Research Station
USDA Forest Service
U.S. Department of Agriculture
Riverside, California

**Lynette Z. Morelan**
Boise National Forest
Boise, Idaho

**Lauri H. Mullen**
Department of Environmental Science
 and Policy
University of California
Davis, California

**Kelly R. Munkittrick**
AECB
National Water Research Institute
Environment Canada
Burlington, Ontario, Canada

**Robert C. Musselman**
Rocky Mountain Research Station
USDA Forest Service
U.S. Department of Agriculture
Fort Collins, Colorado

**Douglas C. Nelson**
Section of Microbiology
Division of Biological Sciences
University of California
Davis, California

**Leon F. Neuenschwander**
College of Forestry, Wildlife and Range
 Sciences
University of Idaho
Moscow, Idaho

**John Nickle**
Department of Zoology
University of Guelph
Guelph, Ontario, Canada

**Oladele A. Ogunseitan**
Department of Environmental Analysis
 and Design
University of California
Irvine, California

**Gerald T. Orlob**
Professor Emeritus
Department of Civil and Environmental Engineering
University of California
Davis, California

**Isaac Oshima**
California Department of Fish and Game
Technical Services Branch
Sacramento, California

**Anitra L. Pawley**
Center for Ecological Health Research
Department of Environmental Science and Policy
University of California
Davis, California

**Cam Portt**
C. Portt & Associates
Guelph, Ontario, Canada

**James Quinn**
Department of Environmental Science and Policy
University of California
Davis, California

**Peter J. Richerson**
Department of Environmental Science and Policy
University of California
Davis, California

**Lisa Ruemper**
Department of Environmental Biology
University of Guelph
Guelph, Ontario, Canada

**William R. Rush**
Boise National Forest
Boise, Idaho

**Camilla M. Saviz**
Department of Civil and Environmental Engineering
University of California
Davis, California

**Susan Schilling**
Pacific Southwest Research Station
USDA Forest Service
U.S. Department of Agriculture
Riverside, California

**Kate M. Scow**
Department of Land, Air and Water Resources
University of California
Davis, California

**Mark R. Servos**
AEPB
National Water Research Institute
Environment Canada
Burlington, Ontario, Canada

**Kevin A. Shuttle**
Environmental Projects Laboratories
Merck & Company, Inc.
Elkton, Virginia

**Darell G. Slotton**
Department of Environmental Science and Policy
University of California
Davis, California

**Thomas H. Suchanek**
Department of Wildlife, Fish and
  Conservation Biology
University of California
Davis, California

**Padma Sudarshana**
Department of Land, Air, and Water
  Resources
University of California
Davis, California

**Brent K. Takemoto**
Research Division
California Air Resources Board
California

**John L. Thornton**
Boise National Forest
Boise, Idaho

**Glen J. Van Der Kraak**
Department of Zoology
University of Guelph
Guelph, Ontario, Canada

**Charles E. Vaughn**
Hopland Research and Extension
  Center
University of California
Hopland, California

**Joshua H. Viers**
Center for Ecological Health Research
Department of Environmental Science
  and Policy
University of California
Davis, California

**Cydney A. Weiland**
Boise National Forest
Boise, Idaho

**Richard A. Williams**
Technical Operations
Merck & Company, Inc.
Albany, Georgia

**Cat E. Woodmansee**
Department of Environmental Science
  and Policy
University of California
Davis, California

**Lee Ann Woodward**
Department of Environmental Science
  and Policy
University of California
Davis, California

**John T. Woolley**
Department of Political Science
University of California
Santa Barbara, California

# Acknowledgment

The information contained in this book comes primarily from presentations associated with the conference: *From Cumulative Impacts Toward Sustainable Solutions: Critical Methodologies for the Study of Ecosystem Health.* Dennis Rolston, the Director of the Center for Ecological Health Research, had the vision for this conference and its tangible product, this book. Without that impetus neither would have happened. We are most grateful to the following organizations for their financial support of the conference:

Center for Ecological Health Research
(Funded by USEPA, R819658)
University of California, Davis
(Dennis Rolston, Director)

Campus Initiative of the Environment
University of California, Davis
(Robert Flocchini, Chair, Executive Steering Committee)

Kearney Foundation for Soil Science
College of Natural Resources
University of California, Berkeley
(Garrison Sposito, Director)

We would like to recognize the extraordinary efforts of Cheryl Smith whose organizational skills, perseverance, and humor made it possible to carry out both the conference and this publication. We would also like to offer a special thank you to Pritam Andreassen, Deborah Caneveri, and Katherine Zimmerman of the Center for Ecological Health Research for all their hard work on the logistics of the conference and the development of this book.

# Table of Contents

**Chapter 1**
The Genesis and Future of the Field of Ecotoxicology ............................................. 1
*John Cairns, Jr.*

## SECTION I: FATE

**Chapter 2**
A Predictor of Seasonal Nitrogenous Dry Deposition in a Mixed Conifer
Forest Stand in the San Bernardino Mountains ....................................................... 17
*Michael J. Arbaugh, Andrzej Bytnerowicz, and Mark E. Fenn*

**Chapter 3**
Integrating Chemical, Water Quality, Habitat, and Fish Assemblage Data
from the San Joaquin River Drainage, California ................................................... 25
*Larry R. Brown, Charles R. Kratzer, and Neil M. Dubrovsky*

**Chapter 4**
Subsurface Contaminant Fate Determination Through Integrated Studies
of Intrinsic Remediation ........................................................................................... 63
*Scott W. Hooper, Kevin J. Daye, Kevin A. Shuttle, and Richard A. Williams*

**Chapter 5**
The Cantara Spill: A Case Study — Pesticide Transport in a Riverine
Environment .............................................................................................................. 71
*Camilla M. Saviz, John F. DeGeorge, Gerald T. Orlob, and Ian P. King*

**Chapter 6**
Distribution and Transport of Air Pollutants to Vulnerable California
Ecosystems ................................................................................................................ 93
*Andrzej Bytnerowicz, John J. Carroll, Brent K. Takemoto, Paul R. Miller, Mark E. Fenn, and Robert C. Musselman*

**Chapter 7**
The History of Human Impacts in the Clear Lake Watershed (California)
as Deduced from Lake Sediment Cores ................................................................. 119
*Peter J. Richerson, Thomas H. Suchanek, Jesse C. Becker, Alan C. Heyvaert, Darell G. Slotton, Jae G. Kim, Xiaoping Li, Laurent M. Meillier, Douglas C. Nelson, and Charles E. Vaughn*

## SECTION II: EFFECTS

**Chapter 8**
The Development of Cumulative Effects Assessment Tools Using Fish Populations ............................................................................................................. 149
*Kelly R. Munkittrick, Mark E. McMaster, Cam Portt, Wade N. Gibbons, Andrea Farwell, Lisa Ruemper, Mark R. Servos, John Nickle, and Glen J. Van Der Kraak*

**Chapter 9**
Air Pollutants and Forests: Effect at the Organismal Scale ................................ 175
*Teresa W.-M. Fan and Richard M. Higashi*

**Chapter 10**
DNA Fingerprinting as a Means to Identify Sources of Soil-Derived Dust: Problems and Potential ........................................................................................ 193
*Mary Ann Bruns and Kate M. Scow*

**Chapter 11**
Microbial Proteins as Biomarkers of Ecosystem Health ..................................... 207
*Oladele A. Ogunseitan*

**Chapter 12**
Application of a Random Amplified Polymorphic DNA (RAPD) Method for Characterization of Microbial Communities in Agricultural Soils ................ 223
*Padma Sudarshana, Jessica R. Hanson, and Kate M. Scow*

**Chapter 13**
Air Pollution and Forests: Effects at the Landscape Level ................................. 233
*Paul R. Miller, John J. Carroll, Susan Schilling, and Raleigh Guthrey*

**Chapter 14**
Mercury in Lower Trophic Levels of the Clear Lake Aquatic Ecosystem, California ............................................................................................................... 249
*Thomas H. Suchanek, Bradley A. Lamphere, Lauri H. Mullen, Cat E. Woodmansee, Peter J. Richerson, Darrell G. Slotton, Lee Ann Woodward, and E. James Harner*

## SECTION III: RISK

**Chapter 15**
Resources at Risk: A Forest Fire-Based Hazard/Risk Assessment ..................... 271
*Timothy A. Burton, Deirdre M. Dether, John R. Erickson, Joseph P. Frost, Lynette Z. Morelan, William R. Rush, John L. Thornton, Cydney A. Weiland, and Leon F. Neuenschwander*

**Chapter 16**
Uncovering Mechanisms of Interannual Variability from Short Ecological Time Series ..................................................................................................... 285
*Alan D. Jassby*

**Chapter 17**
Developing Realistic Air Pollution Exposure/Dose Criteria for Ecological Risk Assessments ............................................................................................ 307
*Allen S. Lefohn*

**Chapter 18**
Survey Methodologies for the Study of Ecosystem Restoration and Management: The Importance of Q-Methodology .............................................. 321
*John T. Woolley, Michael V. McGinnis, and William S. Herms*

**Chapter 19**
The California Water Quality Assessment Spatial Database: A Preliminary Look at Sierra Nevada Riverine Water Quality ..................................................... 333
*Anitra L. Pawley, Joshua H. Viers, Isaac Oshima, and James F. Quinn*

**Index** .................................................................................................................. 349

# 1 The Genesis and Future of the Field of Ecotoxicology

*John Cairns, Jr.*

## CONTENTS

Abstract ............................................................................................................. 1
Introduction ...................................................................................................... 2
An Irreducibly Interdisciplinary Pursuit ........................................................ 2
Divergence Between Questions and Methods ............................................... 3
Future Trends: New Goals and New Tools .................................................... 6
    Sustainability .............................................................................................. 6
    Ecosystem Services .................................................................................... 8
    Ecosystem Health ...................................................................................... 9
Conclusions .................................................................................................... 10
Acknowledgment ........................................................................................... 10
References ...................................................................................................... 11

## ABSTRACT

Some present practices in studying ecosystem health are understood more easily when the genesis of the field of ecotoxicology is examined. However, future trends in the field of ecotoxicology will be most affected by (1) a rapidly expanding interest in the field of sustainable development, (2) ensuring delivery of ecosystem services, and (3) shifting emphasis from avoiding harm in ecosystems to promoting robust health. As the problems that must be addressed grow larger in both temporal and spatial scale and more subtle in the magnitude of impact, the tools to address them must change. Continuing increases in human population and an increased level of affluence for many of the world's people will put unprecedented pressures on natural systems. This will require environmental scientists to adjust to new demands rapidly and policymakers to reduce the lag time between acceptance of an idea by mainstream science and its incorporation into regulatory practices.

**Keywords:** *ecotoxicology — genesis, ecotoxicology — future, ecotoxicology — limitations, sustainability, ecosystem health*

## INTRODUCTION

The field of ecotoxicology has roots in both the study of basic biological processes and practical concern about unintended effects of human-induced changes in the environment. Toxicity tests with animals have probably been used to protect human health as long as poisons have been around, i.e., a very long time. However, published studies linking mining and air pollution to damage in fish or wildlife populations, similar to what is termed "ecotoxicology" today, date to at least the late 1800s.[1-4] Instances of changes to the environment that adversely affected local human populations clearly predate the industrial revolution.[5] As concentrations of human population and the ability of humans to modify their environments increased, the losses of sweet water, soil, air, fish, woodlands, and wildlife became more frequent, longer lasting, and more widespread. The predictable response to these losses was the redirection of some human effort into preventing further losses. Environmental management was born and with it the need for sound information to guide management efforts.

## AN IRREDUCIBLY INTERDISCIPLINARY PURSUIT

The assessment of risk to the environment stemming from a human action is an irreducibly interdisciplinary pursuit. Skills from biology, chemistry, geology, geography, economics, engineering, planning, political science, and applied offshoots of all of these and other disciplines are necessary to the process. This interdisciplinary nature has created turmoil over the years and continues to do so. The environmental practitioner has to have skills drawn from many fields, but, at times, is respected by none of those disciplinary tribes. While these problems with disciplinary boundaries ebb and flow, they have not diminished over the years.

I entered the field of environmental toxicology in a serendipitous fashion. In 1948, I joined one of Ruth Patrick's[6] two river survey teams, totaling 35 personnel, and was 1 of 6 who stayed on at the end of the summer. W. B. Hart of the Atlantic Refining Company (Philadelphia, PA) encouraged Patrick to start the river surveys. The purpose of many of the river surveys was to establish a baseline of environmental conditions before an industry was built and operating, thus providing industrial executives with some estimate of the probable environmental effects of their wastes once the factory went into operation. Although I served as protozoologist for the survey team, Patrick asked me to do fish toxicity testing to complement the river surveys. So, in 1949, all the toxicity testing equipment from the Atlantic Refining Company was moved to the Academy of Natural Sciences, and I was trained in its use by Hart, the senior author of one of the first standard procedures for fish toxicity testing.[7] Not until many years later did I realize the vision of Hart in developing this testing procedure, especially at a time when the attention of the nation was focused on winning World War II. Environmental pollution was certainly not a major concern for industry, government, or citizens at that time. Unintentional damage from discharges of chemicals was quite unimpressive compared to the destruction of warfare.

Despite the clear need for sound scientific information to guide environmental decisions, all the advice given to me at that time, especially from my fellow graduate students, was that studies of environmental pollution were not of particular interest to biologists. The clear corollary was that I was unlikely to get a position in that field at an academic institution once the Ph.D. degree had been completed. Applied biologists were definitely second-class citizens.

This perception was consistent with the experiences I and other members of Patrick's group were having. Many biological journals had little interest in ecological toxicity tests; my manuscripts were regularly returned with comments that such a study was not an area of interest to the journal. As a consequence, many of my early papers on fish toxicity[8] were published in *Notulae Naturae*, a journal of the Academy of Natural Sciences, hithertofore devoted almost entirely to taxonomy.

The assessment of pollution was dominated at that time by what were then called sanitary engineers and chemists. Patrick spent much time espousing the biological assessment of pollution — an uphill battle. Much to my surprise, Don Bloodgood, Clair Sawyer, and other sanitary engineers (now retitled to environmental engineers) had a deep interest in biological assessment of pollution and encouraged me to publish in engineering and other journals outside the standard biological discipline.[9]

As a consequence of the considerable professional barriers to ecotoxicology at that time, I decided to do my Ph.D. dissertation on parasite transfaunation studies[10] in order to be better prepared to meet my obligations to my wife and children. Fortunately, due to the energetic activities of Patrick, extramural funding continued to appear and I, under her tutelage, began to get grants and contracts on my own.

Even today, much work in each constituent discipline is very narrowly focused. For example, Harte et al.[11] surveyed journals in ecology and conservation biology and found that very few articles used chemical and physical characterizations of water, soil, air, climate, and other components of the environment to aid in the interpretation of biological patterns. Only 22% of these articles even mentioned one of these factors in a site description. Harte and colleagues termed this trend "the Pluto Syndrome" — that is, there was little indication in the published papers that the ecological research was not carried out on the planet Pluto. Yet clearly, good risk assessment and subsequent environmental management depends on integrating these interacting factors. New journals, such as *Ecotoxicology, Landscape Ecology, Journal of Ecosystem Health, Ecological Engineering, Ecological Economics, Ecological Applications*, and the like, have been created to facilitate the integration of formerly isolated components of present environmental problems. Since environmental practitioners are often trained and their professional activities are reviewed by institutions still organized along disciplinary boundaries, turf wars continue and impede both scientific and professional progress. It is my hope that increasing interest in sustainable development will result in a more widespread synthesis of previously isolated specialties.

## DIVERGENCE BETWEEN QUESTIONS AND METHODS

About 2 years ago, I had a visitor from a European university who was writing a dissertation in sociology on the boundaries between toxicology and ecology.[2] He interviewed me about the development of toxicity tests and the current practices for using these tests to predict environmental risks. His straightforward questions about the source of various methods served to focus my concerns about a continuing divergence between the important questions in the field of ecotoxicology and the methods that could be used to address these questions. While the problems of concern are now inherently larger in both temporal and spatial scale and more subtle in the magnitude of impact, the predominant tools used in addressing them have not changed. There is always tension between the clearly desirable, but mutually exclusive, goals of precision and accuracy in the design of any toxicity test; the simplest tests are the most reproducible or precise, while the most complex tests are more accurate in that they can monitor the more complex end points that are the object of protection in natural systems. However, because of the regulatory weight many toxicity tests must withstand, a standard tool is often used in place of the most applicable tool. As a consequence, instead of developing new tools designed specifically for the spatial and temporal scale of a problem, old tools (e.g., single species toxicity tests) are combined or retrofitted to address problems on entirely new scales (e.g., landscape ecotoxicology).

The transitional stage I have been recommending for over two decades is to retain a "bottom-up" environmental toxicological capability, but to integrate it with a "top-down" capability. In the bottom-up approach, components (i.e., selected species) are the focus of the testing, and models are developed to predict the response of the entire system from the measured response of these components. The top-down approach attempts to consider the integrity of the entire system, including effects on the many components of an interacting and interdependent system, under realistic conditions. The development of the top-down methods has all the obstacles and difficulties associated with the previous development of bottom-up methods. In addition, much of the information generated is site specific and cannot be easily generalized from one situation to another. However, top-down methods have progressed rapidly in recent years and have demonstrated their usefulness to a great extent.[12]

A key example of methods retrofitted to a new purpose, in contrast to methods specifically developed for new goals, is the use of a collection of toxicity tests with individual species to estimate effects of human actions on communities. Using aquatic systems as an example, a toxicity test monitoring the survival of fish is clearly relevant when a primary management goal is prevention of a fish kill. However, as management goals have become more ambitious by encompassing subtler effects and moving beyond the prevention of gross damage to protection of ecosystem health, test end points should change to suit these new goals. Instead, the same tests and end points are retained and their output is manipulated to provide information upon which environmental management decisions will be made.[13] Retrofitting toxicity tests with individual species to protect entire communities has focused first on "important species," then on the most sensitive species, and now on a collection of representative species.

Important species generally have been operationally defined as fish in aquatic systems.[14]

1. They are vertebrates (as are humans) even though they are cold blooded.
2. Legislators, regulators, and citizens care about fish. Because fish are used for food and recreation, they have social importance; they would be missed immediately in their absence.

3. Fish "integrate" effects on other organisms. Since fish eat organisms lower on the food chain, if fish are present, the lower organisms must also be present.
4. Conditions sufficient to affect fish are likely to be important changes — organisms lower than fish might be affected by lesser changes, but their rates of reproduction, frequency of reproduction, and short life cycles make it easier for them to recover than a whole fish population.
5. The aquarium hobby provided some useful information on how to proceed with laboratory toxicity.

However, if the management goal is fishable waters, not only must fish kills be avoided, but the organisms that the fish eat must also be protected. In order to protect fish, it is necessary to protect the community. Consequently, the focus of ecotoxicology broadened.

As data accumulated, patterns of the toxicity of chemicals to organisms could be found[15,16] especially for organic wastes such as sewage. It was discovered that some organisms were often more sensitive to these conditions than others. However, even if every species in an ecosystem could be cultured and toxicity testing procedures were available for each of them, it would be impossible and impractical to test each one. But if the most sensitive species in the environment could be tested and management decisions could be based on that information, all other species would be protected. Unfortunately, initial patterns in tolerance to organic wastes did not hold true for other chemicals — no one species is most sensitive to all chemicals. After a great deal of effort was expended on identifying the most sensitive species and lingering confusion continued about the value of testing a sensitive rather than a tolerant species, this approach to protecting communities based on information about the response of one species has been largely abandoned.[17]

In its place is an approach that uses the response of a collection or sample of organisms to predict the response of communities.[18] Some of these approaches call for testing representatives of various taxonomic, trophic, or lifestyle groups, i.e., a benthic detritivorous insect, a picivorous fish, an alga. Others call for tests on a random sample of organisms and assume that the pattern of tolerance for a chemical across different organisms will have a certain shape.

The process of evolution often involves gradual modifications of an already existing structure that improves the functioning of a species. However, is this the best way to design tests for risk assessment? Assessment of human health effects involves the prediction of effects on just one species, *Homo sapiens*, whereas assessment of ecological effects must encompass thousands of interacting species. Yet, the most common type of test used in human health effects or ecological risk assessments is largely identical, i.e., the exposure of individual animals of a single species to chemicals. It is still rare to base ecological decisions on direct testing of end points that are the object of protection.

Ideally, the field of ecotoxicology should have well-developed techniques for every scale: local, ecosystem, landscape, and ultimately continental and global levels. The scale of the problem and the management goals should determine the techniques used in a study.[19] The mechanics of change should be determined by studying changes at a scale one level lower than that of the problem, possibly providing subsequent predictive capabilities. Where possible, predictive models would be validated by comparing predictions from the model to observation of effects in the system of interest. Where the scale of predictions make this impossible, individual components of the models should be validated to the degree possible, and peer review enhances assurance that the best consensual understanding of the relevant processes are accurately represented in the model.

For example, Cairns and Niederlehner[20] review methods used in ecotoxicology at the landscape level. Landscape ecotoxicology addresses the problems of toxicants spread over large spatial areas and having effects at that same level of ecological organization. The approach is characterized by the use of end points appropriate to the spatial scale across which a toxicant is dispersed, attention to the interactions between physical and temporal patterns and a process of ecological impairment, and integration of multiple lines of evidence for toxicity at various scales. Tools such as remote sensing and spatially explicit simulation models are characteristic of landscape ecotoxicology. Integrating information from damaged systems, toxicity tests, simulation models, and biomonitoring of both healthy and damaged systems is a desirable approach at this level and has been advocated for a number of years.[21-24] Close connections between designed experiments and surveys of natural systems must be maintained as much as possible with the acknowledgment that, at higher levels of organization, no experiments can be devised to determine thresholds. As the ability to experiment and predict over large spatial and temporal scales breaks down, decisionmakers will be faced with higher degrees of uncertainty than in the past.

## FUTURE TRENDS: NEW GOALS AND NEW TOOLS

New goals must lead to new techniques. Cairns and colleagues[25] have suggested that the field of ecotoxicology is in the early stages of a major paradigm shift that is driven by three interrelated forces: (1) heightened interest in sustainable use of the planet (referred to most commonly as sustainable development), (2) protection of ecosystems services (those functional attributes of ecosystems perceived as useful by human society), and (3) the shift from emphasis on avoiding deleterious effects on organisms and ecosystems to maintaining them in robust health. These factors will shape the future of ecotoxicology.

### SUSTAINABILITY

Malone[26] concluded that management of the environment to assure sustainability of its capacity to support human life is essentially a matter of managing human affairs in a manner that fosters sustainable human development. Anything written on sustainable development must acknowledge the pivotal role of the report from the World Commission on Environment and Development of the United Nations[27] titled *Our Common Future* (which is most commonly referred to as the Brundtland Report after Dr. Gro Harlem Brundtland, then Prime Minister of Norway, the chair of the group that produced the report). That commission defined sustainable development as "...development that meets the needs of the present without compromising the ability of future generations to meet their own needs." Weston[28] notes:

> The Commission integrated sustainable development into the world's economy as follows: Sustainable development is... "a process of change in which the exploitation of resources, the direction of investments, the orientation of technological development ... institutional change and the ability of the biosphere to absorb the effects of human activities are consistent with future as well as present needs." (Reference 28, p.5)

Ecotoxicologists must address the risks of human actions to sustainability, especially the ability of the biosphere to absorb the effects of human activities! Thompson[29] suggests:

> ...nature is the quintessential supplysider, with its resource reserves not readily available on demand. Through systematic recycling and reuse, moreover, nature doesn't push its inventory of renewable and nonrenewable resources beyond critical limits for sustainability. It's a natural economic system with rules of the game that govern the interrelationships among all things — energy, matter, space, time and life — and their most efficient and effective use. (Reference 29, p. 3)

Robèrt and colleagues[30] provide four system conditions deemed necessary to achieve sustainability: (1) substances from the Earth's crust must not systematically increase in the ecosphere, (2) substances produced by society must not systematically increase in the ecosphere, (3) the physical basis for productivity and diversity of nature must not be systematically diminished, and (4) fair and efficient use of resources with respect to meeting human needs. Cairns[31] restated and expanded these conditions as a result of correspondence about "The Natural Step USA Program."

1. Artifacts created by human society may not systematically increase on the planet. This restriction includes persistent toxic chemicals, impervious and heat-holding surfaces which affect hydrologic cycles and climate and fragment ecosystems, and solid wastes (a few illustrative examples).
2. Substances extracted from Earth's crust must not be concentrated or dispersed in ways harmful to the biosphere (e.g., greenhouse gases, heavy metals, or oil).
3. The physical and biological basis for the services provided by nature (i.e., ecosystem services) shall not be diminished.
4. Short-term human "needs" must be reevaluated if meeting them endangers the ecological life support system of the planet.
5. Unless society is willing to accept an ecologically impoverished world, ecological restoration must equal ecological destruction if sustainability is to be achieved.

Subsequently, Cairns[32] added three more conditions to this list.

6. Yet undeveloped technologies should not be depended upon to save society from current problems.

7. Human society should not co-opt so much of the energy captured by photosynthesis that the machinery of nature cannot work.
8. Human society should not co-opt more than a fixed percentage of land surface, i.e., if a world food shortage develops, agricultural land use should remain stable rather than converting more natural systems to agriculture. Instead, grains should be shifted from domesticated animals to direct human use.

Since this manuscript was produced, this list has been greatly expanded.[33]

More specific ways in which these goals of sustainability will influence the future goals and tools in ecotoxicology may be clearer if further consideration is given to the two key concepts of ecosystem services and ecosystem health.

## ECOSYSTEM SERVICES

Sustainability requires the maintenance of ecosystem services per capita. Ecosystem services are those attributes or functions of ecosystems perceived as beneficial to human society.[34] They include the capture of energy from sunlight and conversion into food, fiber, and pharmaceuticals; the regeneration of breathable air, drinkable water, and arable soils; the breakdown of wastes; buffering natural disasters; and many others. Costanza et al.[35] have estimated the value of these services provided to human society by natural systems at little or no cost to be more than $33 trillion per year (U.S. dollars).

One example of an ecosystem service is described by Cairns and Bidwell[36] as they discuss the findings of Professor J. R. Stauffer working on Lake Malawi.[37] Stauffer found that schistosomiasis, a disease for which snails are the secondary hosts, had been kept under control by species of fish that consumed the infected snails. When these fishes were overharvested, the ecosystem service of pestilence control diminished, and the percentage of infected humans rose dramatically. This ecosystem service was not appreciated until Stauffer observed it. Cairns[34] also mentions the research of John Harte of the University of California at Berkeley carried out at Rocky Mountain Biological Laboratory (RMBL), a high-altitude research station on the western slopes of the Rockies in Colorado. Harte found that the soils at 9500 feet in Gunnison County, CO, where RMBL is located, were methane sinks. Methane is a much more effective greenhouse gas than carbon dioxide and is produced by the cattle in Gunnison County. Greenhouse gases, which might raise the Earth's temperature, are important to another major industry in Gunnison County, namely skiing, especially the resort at Mount Crested Butte, where artificial snow must be produced often at significant costs. This ecosystem service would have been unrecognized until Harte found evidence for it. Such situations are likely to increase dramatically as research focuses attention more strongly on ecosystem services.

Ecotoxicology at large scales must evaluate the risks of human activities on the ecosystem services upon which human society clearly depends. The effect of individually small, but collectively significant, cumulative impacts on the provision of ecosystem services is of special concern, as is the effect of population pressure on the diminishing areas providing ecosystem services.

## Ecosystem Health

With continued habitat destruction and increasing population pressures, the maintenance of ecosystem services per capita will depend not only on conservation and restoration, but also on the maintenance of peak ecosystem health in the remaining minimally managed systems so that they can provide ecosystem services at a maximum rate.

In relatively recent times, a shift has occurred in the interpretation of the term "human health." Previously, the concept was widely regarded as the absence of disease. Now, with society's preoccupation with fitness, the term has come to mean much more than the absence of disease. It connotes robust well-being — vitality, energy, resistance to disease, and enjoyment of life. Similarly, regulatory agencies such as the U.S. Environmental Protection Agency (EPA) have focused on the protection of ecosystems, which is judged, then and even now, by the absence of observable deleterious effects. In short, if no harm was evident, the condition of the ecosystem was acceptable to the regulatory agency.

However, the absence of deleterious effects may be due to the inability to measure them. "Ecosystem health" (or other terms such as "ecological integrity," "biological integrity," and "ecosystem condition" all used to describe essentially the same concept) is a more ambitious goal than the absence of detectable harm. Toxicity tests identify three critical responses: a response threshold concentration below which there is no evidence of harm; a zone of greater response where increasing the concentration increases the response; and a threshold beyond which the system stops responding, either because all the organisms are dead or because no further response under these conditions is possible because the system has "peaked out." However, different levels of fitness may exist in the range of concentrations that produce a no-observable response.

There is one marked disadvantage to comparing human health and ecosystem health — human physiology has homeostatic regulatory mechanisms that keep temperature, blood chemistry, pulse rate, and other important life functions within preset limits under "normal operating conditions." But communities, ecosystems, and ecological landscapes do not have preset optimal levels to which they invariably return. These composite biological structures will return to a previous state if a stress does not exceed a limit of tolerance, but feedback loops are different from the homeostatic mechanisms operating in organismal physiology. When a limit of tolerance is exceeded, an ecosystem does not die. Instead, it moves to a different state with different normal operating conditions. E. P. Odum (personal communication) and others have called the feedback loops that do function in ecological systems "homeorhesis" in contrast to the homeostasis that occurs in the physiology of individual organisms. While the set points for human health are intrinsic, the set points for ecosystem health are variable and can be substantially changed by goals externally set by humans.

Cairns and colleagues[19] proposed a framework for developing indicators of ecosystem health similar in concept to the use of "leading economic indicators." Consideration was given to linkages between human activities and the well-being or state of the environment, since environmental degradation affects both biogeochemical and socioeconomic indicators. The important thrust of this concept, however, is that desirable properties in indicators of environmental health vary with the specific management use. For instance, different indicators are required when collecting data to assess the condition of the environment, to monitor trends over time, to provide early warning of environmental harm, or to diagnose the cause of an existing problem. Trade-offs between desirable characteristics, cost, and quantity and quality of information are inevitable when selecting indicators for management use. The decision about what information to collect for various purposes can be made more rationally when available indicators are characterized and matched to management goals. Different circumstances require different evidence, and the same suite of methods or information is not equally suited to all problems. This variety will inevitably cause problems for both regulators and industry (not to mention municipalities) because prescriptive requirements are more suited to courts of law than to ecosystems.

## CONCLUSIONS

Despite contrarian views that have received much publicity, the clear consensus of scientists is that important changes in the environment, both locally and globally, have occurred and these changes have consequences for humans. Ecotoxicologists have an important role to play in dealing with these challenges. Although ecological responses to toxic materials are undeniably a continuing problem, habitat destruction is probably more serious cumulatively and less tractable. With continued habitat destruction, the maintenance of ecosystem services per capita will depend on conservation, restoration, and maintenance of peak ecosystem health in the remaining minimally managed systems so that they can provide ecosystem services at a maximum rate. The continuing increases in human population and increased level of affluence for many of the world's people will put unprecedented pressures on natural systems during the first half of the next century. This will require environmental scientists to adjust to new demands rapidly and policymakers to reduce the lag time between acceptance of an idea by mainstream science and its incorporation into regulatory practices.

## ACKNOWLEDGMENT

I am indebted to Lisa Maddox for transcribing the dictation of the first part of this manuscript and to Eva Call for transcription of the last part and revisions of the entire manuscript. I am also much indebted to Darla Donald for editing the manuscript to meet the requirements of the journal and other editorial effort. Finally, B. R. Niederlehner and Alan Heath provided useful suggestions on an early draft of this manuscript.

For questions or further information regarding the work described in this chapter, please contact:

**Dr. John Cairns, Jr.**
Dept. of Biology
1020 Derring Hall
Virginia Tech
Blacksburg, VA 24061
e-mail: cairnsb@vt.edu

## REFERENCES

1. Adams, W.J. 1995. Aquatic toxicology testing methods. In D.J. Hoffman, B.A. Rattner, G. A. Burton, Jr. and J. Cairns, Jr., eds., *Handbook of Ecotoxicology*. Lewis Publishers, Boca Raton, FL, pp. 25–46.
2. Halffman, W. 1995. The boundary between ecology and toxicology: a sociologist's perspective. In J. Cairns, Jr. and B.R. Niederlehner, eds., *Ecological Toxicity Testing: Scale, Complexity, and Relevance*. Lewis Publishers, Boca Raton, FL, pp. 11–34.
3. Hoffman, D.J. 1995. Wildlife toxicity testing. In D.J. Hoffman, B.A. Rattner, G. A. Burton, Jr. and J. Cairns, Jr., eds., *Handbook of Ecotoxicology*. Lewis Publishers, Boca Raton, FL, pp. 47–69.
4. Rand, G.M, P.G. Wells and L.S. McCarty. 1995. Introduction to aquatic toxicology. In G.M. Rand, ed., *Fundamentals of Aquatic Toxicology*, 2nd ed. Taylor and Francis, Washington, D.C., pp. 3–69.
5. Diamond, J. 1992. *The Third Chimpanzee*. HarperCollins, New York.
6. Patrick, R. 1949. A proposed biological measure of stream conditions based on a survey of Conestoga Basin, Lancaster County, Pennsylvania. *Proc. Acad. Nat. Sci. Philadelphia* 101:277–341.
7. Hart, W.B., P. Doudoroff and J. Greenbank. 1945. The evaluation of the toxicity of industrial wastes, chemicals and other substances to fresh-water fishes. The Atlantic Refining Company, Philadelphia, PA, 317 pp.
8. Cairns, J., Jr. and A. Scheier. 1957. The effects of temperature and hardness of water upon the toxicity of zinc to the common bluegill (*Lepomis macrochirus* Raf.). *Not. Nat. Acad. Nat. Sci. Philadelphia* 299:1–12.
9. Cairns, J., Jr. 1956. Effects of increased temperatures on aquatic organisms. *Ind. Wastes* 1:150–152.
10. Cairns, J., Jr. 1953. Transfaunation studies of the host-specificity of the enteric Protozoa of amphibia and various other vertebrates. *Proc. Acad. Nat. Sci. Philadelphia* 105:45–69.
11. Harte, J., M. Torn and D. Jensen. 1992. The nature and consequences of indirect linkages between climate change and biological diversity. In R. Peters and T.E. Lovejoy, eds., *Global Warming and Biodiversity*. Yale University Press, New Haven, CT, pp. 325–342.
12. Cairns, J., Jr. and B.R. Niederlehner, eds. 1995. *Ecological Toxicity Testing: Scale, Complexity, and Relevance*. Lewis Publishers, Boca Raton, FL.
13. Rapport, D. 1998. Need for a new paradigm. In D. Rapport, R. Costanza, P.R. Epstein, C. Gadet and R. Levins, eds., *Ecosystem Health*. Blackwell Science, Malden, MA, pp. 3–17.

14. Edwards, C.J. and R.A. Ryder. 1990. Biological surrogates of mesotrophic ecosystem health in Laurentian Great Lakes. Report to the Great Lakes Science Advisory Board of the International Joint Commission, Windsor, ON.
15. Cairns, J., Jr., A. Scheier and N.E. Hess. 1964. The effects of alkyl benzene sulfonate on aquatic organisms. *Ind. Wastes* 9:22–28.
16. Patrick, R., J. Cairns, Jr. and A. Scheier. 1968. The relative sensitivity of diatoms, snails, and fish to twenty common constituents of industrial wastes. *Prog. Fish Cult.* 30:137–140.
17. Cairns, J., Jr. 1986. The myth of the most sensitive species. *BioScience* 36:670–672.
18. Smith, E.P. and J. Cairns, Jr. 1993. Extrapolation methods for setting ecological standards for water quality: Statistical and ecological concerns. *Ecotoxicology* 2:203–219.
19. Cairns, J., Jr., P.V. McCormick and B.R. Niederlehner. 1993. A proposed framework for developing indicators of ecosystem health. *Hydrobiologia* 263:1–44.
20. Cairns, J., Jr. and B.R. Niederlehner. 1996. Developing a field of landscape ecotoxicology. *Ecol. Appl.* 6:790–796.
21. O'Neill, R.V., D.L. De Angelis, J.B. Waide and T.F.H. Allen. 1986. *A Hierarchical Concept of Ecosystems*. Princeton University Press, Princeton, NJ.
22. Hunsaker, C.R., R.L. Graham, G.W. Suter, II, R.V. O'Neill, L.W. Barnthouse and R.H. Gardner. 1990. Assessing ecological risk on a regional scale. *Environ. Manage.* 14:325–332.
23. Suter, G.W., II. 1990. End points for regional ecological risk assessment. *Environ. Manage.* 14:9–23.
24. Graham, R.L., C.T. Hunsaker, R.V. O'Neill and B.L. Jackson. 1991. Ecological risk assessment at the regional scale. *Ecol. Appl.* 1:196–206.
25. Cairns, J., Jr., J.R. Bidwell and M.E. Arnegard. 1996. Toxicity testing with communities: Microcosms, mesocosms, and whole-system manipulations. *Rev. Environ. Contam. Toxicol.* 147:45–69.
26. Malone, T.F. 1994. Sustainable human development: A paradigm for the 21st century. A white paper for the National Association of State Universities and Land-Grant Colleges, Research Triangle Park, NC.
27. The World Commission on Environment and Development. 1987. *Our Common Future*. Oxford University Press, Oxford, England, 400 pp.
28. Weston, R.F. 1995. Sustainable development: To better understand the concept. *The Weston Way* 21(1):5–16.
29. Thompson, A.F. 1995. Sustainable development: "Depoliticking" the environment and making it a matter of natural economics. *Sustainable Development* 21:2–3.
30. Robèrt, K.-H., J. Holmberg and K.-E. Eriksson. 1994. Socio-ecological principles for a sustainable society. The Natural Step Environmental Institute Ltd., Stockholm, Sweden, 11 pp.
31. Cairns, J., Jr. 1996. Respect for the inherent worth of each individual vs. the web of life: An unresolved conflict affecting sustainable use of planet. UUFNRV home page at http://www.montgomery-floyd.lib.va.us/pub/compages/uufnrv/articles.html.
32. Cairns, J., Jr. Eight ecological conditions for sustainable use of the planet. *Sustain.* in press.
33. Cairns, J., Jr. 1997. Commentary: Defining goals and conditions for a sustainable world. *Environ. Health Perspect.* 1105(11):1164–1170.
34. Cairns, J., Jr. 1996. Determining the balance between technological and ecosystem services. In P.C. Schulze, ed., *Engineering Within Ecological Constraints*. National Academy Press, Washington, D.C., pp. 13–30.

35. Costanza, R., R. d'Arge, R. de Groot, S. Farber, M. Grasso, B. Hannon, K. Limburg, S. Naeem, R.V. O'Neill, J. Paruelo, R.G. Raskin, P. Sutton and M. van den Belt. 1997. The value of the world's ecosystem services and natural capital. *Nature* 387:253–260.
36. Cairns, J., Jr. and J.R. Bidwell. 1996. The modification of human society by natural systems: Discontinuities caused by the exploitation of endemic species and the introduction of exotics. *Environ. Health Perspect.* 104:1142–1145.
37. Stauffer, J.R., Jr., M.E. Arnegard, M. Cetron, J.J. Sullivan, L.A. Chitsulo, G.F. Turner, S. Chiotha and K.R. McKaye. 1997. The use of fish predators to control vectors of parasitic disease: Schistosomiasis in Lake Malawi — A case history. *BioScience* 47:41–49.

# Section I

*Fate*

# 2 A Predictor of Seasonal Nitrogenous Dry Deposition in a Mixed Conifer Forest Stand in the San Bernardino Mountains

*Michael J. Arbaugh, Andrzej Bytnerowicz, and Mark E. Fenn*

## CONTENTS

Abstract ............................................................................................................. 17
Introduction ...................................................................................................... 18
Materials and Methods .................................................................................... 18
    Site Description ......................................................................................... 18
    Model Development .................................................................................. 19
    Model Parameterization for Barton Flats ................................................ 20
    Branch Rinse and Throughfall Comparison Data .................................. 21
Results and Discussion .................................................................................... 21
Conclusions ...................................................................................................... 23
References ........................................................................................................ 23

## ABSTRACT

A predictor was developed to estimate seasonal surface dry deposition of nitrogenous (N) air pollution to a mixed conifer forest in the San Bernardino Mountains. The estimator requires only foliar rinse measurements at the canopy top, and leaf and branch area estimates by species for the site. The model assumes that the forest lacks a vertical atmospheric concentration gradient, and that $NH_4NO_3$ is the dominant form of particulate ammonium and nitrate. The model was applied to a intensive study site, Barton Flats, located on the eastern side of the San Bernardino Mountains. Seasonal dry surface deposition for three tree species were estimated and then compared with foliar rinse and throughfall estimates for the summer of 1993. The model estimate of summertime dry surface deposition (3.44 kg N/ha) falls within the range of the foliar rinse estimate (4.22 kg N/ha) and the throughfall estimate

(2.95 kg N/ha) for the same time and location. This approach has potential for reducing the number of field and laboratory measurements needed to obtain reliable seasonal surface deposition estimates for many dry forest, shrub, and savanna systems.

**Keywords:** *ammonium, nitrate, deposition, model, ponderosa pine*

## INTRODUCTION

Dry deposition is one of the major mechanisms of nitrogenous (N) transfer from the atmosphere to plant surfaces in the arid California mountains.[1,2] High dry deposition fluxes of N pollutants have been reported in the highly exposed areas of the San Gabriel and San Bernardino Mountains,[3-4] and in lesser amounts in the southern and western portions of the Sierra Nevada.[5]

Foliar rinsing techniques[3,6] have been successfully used for determinations of atmospheric deposition of various ions to trees. The technique is especially valuable in dry climates where in the spring-summer-fall period precipitation (which would allow for throughfall collection) is scarce.

Reduction of the number of dry deposition measurements through modeling would allow larger areas and times to be measured. Unfortunately, forest conditions reduce the ability to apply simple inferential models to this system. This open vegetation type provides increased mixing and edge effects[7] which violate the assumptions of most simple inferential models.[8]

In a previous study, we developed a model to estimate dry N surface deposition along a vertical canopy profile for ponderosa pine trees in the San Bernardino Mountains.[9] In this study we extend this model to estimate summer season dry N deposition to the forest. This requires scaling model estimates from individual branches to the stand level and inclusion of information from several other studies. The accuracy of the new model is evaluated by comparison with empirical foliar rinse and throughfall estimates of seasonal surface N deposition for the same location and time.

## MATERIALS AND METHODS

### SITE DESCRIPTION

The study site was located in a forested area with moderate ozone damage in the vicinity of the Forest Service Visitor Information Center at Barton Flats on State Highway 38.[10] About 1.6 km from the atmospheric monitoring station, three variable area plots were established. In this study only one plot was used, plot 2, which was 0.48 ha in size and 2171 m in elevation. The mixed conifer forest at the study site is comprised of a mixture of ponderosa pine (*Pinus ponderosa*, Laws.) or Jeffrey pine (*P. jeffrey*, Grev. and Balf), white fir (*Abies concolor*, Gord. and Glend.), California black oak (*Quercus kelloggii*, Newb.), and a herbaceous or shrubby understory. Ponderosa and Jeffrey pine comprise 49% of stand basal area and 37%

of the total stems. White fir and black oak are about 24% of the stand basal area each. White fir comprises about 36% of the total stems, while black oak is only 21% of the stems.

## MODEL DEVELOPMENT

In a previous study, a simple model was developed to predict N flux to leaf surfaces at different canopy positions as a function of flux at the top of the tree canopy and tree height.[9] A single model was found to be adequate for predicting N deposition to leaf surfaces due to nitric acid vapor ($HNO_3$) and particulate nitrate ($NO_3^-$), and ammonium ($NH_4^+$):

$$F_{Nh} = 0.79\exp^{-1.26(1 - h/t)} F_{Nt}, \text{ for } h < t \quad (1)$$

$$R^2_n = 0.70, \text{ RMSE} = 1.27$$

where  $F_{Nh}$ = flux to leaves from either particulate nitrate and nitric acid or ammonium at the canopy level h
  $h$ = a vertical height along the tree canopy
  $t$ = max(h) = top of canopy

The model is applicable to canopy heights less than t. At the top of the canopy a discontinuity in the model arises, likely due to increased turbulence at the canopy top relative to turbulence within the canopy.[9]

In this study we used the newly developed model to predict surface N dry deposition for the entire canopy. To do this, we assumed that the site consists of co-dominant trees of approximately equal height and that $NH_3$ is not deposited on surfaces, but only taken up through stomates. The model was then applied to estimate deposition to leaf and branch surface dry deposition at 5% vertical sections of the forest canopy. The modified model form is

$$F_{Nhoi} = 0.79\exp^{-1.26(1 - ho)} F_{nt} (\alpha_{1i}LA_{iho} + \alpha_{2i}BA_{iho})$$

where  $F_{Nhoi}$ = flux to the 5% of canopy centered at height ho for tree species i
  $i$ = tree species i
  $\alpha_{1i}$ = constant parameters reflecting surface deposition differences between ponderosa pine foliage and foliage of other tree species
  $\alpha_{2i}$ = constant parameters reflecting surface deposition differences between ponderosa pine foliage and wood surface area of tree species i
  ho = 5% vertical sections of the forest; ho = 0.025 corresponds to bottom 5% of canopy and ho = 0.975 corresponds to the top 5% of the forest canopy
$LA_{iho}, BA_{iho}$ = leaf area and wood surface areas, respectively, for species i at height h/t

This model scales the canopy foliage deposition model (1) to the stand level using leaf area and wood area information, the differences between deposition to ponderosa pine foliage and foliage and wood surfaces of other tree species in the stand. The total stand washable surface deposition also requires estimates for open ground areas, which need to be measured separately from the canopy.

## MODEL PARAMETERIZATION FOR BARTON FLATS

Parameterization information for fluxes and constants were gathered as part of the large study "Assessment of acidic deposition and ozone effects on conifer forests in the San Bernardino Mountains"[7] at Barton Flats, a mixed conifer stand at the eastern end of the mountains. Deposition of nitrate, ammonium, and sulfate to branches of seedlings and mature trees of ponderosa pine at various levels of the canopy were measured during three photochemical smog seasons (1992, 1993, and 1994). Fluxes of $NO_3^-$ and $NH_4^+$ to branches of ponderosa pine seedlings were determined on a tower located in a stand of mature ponderosa/Jeffrey pines.

Branch wash deposition was estimated from four seedlings located at each of four different levels of the tower (29 m at the top of the canopy; 24 m and 16 m in the middle of the canopy; and 12 m at the canopy bottom). In the beginning of each collection period, branches about 10 cm long were thoroughly rinsed with double deionized water. The branches were rinsed again at the end of the collection period (14 days) with about 100 mL of double deionized water, and the rinses were collected in 250 mL Nalgene bottles. Bottles were placed on ice, then immediately transferred to the laboratory and stored at –18°C. Concentrations of $NO_3^-$ were determined by ion chromatography (Dionex 4000i ion chromatograph), and concentrations of $NH_4^+$ were determined colorimetrically with a Technicon TRAACS Autoanalyzer. Seedling branch rinses were compared with mature ponderosa pine branch rinses taken at the same times and locations, and there were no significant differences between the two measures.[7]

Leaf area ($LA_i$) for each overstory species was estimated using a litterfall method for determining leaf area.[11] This method was modified because litter collectors were placed only under large trees and, thus, did not collect litter from areas representative of the entire plot. Therefore, our strategy was to estimate the total annual litterfall per tree, determine the surface area of the annual litterfall, and estimate the leaf area of the tree by factoring in the number of annual whorls on the tree (the annual turnover rate 't' used from Reference 11). Thus, the average surface area of the litter produced by a tree in one year multiplied by the number of years of foliage left on the tree approximates the leaf area of that tree.

Using this method, the leaf area of ponderosa pine was estimated to be 3800 m²/ha, 9201 m²/ha for white fir, and 5891 m²/ha for black oak. In addition to forest foliage and branch deposition, estimates of dry surface deposition to the ground (356 g N/ha) and wet deposition due to summer precipitation (420 g N/ha) were included from a previous study.[7]

Estimates of deposition differences to ponderosa pine wood surfaces ($\alpha_2$) and foliage ($\alpha_1$) and wood surfaces of other species ($\alpha_2$) were obtained during a period of intense measurements in the summer of 1993. Values were compared with seedling branch washes, and conversion factors were developed to estimate seasonal deposition to these surfaces from seedling branch washes. Surface deposition to branch wood was found to be approximately 7.9 times greater than deposition to foliage for ponderosa pine and white fir and 3.7 times greater for black oak. Using information from a fell tree study,[12] wood surface area ($BA_i$) was found to be 2.9% of leaf area for ponderosa pine. This was assumed to be true for fir as well, while black oak was assumed to be 10.1% of oak leaf area (Dr. M. Krywult, personal communication).

### Branch Rinse and Throughfall Comparison Data

Empirical dry surface N deposition estimates were developed independently of the model, but with similar data used for model development and parameterization.[7] Flux rates to the entire canopy were assumed to be the average of four flux measurements (29, 24, 16, and 12 m) over the season. Differences between seedling and mature-tree branch rinse deposition were included in the calculations, as were all differences between branch rinse deposition between species. Some constants also differed for the empirical estimates, reflecting slightly different sources of supporting data not gathered as part of the assessment study. For instance, canopy leaf area distributions were used for the model, but not in the empirical estimates. A more complete description of the methods used can be found in Reference 7.

Throughfall in this study was considered to be precipitation penetrating the tree canopy then falling to the forest floor, while rainfall was defined to be precipitation falling to the ground between tree canopies. Flip-top throughfall collectors were developed that remained closed until a small amount of precipitation collected in a funnel which directed a small amount of moisture to wet a narrow strip of dissolvable paper. When the paper dissolved, a counterweight caused the lid to flip open. The same collectors were used to measure throughfall and rainfall. Construction and operation of the collectors are described by Glaubig and Gomez.[13] Rainfall was collected in stand openings free of canopy influences, but adjacent to throughfall collector trees. Throughfall and rainfall were collected during the spring and early fall period. Rare summer rain events were also measured. During 1993, fortunate precipitation events in late spring and early fall resulted in comparable wet deposition measurement and dry deposition measurement periods (wet deposition was approximately 2 weeks longer than the dry deposition measurement period for that year.) We assumed that precipitation was sufficient to completely remove washable N dry deposition for both foliage and wood surfaces for the spring and fall precipitation events.

## RESULTS AND DISCUSSION

Model estimates of seasonal dry N surface deposition were similar to that obtained independently using dry deposition data (Table 1). These results indicate that flux at the top of the canopy is comparable to total forest seasonal dry surface deposition, as measured by branch rinsing. The estimate is also comparable to throughfall estimates of N surface deposition, thus model agreement with empirical results are likely not due to using similar data.

## TABLE 1
## Comparison of Summertime Inorganic N ($NO_3^-$ + $NH_4^+$) Deposition (kg/ha) in Plot 2 Based on Model Estimates, Branch Rinse, and Throughfall Data

| Species or Parameter | Model Estimate | Branch Rinses | Throughfall |
|---|---|---|---|
| Fir | 0.87 | 0.99 | 1.10 |
| Oak | 1.24 | 1.70 | 0.45 |
| Pine | 0.44 | 0.64 | 0.51 |
| Ground[a] | 0.36 | 0.36 | 0.36 |
| Rain[b] | 0.53 | 0.53 | 0.53 |
| Total | 3.44 | 4.22 | 2.95 |
| Total (without oak) | 2.20 | 2.52 | 2.50 |

[a] Assumed to be equal to the average of three species foliage deposition multiplied by plot size. Included in all estimates for the purpose of comparison.

[b] Estimated from flip-top rain collectors located in open areas of the plot, and included in all three approaches for the purpose of comparison.

Deposition estimates without the oak data are very similar between the model, branch rinse, and throughfall estimates (Table 1). Similarities between branch rinse and throughfall estimates are expected since both are based on similar processes of dry deposition inputs and aqueous rinsing of the canopy surfaces. The model closely tracks the empirical estimates and therefore may be useful for estimating branch rinse or throughfall estimates.

Estimates of total surface deposition with black oak differed for the three estimates. Throughfall estimates were just 25% of the value estimated from branch rinses. The low throughfall in oak may have been due to stem flow not measured by throughfall collectors. For conifers this is a minor component of deposition, but for deciduous forests as much as 30% of the water input to the forest may be from stem flow.[14] The irregular canopy of oak also makes measurement by throughfall more dependent on the position of collectors. In this study oak throughfall was often similar to rainfall,[15] indicating that some throughfall likely did not intercept the diffuse black oak canopy. In addition, the irregular crowns of black oaks may have resulted in variability in leaf and wood area that was not reflected in the constants used in the empirical and model estimates.

The model's ability to be applied to other locations is presently limited in two general ways. The first is that it has been developed at a single stand, therefore extrapolation to other locations depend on comparability of the vertical deposition gradient as estimated in (1), and that differences between ponderosa pine foliage and other surfaces are comparable between locations. The second limitation is that branch wash estimates from the canopy top, which ultimately control the model estimates, are difficult to obtain. Mixed forest canopies may be over 30 m in height, so some type of scaffold system is required to get canopy top branch washes, which makes the data expensive to obtain. Clearly, some less expensive methods need to be developed before the model can be evaluated or applied to a number of locations.

## CONCLUSIONS

The preliminary model developed in this study has the potential of reducing the necessary work to assess whole canopy flux, as measured by branch rinsing, and is applicable over a range of summer atmospheric deposition concentrations and temperatures. It reduces the amount of data needed to assess summer dry surface N deposition in the mixed conifer forest stand studied to two components: (1) foliar rinse leaf surface deposition at the canopy top and (2) site leaf and branch area by tree species (and ground surface).

Additional locations, especially toward the western side of the San Bernardino Mountains and in the southern Sierra Nevada, need to be examined to determine how different pollutant loads, species mixtures, and meteorological conditions affect the $a_0$ and $a_1$ parameters for the mixed conifer forest.

This approach may have application to many ecosystems lacking vertical concentration gradients of pollutants (those with discontinuous canopies), where the dominant form of aerosol N pollution is $NH_4NO_3$ and aerosol $NO_3^-$ and $NH_4^+$ are in constant proportions. Such systems might include coastal sage scrub; oak-grassland; some chaparral systems; and other mediterranean, arid, and tropical ecosystems. Application of the model to shrub systems may be especially valuable because it is convenient and affordable to estimate deposition at the canopy top.

For questions or further information regarding the work described in this chapter, please contact:

**Michael J. Arbaugh**
4955 Canyon Crest Dr.
Riverside, CA 92507-6099
e-mail: marbaugh@deltanet.com

## REFERENCES

1. Rundell, P.W. and D.J. Parsons. 1977. Montane and subalpine vegetation in the Sierra Nevada and Cascade ranges. In: Barbour, M.J., Major, J. eds., *Terrestrial Vegetation of California*, pp. 559–599, Wiley, New York.
2. Bytnerowicz, A. and M.E. Fenn. 1996. Nitrogen deposition in California forests: a review. *Environ. Pollut.* 2:127–146.
3. Bytnerowicz, A., P.R. Miller, D.M. Olszyk, P.J. Dawson and C.A. Fox. 1987. Gaseous and particulate air pollution in the San Gabriel Mountains of southern California. *Atmos. Environ.* 21:1749–1757.
4. Fenn, M.E. and A. Bytnerowicz. 1993. Dry deposition of nitrogen and sulfur to ponderosa and Jeffrey pine in the San Bernardino National Forest in southern California. *Environ. Pollut.* 81:277–285.
5. Bytnerowicz, A., P.J. Dawson, C.L. Morrison and M.P. Poe. 1991. Deposition of atmospheric ions to pine branches and surrogate surfaces in the vicinity of Emerald Lake watershed, Sequoia National Park. *Atmos. Environ.* 25A:2203–2210.

6. Lindberg, S.E. and G.M. Lovett. 1985. Field measurements of particle dry deposition rates to foliage and inert surfaces in a forest canopy. *Environ. Sci. Technol.* 19:238–244.
7. Bytnerowicz, A., M.E. Fenn and R. Glaubig. 1996. Dry deposition of nitrogen and sulfur to forest canopies at three plots. In: Assessment of Acidic Deposition and Ozone Effects on Conifer Forests in the San Bernardino Mountains. Final Report to California Air Resources Board, Contract No. A032-180, pp. 4–1 through 4–48, Sacramento, CA.
8. Hicks, B.B. and T.P Meyers. 1988. Measuring and modelling dry deposition in mountainous areas. In: Unsworth, M.H., Fowler, B. eds. *Acid Deposition at High Elevation Sites*, pp. 541–552, Amsterdam: Kluwer.
9. Arbaugh, M., A. Bytnerowicz and M.E. Fenn. 1996. Predicting N flux along a vertical canopy gradient in a mixed conifer forest stand of the San Bernardino Mountains. In: *Proceedings*, The effects of air pollution and climate change on forest ecosystems, USDA Forest Service GTR-164, http://www.rfl.pswfs.gov/pubs/gtr-psw-164/index.html. Printed version in process.
10. Miller, P.R., J. Chow, J.G. Watson, A. Bytnerowicz, M.E. Fenn, M. Poth and G. Taylor, Jr. 1996. Assessment of acidic deposition and ozone effects on conifer forests in the San Bernardino Mountains. Final Report to California Air Resources Board, Contract No. A032-180, Sacramento, CA.
11. Marshall, J.D. and R.H. Waring. 1986. Comparison of methods of estimating leaf-area index in old-growth Douglas-fir. *Ecology* 67:975–979.
12. Arbaugh, M. J. 1995. Spatial extrapolation and forecast uncertainty of three simulation models for ponderosa pine along the Front Range of Colorado. USDA Forest Service GTR RM-265.
13. Glaubig, R. and A. Gomez. 1994. A simple, inexpensive rain and canopy throughfall collector. *J. Environ. Qual.* 23:1103–1107.
14. Butler J.T. and G.E. Likens. 1995. A direct comparison of throughfall plus stemflow to estimates of dry and total deposition for sulfur and nitrogen. *Atmos. Environ.* 29:1253–1265.
15. Fenn, M.E., R. Glaubig, D. Jones and S. Schilling. 1996. Wet deposition and throughfall measurments in relation to the forest canopy at 3 plots. In: Assessment of Acidic Deposition and Ozone Effects on Conifer Forests in the San Bernardino Mountains. Final Report to California Air Resources Board, Contract No. A032-180, pp. 7–1 through 7–31, Sacramento, CA.

# 3 Integrating Chemical, Water Quality, Habitat, and Fish Assemblage Data from the San Joaquin River Drainage, California

*Larry R. Brown, Charles R. Kratzer, and Neil M. Dubrovsky*

## CONTENTS

Abstract ............................................................................................................... 26
Introduction ........................................................................................................ 26
Methods .............................................................................................................. 28
    Study Design ............................................................................................... 28
    Sample Collection ....................................................................................... 32
    Analytical Methods ..................................................................................... 33
    Data Analysis .............................................................................................. 33
Results ................................................................................................................ 35
    Pesticide Studies ......................................................................................... 35
    Diazinon Studies ......................................................................................... 37
    Fish Assemblages ........................................................................................ 41
    Data Integration .......................................................................................... 45
Discussion .......................................................................................................... 50
    Pesticide Studies ......................................................................................... 50
    Diazinon Studies ......................................................................................... 52
    Species Assemblages .................................................................................. 53
    Data Integration .......................................................................................... 55
Conclusion ......................................................................................................... 58
References .......................................................................................................... 59

## ABSTRACT

As part of the National Water Quality Assessment of the U.S. Geological Survey, studies of concentrations of dissolved pesticides and other measures of water quality, stream habitat, and fish assemblages were conducted in the San Joaquin River drainage, California, from 1992 to 1994. Forty-nine pesticide compounds were detected in one or more water samples collected for analysis. The ten most frequently occurring compounds were simazine, diazinon, metolachlor, chlorpyrifos, DCPA, EPTC, trifluralin, atrazine, diuron, and cyanazine. Detailed studies at four sites revealed significant seasonal variability in the number of pesticides detected in individual site samples and in the summed concentrations of all pesticides detected in a site sample. Concentrations of diazinon were related with application patterns to different crop types and to timing of rainfall. Significant correlations were found between water quality, habitat, and fish assemblage data. The correlations, however, were dependent on the number and locations of the sites included in the analysis. There was no clear link between pesticide concentrations and fish assemblages. Our results highlighted differences in the temporal and spatial scales of sampling between the water quality and ecological studies. The water quality and pesticide studies emphasized sampling over short periods of time at few sites, but the ecological studies emphasized single samples over a large geographic area because of lack of information on the range of possible ecological conditions in the study area.

**Keywords:** *water quality, pesticides, habitat, fish assemblages, biomonitoring*

## INTRODUCTION

The San Joaquin and Tulare basins of California (Figure 1) encompass an area of about 7.4 million hectares and include about 4 million hectares of irrigated agricultural land in the San Joaquin Valley floor.[1] This intensive agricultural activity, accompanied by increasing urbanization, has affected water quality and aquatic habitats in the region through several mechanisms. Intensive use of pesticides has resulted in the presence of dissolved pesticides in surface water, primarily through rainfall runoff and irrigation return flows in agricultural areas. Similarly, large-scale application of fertilizers have contributed to nutrient loads in many streams. Agricultural return flow has been associated with other changes in water quality, including increased concentrations of dissolved solids (salinity) and trace elements.[2-4] Much of the original wetland and riparian habitat has been lost because of clearing and flood control for urban and agricultural activities.[1,3] Also, the damming and diversion of streams to provide water supply and flood control for municipal and agricultural purposes have resulted in changes to the basin hydrology.[1,5] These changes in water quality and habitat have been accompanied by changes in the fish fauna, including declines or extirpations of native species and the introduction of new species.[1,6,7-10] Introduced species appear to be better adapted for the altered habitat conditions on the valley floor. They also may affect native species through competition and predation. Despite the quantity of work done on individual topics, there have been no interdisciplinary studies to provide an integrate assessment and understanding of the interactions among water quality, habitat quality, and fish assemblages.

Integrating Chemical, Water Quality, Habitat, and Fish Assemblage Data

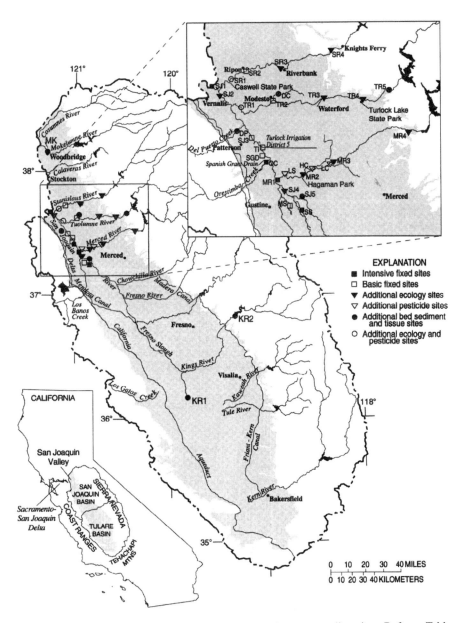

**FIGURE 1** San Joaquin-Tulare basins study unit study area sampling sites. Refer to Table 1 for site names and the activities conducted at each site.

The following work was done as a part of the National Water Quality Assessment (NAWQA) Program of the U.S. Geological Survey (USGS). The purpose of the NAWQA Program is to describe current water quality conditions, define long-term trends in water quality, and elucidate the natural and human-induced processes that affect water quality in the U.S. The methodology of the NAWQA Program is to integrate data collected at the local level of the study area into regional and national analyses. Benthic algae, benthic macroinvertebrates, and fishes are utilized as biomonitors of environmental quality. Studies in the San Joaquin-Tulare basins study area of the NAWQA Program have focused on the lower San Joaquin River and its tributaries, particularly three east-side tributaries: the Stanislaus, Tuolumne, and Merced rivers (Figure 1). The surface water design included collection of physical, chemical, and biological data at varying temporal and spatial scales, ranging from hourly measurements of dissolved pesticide concentrations during a rainstorm to synoptic studies where one-time biological collections were made at sites over a broad geographic area.

The overall purpose of this chapter is to summarize some of the data and analyses conducted to date by the San Joaquin-Tulare basins NAWQA. Within that overall purpose, several specific objectives are addressed. First, data on the number and concentration of dissolved pesticides present in the streams of the study area are summarized. Second, additional studies that focused on variability in concentrations of diazinon are summarized. Third, the results of studies on the structure of fish assemblages at two geographic scales are summarized. Fourth, pesticide, water quality, habitat quality, and fish assemblage data were integrated. In particular, correlations between fish assemblages and the other groups of variables were sought. Some of the challenges that develop when attempting to integrate studies conducted at different temporal and spatial scales also are summarized.

## METHODS

### STUDY DESIGN

The basic NAWQA study design is a nested hierarchy of sites that includes different groups of sites studied with different intensities with regard to the number of variables studied and frequency of sampling.[11] The intent of this design is to maximize information transfer between the disciplines by using the same core group of sites (Table 1). The intensive fixed sites (i.e., geographically fixed, long-term sites) are sampled for dissolved pesticides and related compounds, such as pesticide degradation products (hereafter included with pesticides), nutrients, major ions, hydrophobic organic contaminants and trace elements in bottom sediment, hydrophobic organic contaminants and trace elements in tissues of biota, composition of the benthic algae assemblage, composition of the benthic invertebrate assemblage, composition of the fish assemblage, and habitat characteristics. The San Joaquin-Tulare basins study area includes four intensive fixed sites (Figure 1 and site codes SJ1, MR1, OC, and SS in Table 1). Water samples for analysis of 90 dissolved pesticides (Table 2) were collected at intervals ranging from hourly to monthly (1993). Samples for nutrients and major ions were collected monthly for 2 years (1993 to 1994). Samples for other variables were taken only once. Field measurements of water temperature, specific conductance, pH, alkalinity, and dissolved oxygen were taken during all sampling activities, except during the special diazinon study discussed later. Discharge was monitored continuously.

## TABLE 1
## Site Name, Site Code Used in Figures, and Type of Site for All Sites Sampled During the Study

| Site Name | Site code | Site type[a] |
|---|---|---|
| Del Puerto Creek at Vineyard Road | DP | BST |
| Dry Creek in Modesto | DC | BST |
| Highline Canal | HC | PES |
| Kings River at Empire Weir 2 | KR1 | BST |
| Kings River near Pine Flat Dam | KR2 | BST |
| Livingston Canal | LC | PES |
| Lower Stevinson Lateral | LS | PES |
| Merced River at River Road | MR1 | IFS (BFS), ECO, BST, PES |
| Merced River at Hagamann County Park | MR2 | ECO |
| Merced River at McConnell State Park | MR3 | ECO |
| Merced River near Snelling Diversion Dam | MR4 | ECO |
| Mokelumne River near Woodbridge | MK | BST |
| Mud Slough near Gustine | MS | BFS, ECO, BST |
| Orestimba Creek at River Road | OC | IFS (BFS), ECO, BST |
| Salt Slough at Lander Avenue | SS | IFS (BFS), ECO, BST |
| San Joaquin River near Vernalis | SJ1 | IFS (BFS), ECO, BST, PES |
| San Joaquin River at Maze Road | SJ2 | ECO |
| San Joaquin River near Patterson | SJ3 | BFS, ECO, BST |
| San Joaquin River at Fremont Ford | SJ4 | ECO |
| San Joaquin River at Lander Avenue | SJ5 | BST |
| Spanish Grant Drain | SGD | BFS, ECO, BST |
| Stanislaus River at Caswell State Park | SR1 | ECO, PES |
| Stanislaus River near Ripon | SR2 | BFS, ECO, BST, PES |
| Stanislaus River near Riverbank | SR3 | ECO |
| Stanislaus River near Knights Ferry | SR4 | ECO |
| Tuolumne River at Shiloh Road | TR1 | ECO, PES |
| Tuolumne River at Modesto | TR2 | BFS, ECO, BST, PES |
| Tuolumne River near Waterford | TR3 | ECO |
| Tuolumne River at Turlock State Recreation Area | TR4 | ECO |
| Tuolumne River at Old La Grange Bridge | TR5 | BST |
| Turlock Irrigation District Lateral 5 | TI | BFS,[b]BST |

[a] IFS (BFS) = intensive fixed site (all the same parameters measured at BFS sites are monitored with the addition of pesticides), BFS = basic fixed site, ECO = ecological site, BST = bed sediment and tissue site, PES = diazinon synoptic site.

[b] Water quality data from this site were not included in analyses because ecological data were not collected.

The basic fixed sites[11] were sampled for the same suite of variables, except dissolved pesticides. The San Joaquin-Tulare basins study area includes six basic fixed sites and the four intensive fixed sites, for a total of ten basic fixed sites (Figure 1 and Table 1). Water quality data from the Turlock Irrigation Lateral 5

## TABLE 2
## List of Analytes Included in Pesticide Analyses at the Intensive Fixed Sites

| Compound | MDL | Compound | MDL |
|---|---|---|---|
| **Amide herbicides** | | **Organophosphate insecticides** | |
| Alachlor | 0.002[a] | Azinphos-methyl | 0.001[a] |
| Diethylaniline, 2,6- | 0.003[a] | Chlorpyrifos | 0.004[a] |
| Metolachlor | 0.002[a] | Diazinon | 0.002[a] |
| Napropamide | 0.003[a] | Disulfoton | 0.017 |
| Pronamide | 0.003[a] | Ethoprop | 0.003[a] |
| Propachlor | 0.007[a] | Fonofos | 0.003[a] |
| Propanil | 0.004[a] | Malathion | 0.005[a] |
| | | Parathion | 0.004 |
| **Carbamate herbicides** | | Parathion, methyl- | 0.006 |
| Butylate | 0.002[a] | Phorate | 0.002 |
| EPTC | 0.002[a] | Terbufos | 0.013 |
| Molinate | 0.004[a] | **Pyrethroid insecticides** | |
| Pebulate | 0.004[a] | Esfenvalerate | 0.019 |
| Propham | 0.035 | Permethrin, cis- | 0.005[a] |
| Thiobencarb | 0.002[a] | | |
| Triallate | 0.001[a] | **Triazine herbicides** | |
| | | Atrazine | 0.001[a] |
| **Carbamate insecticides** | | Atrazine, desethyl- | 0.002[a] |
| Aldicarb | 0.016[a] | Cyanazine | 0.004[a] |
| Aldicarb sulfone | 0.016 | Metribuzin | 0.004[a] |
| Aldicarb sulfoxide | 0.021 | Prometon | 0.018[a] |
| Carbaryl | 0.003[a] | Simazine | 0.005[a] |
| Carbofuran | 0.003[a] | | |
| Carbofuran-3HDXY | 0.014 | **Uracil herbicides** | |
| Methiocarb | 0.026 | Bromacil | 0.035 |
| Methomyl | 0.017[a] | Terbacil | 0.007[a] |
| Oxamyl | 0.018 | | |
| Propoxur | 0.035 | **Urea herbicides** | |
| | | Diuron | 0.020[a] |
| **Chlorophenoxy herbicides** | | Fenuron | 0.013 |
| 2,4,5-T | 0.035 | Fluormeturon | 0.035 |
| 2,4-D | 0.035[a] | Linuron | 0.002[a] |
| 2,4-DB | 0.035 | Neburon | 0.015 |
| Dichlorprop | 0.032[a] | Tebuthiuron | 0.010[a] |
| MCPA | 0.050[a] | | |
| MCPB | 0.035 | **Miscellaneous herbicides** | |
| Silvex | 0.021 | Aciflourfen | 0.035 |
| Triclopyr | 0.050[a] | Bentazon | 0.014 |
| | | Bromoxynil | 0.035 |
| **Dinitroaniline herbicides** | | Chloramben | 0.011 |
| Benfluralin | 0.002[a] | Clopyralid | 0.050 |
| Ethafluralin | 0.004[a] | Dicamba | 0.035 |
| Oryzalin | 0.019 | Dinoseb | 0.035 |

## TABLE 2 (CONTINUED)
### List of Analytes Included in Pesticide Analyses at the Intensive Fixed Sites

| Compound | MDL | Compound | MDL |
|---|---|---|---|
| **Dinitroaniline herbicides (cont.)** | | **Miscellaneous herbicides (cont.)** | |
| Pendimethalin | 0.004[a] | Norflurazon | 0.024[a] |
| Trifluralin | 0.002[a] | Picloram | 0.050 |
| **Organochlorine fungicides** | | **Miscellaneous insecticides** | |
| Chlorothalonil | 0.035 | 1-Naphthol | 0.007 |
| | | DNOC | 0.035 |
| **Organochlorine herbicides** | | Propargite | 0.013[a] |
| Dichlobenil | 0.020 | | |
| **Organochlorine insecticides** | | **Fumigants and associated compounds** | |
| Dachthal | 0.002[a] | Bromomethane | 0.200 |
| Dachthal-mono-acid | 0.017 | Dibromo-3-chloropropane, 1,2- | 0.030 |
| DDE, p,p'- (DDT) | 0.006[a] | Dibromoethane, 1,2- | 0.040 |
| Dieldrin (Aldrin) | 0.001[a] | Dichloropropene, $cis$-1,3- | 0.200 |
| HCH, $alpha$ | 0.002[a] | Dichloropropene, $trans$-1,3- | 0.200 |
| HCH, $gamma$ | 0.004[a] | Trichloropropane, 1,2,3- | 0.200 |

*Note:* Degradation products are indented from the parent compound or, when the parent compound is not an analyte, the parent compound is given in parentheses. Method detection limits (MDL) are given in micrograms per liter (pg/L).

[a] Analytes detected during the study.

---

basic fixed site (Site TI in Figure 1) are not included in this chapter because no fish data were collected at the site, so only nine of the basic fixed sites are included in the analyses.

Another component of the NAWQA study design is a bed sediment and tissue survey to assess the occurrence of inorganic and hydrophobic organic contaminants. In the San Joaquin-Tulare basins study area, this study included all the basic fixed sites and an additional seven sites (Figure 1 and Table 1). These sites were sampled in October and November 1992. For the purposes of this chapter, these variables are simply considered as two additional measures of water quality.

Synoptic studies are special studies designed to meet local, regional, or national objectives, and may or may not include the intensive or basic fixed sites. This chapter includes data from two such studies. In 1994, a study was conducted of diazinon transport during storms from the Stanislaus, Tuolumne, and Merced rivers to the San Joaquin River. The purpose of the study was to describe variability in diazinon concentrations and determine the significance of east-side valley sources to total diazinon transport in the San Joaquin River Basin. The diazinon study also utilized diazinon data collected at the San Joaquin River near Vernalis by the Toxic Contaminant Hydrology (TCH) Program of the USGS. During a January storm, four sites were monitored for 54 h. During a February storm, nine sites were monitored for 72 h. The sites included three of the basic fixed sites and one of the intensive fixed sites (Table 1).

An ecological study of benthic algae assemblages, benthic macroinvertebrate assemblages, fish assemblages, and habitat was conducted at 20 sites (4 sites in 1993 and 16 sites in 1994). The study included 9 of the 10 basic fixed sites (Table 1); all four of the intensive fixed sites were included. The purpose of the study was to determine the range of ecological conditions that exist in the lower San Joaquin River drainage and the correlation between water quality variables, habitat variables, algae assemblages, invertebrate assemblages, and fish assemblages. Only the water quality, habitat, and fish assemblage data are discussed in this paper because identifications of algae and invertebrate taxa have not been completed. Four sites were located on each of the Stanislaus, Tuolumne, and Merced rivers (Figure 1), with the most upstream site near the first dam impassible to fish and the most downstream site near the confluence with the San Joaquin River. Four sites were located on the mainstem San Joaquin River, both upstream and downstream of the confluences of the previously mentioned large tributaries. The remaining four sites were located on smaller tributaries entering the San Joaquin River from the south and west.

## Sample Collection

Water samples for pesticide, nutrient, and major ion analyses at the basic and intensive fixed sites were collected as depth- and width-integrated samples as specified in NAWQA guidelines.[12] Water samples collected during the diazinon study were a mixture of depth- and width-integrated, depth-integrated, and grab samples. Water samples collected during the ecological synoptic study for nutrient analysis were grab samples, as were samples collected for field measurements of specific conductance, pH, and alkalinity. Water temperature and dissolved oxygen measurements were taken in the river. Stream discharge was determined from USGS gage records, California Department of Water Resources gage records (San Joaquin River near Patterson [SJ3] only), or measured at ungaged sites.

Standard NAWQA protocols were followed for collection of tissue[13] and sediment[14] during the bed-sediment and tissue contaminant surveys. The fish assemblage and habitat characteristics were sampled using standard USGS methods,[15,16] except that surveys in reaches sampled by backpack electroshocker consisted of one pass through the reach instead of the recommended two. Fish were collected (or observed directly when snorkeling) using a combination of electroshocker (boat or backpack), seine (3.0, 9.1, or 15.2 m with 6-mm mesh), or snorkeling, as appropriate for each site. Captured fish were identified in the field, counted, and at least the first 30 individuals of each species were weighed and measured. The fish observed while snorkeling were identified, counted, and their lengths estimated. Habitat variables were measured at each of six transects within each sampling reach. Fish sampling occurred in August or September of each year. Habitat data and nutrient samples were collected within a month of fish sampling, except at four of the basic fixed sites (SGD, OC, SS, and MS in Table 1). At these sites, data collected as part of the monthly water quality monitoring program were utilized. Data from the date closest in time to the date of fish sampling were used.

## ANALYTICAL METHODS

Most chemical analyses were conducted at the USGS National Water Quality Laboratory (NWQL) in Arvada, CO. Some samples from the diazinon study were analyzed at the USGS District Office in Sacramento, CA. Pesticide samples were processed using the methods of Zaugg et al.[17] and Werner et al.,[18] except for some samples collected during the diazinon study that were processed using the methods of Crepeau et al.[19] All samples were kept cold and then filtered through a baked 0.7-μm glass-fiber filter to remove particulate matter. The pesticides were extracted by solid-phase extraction cartridges, and the adsorbed chemicals were eluted for analysis by capillary-column gas chromatography and mass spectrometry. Nutrient and major ion samples were packaged on ice and sent to the laboratory for analysis. Samples were analyzed using USGS analytical methods.[20-22] Field measurements of specific conductance, pH, water temperature, and dissolved oxygen were made with electronic meters. Alkalinity was determined by titration.

Fish collected for tissue samples were frozen whole in the field; invertebrates were depurated for 24 h prior to freezing. Samples then were sent to the NWQL and analyzed using the methods of Leiker et al.[23] In summary, samples were homogenized, and a subsample was taken for lipid analysis. Hydrophobic organic chemicals were isolated through solvent exchange and gel-permeation chromatography. Chemicals in the extract were analyzed by dual capillary-column gas chromatography and mass spectrometry with electron-capture detection. Sediment samples were frozen in the field, sent to the NWQL, and analyzed using the methods of Foreman et al.[24] Sediment analytical procedures were similar to tissue procedures, except that subsamples were taken to determine moisture content, organic carbon content, and inorganic carbon content.

## DATA ANALYSIS

Pesticide data from each of the intensive fixed sites were summarized as the number of chemicals detected per sampling date and as the summed concentration of all dissolved pesticides detected per sampling date. If a pesticide was not detected, the concentration was assumed to be zero for this analysis. On days when more than one sample was collected, the single sample with the highest total concentration was selected. The overall frequency of occurrence of the ten most frequently detected pesticides was based on all samples collected. Diazinon concentration data from the intensive fixed site studies and diazinon study (including both NAWQA and TCH data) were graphically analyzed in relation to stream discharge and date. Associations between diazinon concentrations and application rates were assessed on the basis of application data collected by the California Department of Pesticide Regulation.[25]

The data from the bed sediment and tissue studies presented in this chapter are the concentration of total-DDT chemicals (sum of DDD, DDE, and DDT) in tissue (μg/kg wet weight) and in sediment, standardized by total organic carbon content (μg/kg of total organic carbon). The association between total-DDT in sediment and in tissue was explored with regression analysis.

The water quality, habitat, and fish data were used in separate analyses of all 20 ecological study sites (9 basic fixed sites and 11 additional sites) and of the 9 basic fixed sites only. Analysis of the data from all 20 sites proceeded as follows. Fish assemblages were analyzed with TWINSPAN (two-way indicator species analysis)[26] and DCA (detrended correspondence analysis).[27-29] Fish data were analyzed as percentages of the total number of fish captured at each site. Numbers of western mosquitofish (*Gambusia affinis*) and lampreys (*Lampetra* sp.) were not included because they were poorly sampled. To reduce the influence of rare species, only species found at three or more sites, and making up at least 5% of the individuals captured at one site, were included in the analyses. TWINSPAN is a divisive classification technique that produces an ordered data matrix of sites and species. The resulting matrix groups sites with similar percentage abundances of species and groups species with similar percentage abundances at different sites. The results were used to define major ecological groups (species assemblages) and major site groupings. The species within any particular TWINSPAN species group can often be found at sites within several TWINSPAN site groups. DCA is a multivariate technique derived from reciprocal averaging that maximizes the correlation between species scores and sample scores along an assumed gradient.[30] Thus, sample scores are constrained by species scores, and species scores are constrained by sample scores in an iterative process. In this study, DCA ordinates species on the basis of their individual percentage abundance at each of the sites and ordinates sites on the basis of percentages of the species at individual sites. By viewing the results of the species and site ordinations simultaneously, sites generally can be interpreted to be characterized by one or more species; however, the results do not necessarily mean that any species is restricted to a particular range of sites.

Water quality data collected during both fish sampling and habitat sampling were analyzed as maximum values of temperature, specific conductance, pH (low pH acid rain is not a problem), and alkalinity. Minimum values of discharge and dissolved oxygen were used. These extreme values would be most stressful to fish and would most likely affect their survival and distribution. The initial long list of water quality and habitat variables was reduced by deleting variables strongly correlated with other more general variables, variables with little variability, or variables with many nondetections. Using the Spearman rank correlation, the remaining variables were correlated with site scores on the first two DCA axes. Variables with significant correlations were classified as habitat variables or as water quality variables. Each variable was examined for normality using a normal probability plot and, when appropriate, $\log_{10}(x + 1)$ transformed.

A reduced set of composite variables were derived from each data set using principal components analysis (PCA). Variables for PCA were standardized to a mean of 0 and a standard deviation of 1. Only principal components (PC) with an eigenvalue greater than 1 were interpreted. Associations between DCA axis site scores and physical variable PC scores were explored using Spearman rank correlations and regression analysis. Associations between habitat and water quality variables were examined with Spearman rank correlations among PCs and canonical correlation analysis of the original variables.

The analytical scheme for the basic fixed sites was basically the same as that described earlier with the following exceptions. For water quality variables, median values from late March through September (n = 7) were used to characterize each site. This period was chosen because, with a few exceptions, the adults of the species captured in the study are tolerant of moderately degraded water quality conditions. Therefore, we targeted the time period when the most sensitive life stages (eggs, larvae, and juveniles) of most of the species are present. Also, these variables were highly variable during the study, but the highest values, presumably most stressful for fish, occurred during this period. Total-DDT in bed sediment and tissue were included as chemical variables. Because of the small number of sites, all species making up at least 5% of the fish at one site were included in the analyses. The results obtained, using all 20 sites and the 9 basic fixed sites, were compared to determine if the data sets differed in identifying correlations between water quality, habitat quality, and fish assemblages.

## RESULTS

### PESTICIDE STUDIES

The intensive fixed-site data (Figure 2) indicate significant variability on the scale of days to weeks for the total concentration of pesticides. Total dissolved pesticide concentrations generally were less than 1 µg/L at all sites from January through April, with the exception of short-term increases to over 4 µg/L in February at the Merced River and Orestimba Creek sites. From the end of April through September, concentrations were less than 1 µg/L at the Merced River and San Joaquin River sites, except for July at the Merced River site and September at the San Joaquin River site. Concentrations were more variable at the Orestimba Creek and Salt Slough sites, with many values greater than 1 µg/L. The extremely high concentration (21.266 µg/L) detected at Orestimba Creek in August was attributed to elevated levels of propargite (20 µg/L). Concentrations were less than 1 µg/L through the end of the year at all sites, except Salt Slough. The Merced River site generally had the lowest concentrations of pesticides. The San Joaquin River site generally had lower concentrations than the Salt Slough or Orestimba Creek sites.

The number of analytes detected in each sample also varied among the sites and time periods (Figure 2). The Merced River generally had the fewest compounds detected — a maximum of ten and a median of four. Orestimba Creek showed a consistently higher number of analytes detected from May through September, with most samples containing 15 or more analytes. However, the median number of chemicals detected at Orestimba Creek (median = 8) was similar to the medians for the Salt Slough and the San Joaquin River (median = 10 and 8, respectively). The number of analytes detected at all sites appeared to be most variable from January through April, generally with a higher number of compounds present from May through September. The number of analytes then declined and stayed relatively low for the rest of the year, except at the Salt Slough site.

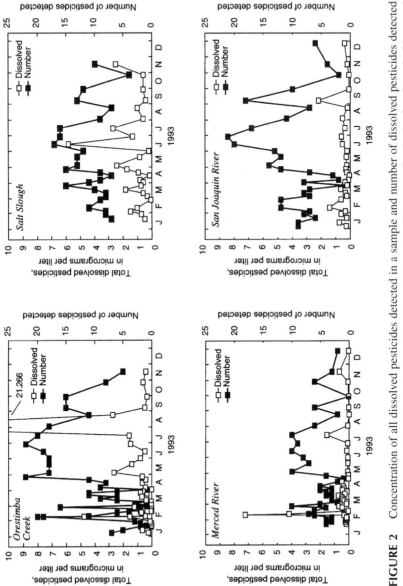

**FIGURE 2** Concentration of all dissolved pesticides detected in a sample and number of dissolved pesticides detected in a sample in 1993 for the intensive fixed sites at the Merced River, the Orestimba Creek, the San Joaquin River, and the Salt Slough.

Forty-nine pesticides were detected in one or more water samples (Table 2). The ten pesticides with the highest frequency of occurrence included eight herbicides and two insecticides. Of the herbicides, simazine was detected in 133 of 142 samples (93.7%), metolachlor in 101 of 142 samples (71%), DCPA in 91 of 142 samples (64%), EPTC in 77 of 142 samples (54%), trifluralin in 63 of 142 samples (44%), atrazine in 57 of 142 samples (40%), diuron in 28 of 76 samples (36.8%), and cyanazine in 49 of 142 samples (34.5%). Diazinon, the most frequently detected insecticide, was followed by chlorpyrifos. These insecticides were detected in 108 (76%) and 91 (64%) of 142 samples, respectively.

## Diazinon Studies

There was substantial variability in concentrations of diazinon among different rivers and temporal variation within those rivers. A detailed analysis of the diazinon data from the intensive fixed sites indicates that concentrations were related to discharge and application patterns. In the Orestimba Creek drainage basin, peak application occurred during August 1993,[25] but peak concentrations occurred during January and February 1993, which coincided with a less intensive application in January. The August application to a wide variety of crops,[25] occurred during low stream discharge. Apparently, conditions were not favorable for transport of diazinon to surface water because concentrations never exceeded 0.6 µg/L. The January application was largely done on almond trees when rainstorms were still occurring and when the probability of transport to surface water was high, resulting in concentrations of over 3.5 µg/L. In the Merced River drainage, the application was mostly done on almond trees during January and February 1993, and only small amounts were applied to other crops throughout the remainder of the year.[25] Consequently, most detections and highest concentrations were associated with winter storm runoff in February. The Salt Slough drainage showed a different pattern, with maximum applications in March, August, and September to crops, excluding orchards.[25] Concentrations were variable but low, never exceeding 0.3 µg/L. The site at San Joaquin River near Vernalis integrates all upstream uses on the tributaries. Application rates were highest in January and February 1993, and application in those months was primarily done on almond orchards. The highest concentrations (from 0.1 to over 0.6 µg/L) occurred in January and February 1993, indicating that orchard applications, combined with rainstorms, were the most important factors in determining dissolved diazinon concentrations in the San Joaquin River.

The diazinon synoptic study documented spatial variability in diazinon concentrations of rivers and temporal variation within the large east-side tributary rivers, both between storms and during single storms (Figure 3). The Tuolumne River had the highest concentrations in both storms, followed by the Merced and Stanislaus rivers. Concentrations tended to be higher during the January storm, but stream discharge was lower. Concentrations were highly variable over the course of a storm. Diazinon concentrations at the San Joaquin River near the Vernalis site reflected the source of storm runoff and transport time. The highest concentrations during both storms closely coincided with the arrival of runoff from the Tuolumne River. Load calculations indicated that the January storm transported more than twice as much

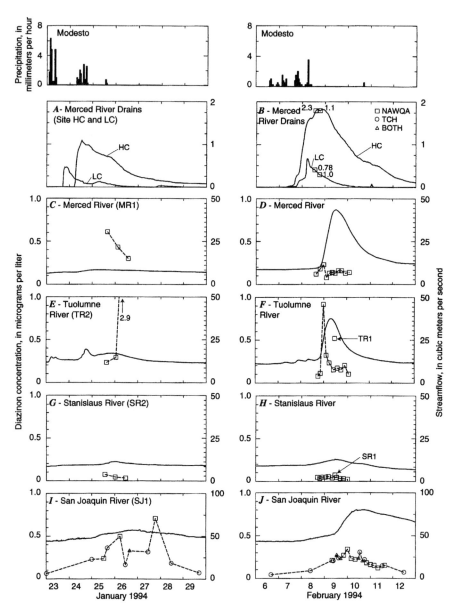

**FIGURE 3** Precipitation in Modesto, stream discharge (solid lines), and concentrations of dissolved diazinon (boxes and dotted lines) at sampling sites in the Merced River drains (A, B), Merced River (C, D), Tuolumne River (E, F), Stanislaus River (G, H), and San Joaquin River (I, J) drainages during storms in January and February of 1994. See Table 1 for site codes. Data are shown for samples collected by NAWQA, the Toxic Contaminant Hydrology (TCH) Program, and one joint sample.

Integrating Chemical, Water Quality, Habitat, and Fish Assemblage Data 39

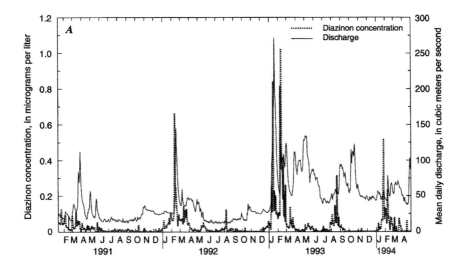

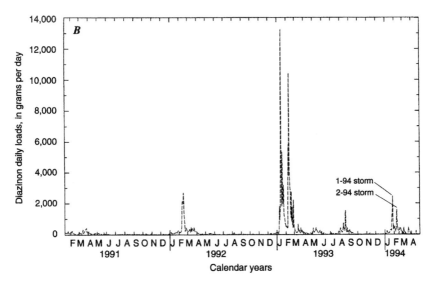

**FIGURE 4** (A) Daily dissolved diazinon concentrations and (B) daily loads of dissolved diazinon at the San Joaquin River near Vernalis from 1991 to 1994.

diazinon as did the February storm, despite the lower runoff in January (Figure 4). The total transport of diazinon by the January and February storms was estimated to be only about 0.05% of the total amount applied during that period. Load estimates indicate that about 74% of the annual diazinon load in the San Joaquin River occurred in January and February during the years 1991 to 1993 (Figure 4).

## TABLE 3
## Common and Scientific Names of Species Collected

| Species Collected | Origin | Species Code | Percent of ECO Sites | Percent of BFS Sites |
|---|---|---|---|---|
| **Petromyzontidae (Lampreys)** | | | | |
| Unknown lampreys, *Lampetra* spp. | N | —[a] | 15 | 11 |
| **Clupeidae (Shad and Herring)** | | | | |
| Threadfin shad, *Dorosoma petenense* | I | TFS | 30 | 44 |
| **Salmonidae (Salmon and Trout)** | | | | |
| Rainbow trout, *Oncorhynchus mykiss* | N | — | 5 | 0 |
| **Cyprinidae (Minnows)** | | | | |
| Common carp, *Cyprinus carpio* | I | CP | 90 | 100 |
| Fathead minnow, *Pimephales promelas* | I | FHM | 40 | 67 |
| Goldfish, *Carassius auratus* | I | GF | 50 | 67 |
| Hardhead, *Mylopharodon conocephalus* | N | HH | 25 | 0 |
| Hitch, *Lavinia exilicauda* | N | — | 15 | 0 |
| Red shiner, *Cyprinella lutrensis* | I | RSH | 45 | 78 |
| Sacramento blackfish, *Orthodon microlepidotus* | N | — | 5 | 11 |
| Sacramento squawfish, *Ptychocheilus grandis* | N | SQ | 25 | 0 |
| **Catostomidae (Suckers)** | | | | |
| Sacramento sucker, *Catostomus occidentalis* | N | SKR | 45 | 11 |
| **Ictaluridae (Catfish)** | | | | |
| Black bullhead, *Ameiurus melas* | I | BLBH | 40 | 44 |
| Brown bullhead, *A. nebulosus* | I | — | 15 | 0 |
| Channel catfish, *Ictalurus punctatus* | I | CCF | 55 | 56 |
| White catfish, *A. catus* | I | WCF | 75 | 89 |
| **Poeciliidae (Livebearers)** | | | | |
| Western mosquitofish, *Gambusia affinis* | I | — | 75 | 89 |
| **Atherinidae (Silversides)** | | | | |
| Inland silverside, *Menidia beryllina* | I | ISS | 30 | 44 |
| **Percichthyidae (Temperate Basses)** | | | | |
| Striped bass, *Morone saxatilis* | I | — | 20 | 33 |
| **Centrarchidae (Sunfish)** | | | | |
| Black crappie, *Pomoxis nigromaculatus* | I | — | 15 | 22 |
| Bluegill, *Lepomis macrochirus* | I | BG | 90 | 89 |
| Green sunfish, *L. cyanellus* | I | GSF | 90 | 89 |
| Largemouth bass, *Micropterus salmoides* | I | LMB | 75 | 67 |
| Redear sunfish, *L. microlophus* | I | RSF | 60 | 56 |
| Smallmouth bass, *M. dolomieu* | I | SMB | 60 | 33 |
| White crappie, *P. annularis* | I | — | 10 | 22 |

## TABLE 3 (CONTINUED)
## Common and Scientific Names of Species Collected

| Species Collected | Origin | Species Code | Percent of ECO Sites | Percent of BFS Sites |
|---|---|---|---|---|
| **Percidae (Perch)** | | | | |
| Bigscale logperch, *Percina macrolepida* | I | — | 5 | 0 |
| **Embiotocidae (Surf Perch)** | | | | |
| Tule perch, *Hysterocarpus traski* | N | TP | 25 | 11 |
| **Cottidae (Sculpin)** | | | | |
| Prickly sculpin, *Cottus asper* | N | PSCP | 35 | 11 |

*Note:* Shown are origin (N = native to California and I = introduced to California), species code used in figures, percentage of the 20 ecological sites (ECO) where each species was found, and percentage of the 9 basic fixed sites (BFS) where each species was found. Species are arranged by family with the common family name in parentheses.

[a] — indicates species not included in statistical analyses of all 20 ecological sites because of rarity (captured at 3 or fewer site and never more than 5% of the fish caught at any site) or because they were not sampled consistently with the methods used.

### FISH ASSEMBLAGES

A total of 29 species of fish were collected during the sampling of all 20 sites: 9 species are native to California and 20 species have been introduced (Table 3); 23 species were present at the 9 basic fixed sites, of which 5 are native and 18 introduced (Table 3). There were no strong patterns in number of species captured at a site either within or among streams. The mainstem San Joaquin River sites tended to have the most species (12 to 18), but this range overlapped with that of the Tuolumne River (10 to 15) and the small southern and western tributaries (8 to 13). The Stanislaus River had somewhat fewer species (8 to 11). This range was similar to that of the three most downstream sites on the Merced River (10 to 11). The site located farthest upstream on the Merced River had only five species present, and all of them were native species.

On the basis of data from all 20 of the ecological sites, TWINSPAN defined six groups of species (Figure 5A) and six groups of sites (Figure 5B). Two of the species groups consisted of single species and one of the site groups consisted of a single site. DCA axis 1 summarizes a range of conditions from San Joaquin River mainstem sites with high percentages of introduced species, particularly red shiner, fathead minnow, inland silverside, and threadfin shad, to upstream tributary sites with high percentages of native species.

The San Joaquin River mainstem site group (Figure 5B), including the small tributary streams and drains, was characterized by high percentages of red shiner, fathead minnow, inland silverside, and threadfin shad (San Joaquin mainstem species

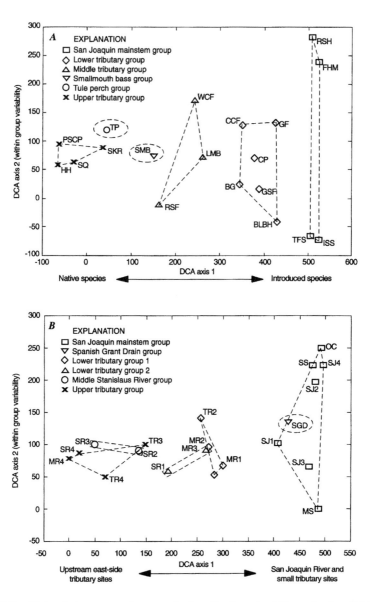

**FIGURE 5** Detrended correspondence analysis ordinations for (A) species (see Table 3 for species codes) and (B) sampling sites (see Table 1 for site codes) using data from all 20 ecological sites. Similar symbols (enclosed by dotted lines) indicate groups of similar species or sites as determined by TWINSPAN (two-way indicator species analysis). Analysis was based on percentage abundances of species at a site.

group, Figure 5A). Native species were largely absent from the mainstem San Joaquin River. All four of the characteristic species were found at the sites on the mainstem San Joaquin River, Salt Slough and Mud Slough. The only species of the San Joaquin mainstem group captured at sites on the Stanislaus, Tuolumne, or

Merced rivers was the red shiner, which was present at the Merced River site closest to the San Joaquin River. The species in this group were never found in isolation, but were always found in association with species within two other species groups that were characteristic of the lower (lower tributary species group) and middle (middle tributary site group) reaches of the large east-side tributaries. The Spanish Grant Drain site was separated from the other mainstem group sites, apparently because of higher percentages of species associated with the lower tributary species group.

An upper tributary site group was separated from the other sites on the basis of high percentages of the native species: Sacramento squawfish, hardhead, Sacramento sucker, and prickly sculpin (upper tributary group, Figure 5A). The two most upstream sites on both the Stanislaus and Tuolumne rivers and the most upstream site on the Merced River were unique because of the presence of all four of the native species. Though high percentages of native species were characteristic of this site group, one or more introduced species also were present, except on the Merced River near the Snelling Diversion Dam, where only native species were present. An associated group of middle Stanislaus River sites (Figure 5B) was distinguished by the presence of tule perch, a native species which was abundant only in the Stanislaus River and was found only at one other site in the drainage (SJ2 in Table 1). Similarly, smallmouth bass was found at high percentage abundances in the Stanislaus River; however, smallmouth bass was found at a wider range of sites than tule perch.

The remaining sites were organized into two groups of lower and middle tributary sites that were not characterized by a particular group of species (lower tributary groups 1 and 2, Figure 5B). Species with high percentage abundances at these sites formed two groups. There was a large group of introduced species that included channel catfish, goldfish, common carp, green sunfish, bluegill, and black bullhead (lower tributary group, Figure 5A). There was also a smaller group of introduced species that included largemouth bass, redear sunfish, and white catfish (middle tributary group, Figure 5A). The species in the larger group generally had higher percentage abundances at sites in the San Joaquin River mainstem group. The species in the smaller group generally had higher percentage abundances at the middle and lower tributary sites.

Variation within the major species and site groupings was summarized by DCA axis 2 (Figure 5A and B). Variation was particularly high in the mainstem San Joaquin River site group because there was a great deal of variability at specific sites in percentages of the four characteristic species.

Using data only from the basic fixed sites, TWINSPAN defined six groups of species (Figure 6A) and four groups of sites (Figure 6B). Two species groups were composed of single species, and one group was composed of two species. Two of the site groups consisted of single sites. In contrast to the analysis of all 20 sites, DCA axis 1 (Figure 6A and B) did not spread the sites out smoothly, but instead grouped most of the sites in the middle with two sites with unique characteristics on either side. The Mud Slough site was separated from the other sites primarily because of its high percentages of threadfin shad and black bullhead. The Stanislaus River was separated from the other sites primarily because it was the only site with tule perch and Sacramento sucker and, to a lesser extent, because of a high percentage

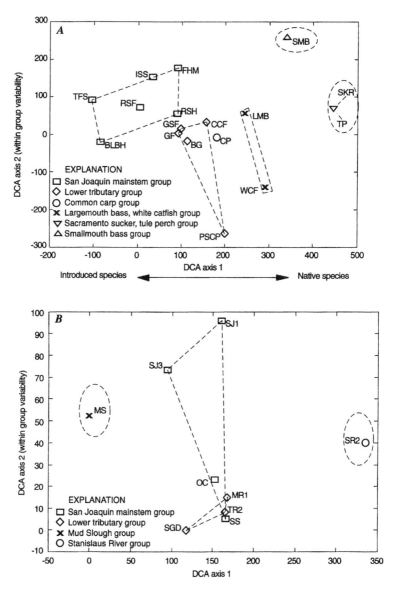

**FIGURE 6** Detrended correspondence analysis ordinations for (A) species (see Table 3 for species codes) and (B) sampling sites (see Table 1 for site codes) using data from the nine basic fixed sites. Similar symbols (enclosed by dotted lines) indicate groups of similar species or sites as determined by TWINSPAN (two-way indicator species analysis). Analysis was based on percentage abundances of species at a site.

of smallmouth bass. The separation of the Tuolumne River, Merced River, and Spanish Grant Drain sites from the mainstem San Joaquin sites near Vernalis and Patterson, the Salt Slough site, and the Orestimba Creek site was based primarily on the predominance of red shiner, fathead minnow, inland silverside, and threadfin shad at the latter sites. However, on the basis of DCA axis 1 scores, these sites were more similar to each other than to the Stanislaus River and Mud Slough sites. DCA axis 2 (Figure 6A and B) summarizes variation within the site and species groups.

## Data Integration

The sites sampled for the fish study varied widely in water quality (Table 4) and habitat characteristics (Table 5). Much of the variability was associated with differences among streams. Specific conductance and alkalinity were low and relatively constant in the Stanislaus River, but increased from upstream to downstream sites in the Merced and Tuolumne rivers. Specific conductance and alkalinity were lowest in these three large east-side tributary streams and highest in the largest stream, the mainstem San Joaquin River, and the smallest streams, the southern and western tributaries. Specific conductance and alkalinity declined upstream to downstream in the mainstem San Joaquin River, an opposite trend to that observed in the east-side tributary rivers. Patterns in nutrient levels were generally similar to those of specific conductance and alkalinity.

The Stanislaus River had the largest and most consistent discharge of the three large east-side tributaries (Table 5). Discharge increased only 38% from the most upstream site to the most downstream. In contrast, the Tuolumne River increased 392% from the farthest upstream site to the site with the highest discharge, the second downstream site (TR2) rather than the most downstream site (TR1). Discharge in the Merced River declined 42% from the most upstream site to the next site downstream. The Stanislaus River was deeper and had higher water velocities than the other two rivers. The Stanislaus River also was deeper than the San Joaquin River and the smaller tributaries, but velocities were similar to those measured in the mainstem San Joaquin River. Discharge and width of the mainstem San Joaquin River sites increased from upstream to downstream sites. The only sites with mean dominant substrate sizes larger than sand were located in the upper reaches of the three large east-side tributaries.

Principal components analysis (see Table 6) of the physical data collected at all 20 ecological sites produced two composite variables for the water quality data alone (WQ1 and WQ2), only one variable for the habitat data set (HAB1), and two composite variables for the combined habitat and water quality data set (ALL1 and ALL2). Five water quality variables and four habitat variables were included in the analysis. The high variable loadings for all the water quality variables on WQ1 indicate that they were highly correlated. WQ2 primarily represents a gradient in alkalinity independent of the other variables. HAB1 summarizes a gradient on the basis of instream cover for fish, substrate type, and stream gradient. ALL1 basically describes a gradient from sites having low values for nutrients and other water quality measures, coarse substrate, abundant instream cover, and high gradient to sites with high values for nutrient and other water quality measures, fine substrates, low gradient, and little instream cover. ALL2 stresses mean water velocity independent of stream width, depth, or discharge. A number of the variables measured were omitted from further analysis during the data screening process. For example, the original data included three measures of phosphorus, but only orthophosphate was

## TABLE 4
### Values for Selected Water Quality Variables at Sites Included in the Fish Study

| Location | Specific Conductance (mmhos/cm) | pH | Alkalinity (mg CaCO$_3$/L) | Nitrite + Nitrate (mg N/L) | Orthophosphate (mg P/L) |
|---|---|---|---|---|---|
| Tuolumne River at Shiloh Road (TR1) | 418 | 8.1 | 128 | 1.8 | 0.23 |
| Tuolumne River at Modesto (TR2) | 278 | 8.6 | 89 | 1.0 | 0.34 |
| Tuolumne River near Waterford (TR3) | 213 | 7.9 | 72 | 0.12 | 0.04 |
| Tuolumne River at Turlock State Recreation Area (TR4) | 93 | 8.1 | 41 | 0.025 | 0.02 |
| Stanislaus River at Caswell State Park (SR1) | 97 | 7.6 | 39 | 0.18 | 0.03 |
| Stanislaus River near Ripon (SR2) | 80 | 7.8 | 36 | 0.12 | 0.02 |
| Stanislaus River near Riverbank (SR3) | 76 | 7.9 | 34 | 0.15 | 0.02 |
| Stanislaus River near Knights Ferry (SR4) | 62 | 7.6 | 28 | 0.05 | 0.005 |
| San Joaquin River near Vernalis (SJ1) | 909 | 8.3 | 138 | 1.7 | 0.11 |
| San Joaquin River at Maze Road (SJ2) | 1255 | 8.5 | 151 | 3.1 | 0.13 |
| San Joaquin River near Patterson (SJ3) | 1490 | 8.1 | 165 | 3.3 | 0.29 |
| San Joaquin River at Fremont Ford (SJ4) | 2300 | 8.1 | 154 | 3.9 | 0.06 |
| Merced River at River Road (MR1) | 355 | 8.0 | 90 | 3.1 | 0.04 |
| Merced River at Hagamann County Park (MR2) | 206 | 8.0 | 57 | 2.5 | 0.02 |
| Merced River at McConnell State Park (MR3) | 74 | 7.9 | 30 | 0.052 | 0.03 |
| Merced River near Snelling Diversion Dam (MR4) | 42 | 7.3 | 18 | 0.025 | 0.005 |
| Mud Slough near Gustine (MS) | 4670 | 7.8 | 389 | 0.025 | 0.05 |
| Orestimba Creek at River Road (OC) | 492 | 8.0 | 125 | 1.4 | 0.07 |
| Salt Slough at Lander Avenue (SS) | 1540 | 8.6 | 176 | 4.0 | 0.13 |
| Spanish Grant Drain (SGD) | 529 | 7.7 | 72 | 1.5 | 0.19 |

# TABLE 5
## Values for Selected Habitat Variables at Sites Included in the Fish Study

| Location | Discharge (m³/s) | Mean Depth (m) | Mean Velocity (m/s) | Mean Dominant Substrate[b] | Width (m) | Open Canopy[a] (degrees) | Instream Cover (% of area) |
|---|---|---|---|---|---|---|---|
| Tuolumne River at Shiloh Road (TR1) | 2.72 | 0.48 | 0.26 | 4.0 | 37.6 | 146 | 19 |
| Tuolumne River at Modesto (TR2) | 3.74 | 0.42 | 0.30 | 4.0 | 38.9 | 129 | 14 |
| Tuolumne River near Waterford (TR3) | 1.17 | 0.61 | 0.15 | 5.9 | 26.9 | 133 | 28 |
| Tuolumne River at Turlock State Recreation Area (TR4) | 0.76 | 0.39 | 0.18 | 6.0 | 40.6 | 117 | 28 |
| Stanislaus River at Caswell State Park (SR1) | 10.75 | 1.17 | 0.39 | 4.0 | 30.5 | 117 | 9 |
| Stanislaus River near Ripon (SR2) | 9.49 | 0.97 | 0.42 | 4.0 | 26.8 | 95 | 18 |
| Stanislaus River near Riverbank (SR3) | 9.95 | 1.51 | 0.30 | 4.2 | 34.2 | 114 | 62 |
| Stanislaus River near Knights Ferry (SR4) | 7.79 | 1.69 | 0.41 | 6.3 | 31.1 | 114 | 12 |
| San Joaquin River near Vernalis (SJ1) | 22.60 | 0.95 | 0.41 | 4.0 | 93.2 | 166 | 6 |
| San Joaquin River at Maze Road (SJ2) | 15.09 | 0.82 | 0.36 | 4.0 | 60.8 | 159 | 2 |
| San Joaquin River near Patterson (SJ3) | 10.53 | 0.69 | 0.39 | 4.0 | 52.7 | 158 | 4 |
| San Joaquin River at Fremont Ford (SJ4) | 4.68 | 0.93 | 0.33 | 4.0 | 21.9 | 156 | 3 |
| Merced River at River Road (MR1) | 1.65 | 0.37 | 0.31 | 4.0 | 21.4 | 143 | 7 |
| Merced River at Hagamann County Park (MR2) | 1.79 | 0.49 | 0.23 | 3.9 | 22.0 | 116 | 9 |
| Merced River at McConnell State Park (MR3) | 1.38 | 0.82 | 0.19 | 4.3 | 21.2 | 133 | 31 |
| Merced River near Snelling Diversion Dam (MR4) | 2.38 | 0.85 | 0.13 | 6.8 | 51.7 | 137 | 25 |
| Mud Slough near Gustine (MS) | 0.06 | 0.52 | 0.08 | 3.0 | 8.5 | 162 | 2 |
| Orestimba Creek at River Road (OC) | 0.79 | 0.81 | 0.55 | 3.7 | 5.7 | 51 | 11 |
| Salt Slough at Lander Avenue (SS) | 5.64 | 0.80 | 0.33 | 3.0 | 17.0 | 129 | 4 |
| Spanish Grant Drain (SGD) | 0.16 | 0.55 | 0.22 | 3.0 | 3.8 | 68 | 7 |

[a] Open canopy is the number of degrees of sky uncovered by trees or other objects. The maximum value is 180°. The measurement is taken from midstream by determining the number of degrees covered by trees and other objects on each streambank while sighting along a transect line positioned perpendicular to stream flow. The two readings are summed and then subtracted from 180°. Measurements were taken at each of the six habitat transects, and the values were averaged.

[b] Dominant substrate size was categorized as 1 = detritus (unconsolidated plant material), 2 = mud, 3 = silt, 4 = sand (0.02–2.00 mm in diameter), 5 = gravel (2–64 mm in diameter), 6 = cobble (64–256 mm in diameter), 7 = boulder (>256 mm in diameter), 8 = bedrock or hardpan (solid rock or clay forming a continuous surface).

### TABLE 6
### Principal Component Loadings for Environmental (Habitat and Water Quality) Variables

| | Principal Component | | | | | |
|---|---|---|---|---|---|---|
| | Water Quality | | Habitat | | Combined | |
| Environmental Variable | WQ1 | WQ2 | HAB1 | HAB2 | ALL1 | ALL2 |
| | Ecological Sites (n = 20) | | | | | |
| Specific conductance (μmhos/cm)[a] | 0.91 | –0.37 | —[b] | — | –0.92 | –0.29 |
| Alkalinity (mg $CaCO_3$/L) | 0.70 | –0.70 | — | — | –0.74 | –0.60 |
| pH[a] | 0.78 | 0.37 | — | — | –0.71 | 0.33 |
| Nitrate + nitrite (mg N/L)[a] | 0.77 | 0.47 | — | — | –0.74 | 0.51 |
| Orthophosphate (mg P/L)[a] | 0.88 | 0.20 | — | — | –0.81 | 0.20 |
| Mean velocity (m/s) | — | — | –0.42 | — | –0.28 | 0.73 |
| Mean dominant substrate | — | — | 0.80 | — | 0.79 | –0.06 |
| Cover (% area)[a] | — | — | 0.83 | — | 0.80 | 0.27 |
| Gradient (m/km)[a] | — | — | 0.81 | — | 0.73 | 0.13 |
| Proportion of variance explained | 0.66 | 0.21 | 0.54 | — | 0.55 | 0.17 |
| | Basic fixed sites (n = 9) | | | | | |
| Hardness (mg $CaCO_3$/L) | –0.92 | — | — | — | 0.92 | –0.29 |
| Alkalinity (mg $CaCO_3$/L) | –0.95 | — | — | — | 0.96 | –0.13 |
| Specific conductance (μmhos/cm)[a] | –0.98 | — | — | — | 0.96 | –0.25 |
| Total dissolved solids (mg/L)[a] | –0.98 | — | — | — | 0.95 | –0.26 |
| Orthophosphate (mg/L) | –0.61 | — | — | — | 0.61 | 0.37 |
| Iron (μg/L)[a] | 0.89 | — | — | — | –0.86 | 0.12 |
| Total-DDT (μg/kg tissue wet weight)[a] | –0.70 | — | — | — | 0.70 | 0.40 |
| Mean daily discharge (m³/s)[a] | — | — | –0.89 | 0.43 | –0.53 | –0.74 |
| Mean width (m)[a] | — | — | –0.91 | 0.38 | –0.48 | –0.76 |
| Open canopy (degrees) | — | — | –0.90 | –0.37 | 0.03 | –0.95 |
| Instream cover (% area)[a] | — | — | 0.27 | 0.95 | –0.78 | 0.51 |
| Stream gradient (m/km)[a] | — | — | 0.93 | 0.14 | 0.15 | 0.95 |
| Proportion of variance explained | 0.76 | — | 0.67 | 0.27 | 0.53 | 0.31 |

[a] Variables were $\log_{10}(x + 1)$ transformed for analysis.
[b] No data indicated by —.

included in the analysis because the measures were so closely correlated. Most of the habitat variables included in HAB1 were correlated with altitude.

The analysis of the physical data collected at the nine basic fixed sites included more variables and produced somewhat different results (Table 6). The basic fixed-site analysis produced only one composite variable for the water quality data (WQ1), two composite variables for the habitat data set (HAB1 and HAB2), and two composite variables for the combined habitat and water quality data set (ALL1 and ALL2). WQ1 illustrates the same high correlation of general water quality variables as was found using all 20 ecological sites, except for iron, which showed a negative correlation with the other variables. HAB1 summarizes a stream-size gradient. HAB2 is an instream cover variable. Substrate size was not included in the analysis because substrate consisted of sand and silt at all the basic fixed sites. ALL1 mainly is a water quality component similar to the previous data set, but instream cover also loads heavily with moderate importance for discharge. High concentration of dissolved iron appears to be an indicator of good water quality. Total-DDT in tissue was positively correlated with the other water quality measures, except iron. Thus, it did not add much beyond the other water quality data. ALL2 summarized a gradient in stream size, as measured by discharge, width, and degrees of open canopy. These variables were negatively correlated with stream gradient. The habitat variables were not associated with altitude in this analysis.

Nutrient variables were represented by single measures of phosophorus (orthophosphate) and nitrogen (the sum of nitrite and nitrate) because of high correlations among different measures of the same nutrient. The major ions and metals, except for iron, were dropped because they were closely correlated with total dissolved solids and specific conductance. These constituents ranged in value as follows (in mg/L): boron (5.0 to 3000.0), calcium (7.4 to 100.0), chloride (1.4 to 400.0), fluoride (0.05 to 0.3), iron (12.0 to 150.0), magnesium (2.6 to 62.0), manganese (5.0 to 660.0), molybdenum (0.5 to 9.5), potassium (1.0 to 4.5), selenium (0.5 to 21.5), silica (8.6 to 24.0), sodium (2.8 to 390.0), and sulfate (3.0 to 570.0).

Only total-DDT in tissue was used in the previous analysis because there was a significant positive correlation between total-DDT concentration in tissue and total-DDT concentration in total organic carbon content of sediment (Figure 7). Rather than enter two closely correlated variables into the analysis, total-DDT in tissue was chosen as the more direct measure of organism exposure. Asian clams (*Corbicula fluminea*) were collected at 11 sites; common carp (*Cyprinus carpio*) at 3 sites; and bluegill (*Lepomis macrochirus*), channel catfish (*Ictalurus punctatus*), and crayfish (*Procambarus clarki*) at 1 site each. The tissue values for the fish species fell above the regression line; however, recalculation of the regression using only *Corbicula* data did not significantly affect the results, as shown by analysis of covariance.

Spearman rank correlations (Table 7) among the DCA axes and PCs, derived from the data for all 20 ecological sites, indicated significant correlations between DCA axis 1, WQ1, HAB1, and ALL1. The variability, primarily in San Joaquin River mainstem sites, represented by DCA axis 2, was not significantly correlated with either habitat or water quality. HAB1 and WQ1 were significantly correlated with each other (Table 7), as were the original sets of variables (canonical correlation = 0.91, $p$ <0.01). Regressions of DCA axis 1 as a function of the PCs indicated that water quality, habitat, or the combination of the two were good predictors of the fish assemblage (all $p$ <0.001, $r^2$ = 0.81, 0.68, and 0.83, respectively).

Using only the basic fixed-site data, DCA axis 1 was related to water quality, but not to habitat (Table 7). DCA axis 2 was related to habitat, suggesting that the variability in the mainstem group was due to habitat differences among sites. A plot (not shown) of HAB1 and WQ1 grouped sites of similar size and chemistry, but the relationship was not linear, resulting in a nonsignificant correlation. There were insufficient data for canonical correlation. HAB2 was correlated with WQ1, suggesting that cover for fish was limited at sites with poor water quality. Regressions

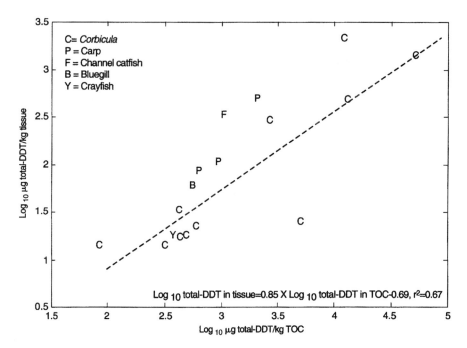

**FIGURE 7** Regression of total-DDT in tissue (wet weight) as a function of total-DDT in sediment (dry weight), normalized to total organic carbon content (TOC) for data collected in 1992.

of DCA axis 1 as a function of the PCs indicated that water quality, habitat, or the combination of the two were adequate predictors of the fish assemblage ($r^2 = 0.59$, 0.71, and 0.66, respectively). Water quality and combined water quality and habitat explained 22 and 17% less of the variance in DCA axis 1, compared to the regressions for all 20 of the ecological sites. The variance explained by habitat essentially was unchanged. This was unexpected because neither habitat PC, when considered alone, was significantly correlated with DCA axis 1.

Insufficient data were available to analyze water quality, habitat, and fish data from the intensive fixed sites alone. DCA axis and PC scores for the four intensive fixed sites were extracted from the basic fixed site scores for graphical analysis. Plots of pesticide concentration and the PC and DCA axes scores (not shown) suggested a positive correlation between general water quality and total dissolved pesticides; however, neither variable appeared related to fish assemblages.

## DISCUSSION

### PESTICIDE STUDIES

The general implication of the San Joaquin-Tulare basins NAWQA pesticide studies is that frequent sampling is needed to understand pesticide transport in surface water.[31] The results of the pesticide studies also suggest that dilution effects and

## TABLE 7
Spearman Rank Correlations Between Detrended Correspondence Analysis (DCA) Axes Site Scores and Principal Components Derived from Water Quality (WQ) Data, Habitat (HAB) Data, and a Combined Data set Including Both Sets of Variables (ALL)

| DCA Axis or Principal Component | Principal Component | | | | | |
|---|---|---|---|---|---|---|
| | Water quality | | Habitat | | Combined | |
| | WQ1 | WQ2 | HAB1 | HAB2 | ALL1 | ALL2 |
| | Ecological Sites (n = 20) | | | | | |
| DCA axis 1 | 0.82 | 0.01 | −0.76 | —[a] | −0.86 | 0.06 |
| DCA axis 2 | 0.43 | 0.22 | −0.17 | — | −0.31 | 0.36 |
| WQ1 | — | 0.23 | −0.75 | — | −0.94 | 0.18 |
| WQ2 | — | — | −0.06 | — | −0.21 | 0.56 |
| HAB1 | — | — | — | — | 0.90 | −0.26 |
| ALL1 | — | — | — | — | — | −0.21 |
| | Basic Fixed Sites (n = 9) | | | | | |
| DCA axis 1 | 0.75 | — | 0.13 | 0.57 | −0.73 | 0.17 |
| DCA axis 2 | 0.00 | — | −0.73 | 0.13 | −0.12 | −0.75 |
| WQ1 | — | — | −0.07 | 0.87 | −0.98 | — |
| WQ2 | — | — | — | — | — | — |
| HAB1 | — | — | — | −0.12 | 0.18 | 0.78 |
| ALL1 | — | — | — | — | — | −0.17 |

*Note:* DCA axes and principal components are numbered according to the proportion of variance explained. Underlined correlations are statistically significant ($p < 0.05$).

[a] No data indicated by —.

unmonitored pesticide sources can affect the interpretation of the importance of various inputs within and exports from a particular drainage basin. For example, concentrations of total dissolved pesticides were consistently high in Orestimba Creek and Salt Slough during the summer period, but elevated concentrations were not apparent at the San Joaquin River site (Figure 2). This lack of response at the San Joaquin River site likely was the result of dilution from the Stanislaus, Tuolumne, and Merced rivers, assuming that concentrations in the former two rivers were as low as those in the Merced River. This would be expected on the basis of similar land uses in each basin. Also, these three large, east-side tributaries account for most of the water flowing through the system. Conversely, the September peak in total pesticide concentration at the San Joaquin River site did not correspond with large concentrations at any of the other three intensive fixed sites, implicating an unmonitored source.

Similar processes also can affect interpretation of the occurrence of pesticides when dilution reduces the concentration of a chemical below the detection limit of the analytical method. For example, only two compounds were detected in the San

Joaquin River in October, despite the detection of eight compounds in Orestimba Creek and four compounds at Salt Slough. Dilution likely is caused by the east-side rivers, including the Merced River, where no pesticides were detected in October. It is notable that total pesticide concentrations in Orestimba Creek and Salt Slough were low in September (<1 µg/L), suggesting that the concentrations of individual compounds were low enough that dilution is a likely explanation. Other processes, however, could be partly or completely responsible. Chemical degradation of pesticide compounds may have occurred, resulting in the conversion of pesticides into compounds that the analytical procedures could not detect. Phase changes may have taken place, with volatile chemicals moving into the atmosphere and hydrophobic compounds adsorbing to sediment particles.

The implications of these data for ecosystem effects of pesticides are difficult to assess because we have not presented results for individual chemicals; however, it seems unlikely that the total concentrations measured would be toxic to adult fish. The $LC_{50}$s (concentration causing 50% mortality of a test group in a given time) for fish species generally are fairly high for pesticides; for example, 60 to 46,000 µg/L for atrazine,[32] 130 to 1420 µg/L for carbofuran,[33] 0.13 to 520 µg/L for chlorpyrifos,[34] 90 to 15,000 µg/L for diazinon,[35] and 1000 to 100,000 µg/L for paraquat.[36] It seems more likely that any direct effects on fish would occur during the egg and larval stages. Most of the species in the San Joaquin River and Sacramento-San Joaquin Delta spawn during the spring, roughly from March through May.[37] The major exceptions are prickly sculpin, which can begin spawning in February, and longfin smelt (*Spirinchus thaleichthys*), which spawns from December through February. Larvae and eggs of these species could be exposed to the winter peak concentrations. Any species could be exposed to transient events, such as the extremely high concentrations detected at Orestimba Creek in August (Figure 2B). Ecosystem effects of pesticides at the concentrations observed are more likely to occur at lower trophic levels such as invertebrates. Bioassays have indicated that water from the San Joaquin River drainage is often toxic to bioassay invertebrates, particulary *Ceriodaphnia dubia*.[38,39] The ultimate effect of this toxicity on the ecosystem, and fish populations and assemblages in particular, remains a matter of speculation.

## DIAZINON STUDIES

Consideration of dissolved concentrations of diazinon in relation to available information on temporal and geographic variability in diazinon application rates leads to a better understanding of transport processes resulting from agricultural practices and land use. In the San Joaquin River drainage, high concentrations of dissolved diazinon in surface water clearly were associated with winter applications to orchards followed by rainfall and runoff. Large summertime applications were not correlated with high concentrations in surface water, suggesting that potential effects of such applications on aquatic ecosystems are minimal. Similar analyses are being conducted for other chemicals detected in the pesticide studies.

Understanding mechanisms of transport of chemicals to surface water is important because of possible effects of the compounds on the ecosystems of surface waters. Diazinon has been of particular interest recently because some samples of surface water from the San Joaquin River drainage have been toxic to bioassay organisms, particularly *Ceriodaphnia dubia*.[38,39] Measured levels of diazinon in some of the water samples exceeded concentrations found to be toxic to *Ceriodaphnia* in laboratory experiments.[40] The contribution of other compounds to the observed toxicity was unknown, as was the overall effect on the resident communities of biota both within the river and in the Sacramento-San Joaquin Delta. Determining the overall ecosystem effects of diazinon is challenging because pulses of diazinon at toxic levels occur only sporadically, usually in January and February during periods of rainfall (Figure 4). As noted earlier, larvae of only two species of fish, prickly sculpin and longfin smelt, are present in the San Joaquin River or the Sacramento-San Joaquin Delta during this time period, and their sensitivity to the chemical is unknown. A recent risk assessment indicated that concentrations of diazinon in the San Joaquin River are likely toxic to the most sensitive 10% of the arthropod species at times, especially in January and February, but that the probability of direct effects on other organisms, particularly fish, is low.[41] However, without knowledge of which native invertebrates are sensitive, their importance to ecosystem processes, and the additional effects diazinon may have in combination with other pesticides, the effect on the ecosystem remains unknown.

## SPECIES ASSEMBLAGES

Analysis of the ecological data from all 20 sites located on the valley floor area of the San Joaquin River drainage was successful at grouping fish species on the basis of similar percentage abundances and grouping sites on the basis of similar fish species assemblages. Saiki's[8] results from the lower mainstem San Joaquin River and tributaries, primarily smaller southern tributaries, indicated that such groupings might exist, but his work did not include extensive coverage of the large east-side tributaries. Recent sampling efforts have emphasized higher altitudes. Studies in the Sierra Nevada foothills, have shown that native fish assemblages are still common, but are threatened by introduced species and environmental degradation.[6,7,10] The results of this study show that populations of native stream fishes have persisted in the highly modified stream reaches downstream of major dams.

Most of the species within each species group appear to share general ecological requirements and life history characteristics. Sacramento squawfish, hardhead, and Sacramento sucker are all native species which are characteristic of Sierra Nevada foothill streams.[1,6,10] These species are large, long-lived (greater than 10 years), and require several years to mature. They spawn in gravelly areas, generally stream riffles, and simply deposit the eggs on the substrate. The eggs then fall into interstices between substrate particles. Spawning occurs in the spring or early summer. Prickly sculpin is common in foothill streams, but is also common in less streamlike habitats, including large rivers, the Sacramento-San Joaquin Delta, lakes, and reservoirs. Prickly sculpin do not reach the same size as the minnows or live as long, but do require at least 1 year to mature and can live over 5 years. In all habitats, the species is associated with cover, most commonly rocky substrates in streams. Spawning takes place in a nest site located under a rock or other hard substrate. Spawning can begin as early as February.

The members of the two groups of introduced species characteristic of tributary sites generally do well in warm-water lakes and reservoirs, but are also common in streams and rivers.[37] All of the species, except carp and goldfish, are nest builders, with the nest consisting of a depression in the bottom substrate. Gravel is not required for successful spawning. Carp and goldfish spawn on submerged aquatic vegetation. Similar to the native species, all the species in this group require more than 1 year to reach maturity in California, and all live for 5 years or more. Spawning occurs in the spring or early summer.

The introduced species characteristic of the mainstem San Joaquin River and small tributary sites have some differences in life history characteristics, but there are also some important similarities. Inland silverside and threadfin shad are common in large warm-water lakes and reservoirs, but are also common in the lower reaches of large rivers.[37] Both fathead minnow and red shiner are characteristic of unstable habitats, such as intermittent streams, where few other species can survive the harsh physical environment.[37] Both are also commonly associated with warm temperatures and high turbidity. Despite these somewhat different habitat associations, the four species share a number of characteristics. Individuals of all species can live up to 3 years, but most survive only 1 or 2 years. They all mature quickly, and most spawning fish are 1 year old. Fathead minnows and threadfin shad can mature in the same summer they are born. All of the species have extended spawning periods and produce large numbers of eggs for their size. Fathead minnows, red shiners, and inland silversides can spawn multiple times throughout the summer. Fathead minnow and red shiner populations generally exhibit a single summertime peak in spawning activity, but inland silverside populations can have several peak periods within the summer. Individual threadfin shad only spawn once, but the spawning period extends through the summer, peaking in June.

The smallmouth bass is similar to the largemouth bass in most respects, but prefers clear, cool water and has a greater affinity for streams and rivers.[37] This difference in habitat preferences is the likely reason for it being separated from the other introduced species groups. The tule perch is unique from all the other species because it is a live bearer. The young tule perch usually are born in May and June. Tule perch are usually associated with cover of some sort, including tules, fallen trees, overhanging vegetation, large rocks, or dense beds of aquatic vegetation.[37] Cover is especially important as refuge from predators for near-term pregnant females and newly born young. Tule perch tend to disappear from streams with reduced flows, increased turbidity, heavy pollution, or reduced cover, particularly reduced emergent vegetation.

The interpretation of the species groups defined, using only the data from the nine basic fixed sites, is similar in most respects. One major difference is the absence of the upper tributary species group of native species. Two native species, Sacramento squawfish and hardhead, were not captured at the basic fixed sites. Sacramento sucker and tule perch did form a separate group of native species because these species occurred only at the Stanislaus River site. Prickly sculpin moved into one of the tributary groups because it was captured only at the Merced River site. The other major difference is that the San Joaquin mainstem group, including fathead minnow, red shiner, threadfin shad, and inland silverside, was no longer identified as a separate group. The absence of the additional mainstem and small tributary sites apparently reduced the power of the analysis to identify the unique association among these species.

## DATA INTEGRATION

The differences in habitat requirements among the species groups associated with the different site groups suggests that habitat availability is likely an important variable explaining the site groupings. For example, it seems likely that the native species should be most common at the upstream sites that have appropriate spawning habitat. Analysis of the data from all 20 ecological sites with a combination of univariate and multivariate statistical techniques demonstrated correlations between fish assemblages, water quality, and habitat; however, it could not be determined whether fish assemblages were responding to water quality and habitat separately or in combination. Thus, fish assemblages provided a good indication of overall environmental conditions, but differences in fish assemblages could not be attributed to any particular variable or suite of variables.

Similar, but less compelling, results were obtained when the data from only the nine basic fixed sites were analyzed. The association between habitat and fish assemblages was still found, but the association was not simple enough to show up in a simple correlation analysis. A multiple regression analysis was required. The main reason the results were less compelling is that two sites (MS and SR2 in Figure 6) in the DCA ordination assume substantially more importance than the other sites. The status of these sites as outliers gives them substantially more weight in statistical analyses. If the study design had not included one or both of the sites, the results of the statistical tests could have been very different.

The reasons for the negative correlation between dissolved iron concentration and the other water quality variables is unknown. Most iron occurs in the waters of the study area in the particulate phase rather than the dissolved phase.[4] The median percentage of iron in the dissolved phase ranged from 0.3 to 13.3% from 1985 to 1988 during a previous study of water quality in the area.[4] In this study, the observed negative correlation of dissolved iron with other water quality parameters was due primarily to high concentrations in the three east-side tributaries (57 to 150 µg/L), compared to the other sites (12 to 25 µg/L). The difference could be accounted for by a number of factors, including differences in geology between the Sierra Nevada and the rest of the area, some unknown source of dissolved iron related to human activity on the east side of the valley floor, or some unknown process in the large reservoirs on each of the east-side rivers.

The limited geographic scale of pesticide data collected made comparisons with other data sets suggestive at best; however, it is notable that the species group of red shiner, fathead minnow, inland silverside, and threadfin shad was characteristic of the sites with the poorest water quality and possibly temporally variable high concentrations of pesticides. Species with extended spawning seasons would be expected to do well under such conditions because there is a high probability they will be spawning at some times when toxic pesticide concentrations are low. All of the species are also good colonizers, so they would be able to recolonize areas affected by intermittent pulses of pesticides at toxic levels. All of the species are omnivorous, so they would be able to utilize whatever food sources were available.

The weak connection between the pesticide and ecological data highlight some differences in how water quality studies and ecological studies are designed, particularly in regard to spatial and temporal scale. Water quality criteria are concentration based, and the probability of exceeding these criteria is greatest at the most contaminated sites. Many water quality studies focus monitoring efforts on the most contaminated sites. In addition, frequent sampling is usually required to account for temporal variability in concentrations. Chemical sampling is nonintrusive and has no effect on subsequent samples; however, in areas where many chemicals are used, researchers may not always know what compounds to target. Finally, pesticide monitoring studies have the advantage of being able to use pesticide-free sites as references. Because most pesticides are not naturally occurring chemicals, detection of any pesticide indicates an anthropogenic source. In practice, however, detection of chemicals at low concentrations often can be complicated by the physical matrix of the sample being analyzed and the sensitivity of analytical techniques.

In contrast, ecological studies that focused on relating community health to environmental conditions generally do not have either well-defined reference conditions for comparison purposes or fixed standards for comparison of sites, unless there is a large body of existing data that can be used to formulate such standards. In areas without an existing database, such as the San Joaquin River drainage, studies tend to focus on gradients in biological communities and correlations with environmental variables.[6,8-10] Such studies are required to establish the range of fish assemblages and environmental conditions that exist in a particular geographic area and to determine any correlations between them. Defining environmental quality at a particular site as "excellent" or "poor," however, requires formulation of meaningful criteria for identification of excellent or poor fish assemblages or environmental conditions.

A comparison of the DCA ordinations obtained with the two sets of data shows clear differences in the ability to define the range of conditions possible in the area (Figures 5 and 6). Obviously, because the basic fixed sites are concentrated in the lower part of the drainage, the species assemblages and environmental conditions present at the upper tributary sites were missed completely. Because these fish assemblages have not been extensively studied, there are no background data for comparing the lower tributary sites to upper tributary sites where the effects of agricultural and urban development are less severe. In contrast, it could be stated with a fair amount of confidence that the concentrations of dissolved constituents and pesticides should be low at these sites because various agencies, including the USGS, have collected water quality data as part of ongoing and historical studies. The basic fixed-site sampling completely missed two of the native species, the Sacramento squawfish and the hardhead. Tule perch and Sacramento sucker were associated only with the Stanislaus River. The outlying position of the Stanislaus River site in the basic fixed-site ordination is expected on the basis of these results. However, the basic fixed-site data does not show that the Stanislaus River is different from the other tributaries for its entire length. In the 20-site ordination, all of the

Stanislaus River sites appear to the left side of the figure when compared to the other tributary sites. This shows that the Stanislaus River is distinct from the Tuolumne and Merced rivers with respect to fish species. The positions of the sites, other than the Stanislaus River site, do not compare well between analyses. The separation between the lower tributary sites and the mainstem San Joaquin River and small tributary sites was lost in the basic fixed-site DCA ordination. The Mud Slough site also was identified as clearly distinctive by TWINSPAN and in the basic fixed-site DCA ordination, but in the analysis of the data from all 20 ecological sites, the Spanish Grant Drain site was identified as unique by TWINSPAN. In addition, the DCA ordination indicated that the Spanish Grant Drain site was closely allied to the mainstem San Joaquin site group. These differences in results obtained with the data sets are especially important when considering the objectives of the NAWQA Program.

Documentation of long-term trends is an important objective of the NAWQA study design; however, without the data on the range of ecological conditions collected in the drainage, few conclusions would have been possible concerning the meaning of future fish surveys. For example, three tributary sites (SR2, TR2, and MR1 in Table 1) were sampled again in 1995, a high discharge year, as part of an assessment of spatial and temporal variability in fish assemblages. At the Tuolumne and Merced river sites, four species were captured in 1995 that were not captured in 1994, including juveniles of Sacramento squawfish, hardhead, Sacramento sucker, and Sacramento splittail, *Pogonichthys macrolepidotus*. Where did these species come from? On the basis of data from all 20 ecological sites, it can be assumed that all but the splittail either migrated downstream or were washed downstream with high discharges. Because the surveys did not document the presence of splittail in the drainage, it also can be assumed that adults of this species migrated into drainage from the Sacramento-San Joaquin Delta, where the species is relatively common, and spawned, explaining the presence of juveniles in the system. Clearly, a detailed knowledge of the range of ecological conditions in an area is needed to successfully use fish assemblages as biomonitors of environmental change.

The primary reason that bioassays were not included as one of the standard methodologies of the NAWQA Program is because no effective approach was available to determine the exact cause of toxicity in bioassays.[42] Instead, community composition of benthic algae, benthic macroinvertebrates, and fish were targeted as monitors of ecosystem health.[43] Biological communities integrate the effects of water quality and habitat over time, with each biotic component responding to environmental disturbances over different periods of time (generally, algae in weeks, invertebrates in months to years, and fish in years). Fish have been recommended as monitors of environmental quality[44] and fish community structure has been the basis for the development of various Indexes of Biotic Integrity (IBI).[45-50] A basic finding during the development of most IBIs is that scoring systems must be modified for each geographic region and that fish assemblage data from a large number of sites is needed to define, with accuracy, fish assemblage characteristics expected at sites with different environmental quality.[50] These studies reinforce the observation that adequate background information is necessary for biomonitoring to be a useful tool for the assessment of environmental quality.

A perception exists that biomonitoring is difficult because of logistical problems with sampling. Ecological sampling tends to be more physically disruptive and often more labor and time intensive than chemical sampling. Sampling at frequent intervals often is not possible because it physically disrupts the site or causes organisms to leave the site. However, with adequate time and resources to characterize gradients in assemblages, standards can be developed. Because fish assemblages integrate the effects of environmental conditions over time, a single visit to a site can take the place of many water quality samples and can detect responses to habitat conditions, which would not be detected in a water quality study. Most methods of fish sampling also provide the possibility for assessing the condition of individual fish, providing additional data on environmental quality. The lack of a specific response of fish communities to a particular physical or chemical attribute of a site, such as water quality, dissolved pesticides, or habitat quality, might be seen as a shortcoming in a monitoring program. However, once scoring systems and standards can be established for characterizing healthy and unhealthy fish communities, fish surveys can be a fast and inexpensive method to identify sites with degraded physical or chemical conditions. When such sites are identified, detailed studies of water chemistry and physical conditions then can be designed to identify the specific problem.

## CONCLUSION

The results of the NAWQA studies have several implications for future monitoring in the San Joaquin River drainage. The primary result of the pesticide studies was that pesticide concentrations are highly variable geographically and temporally. Consequently, frequent sampling is required to understand the variability of concentrations in surface water. Consideration of data on land use and pesticide application, in conjunction with the results of the diazinon studies, identified the major mechanisms of diazinon transport to surface water. Further analyses should result in similar identification of mechanisms for other compounds; however, it is unclear whether monitoring additional waterways with the same temporal intensity is logistically feasible for the regulatory agencies that would use such data. At present, concentration data are useful only for comparison to regulatory standards for single chemicals. Little is known about the antagonistic and synergistic links among pesticides and the effects of such mixtures on organisms. Without additional interdisciplinary studies that include chemistry, toxicology, ecology, and perhaps other disciplines, it seems unlikely that the ecosystem effects of dissolved pesticides and other chemicals will ever be understood.

The analysis of water quality data suggests that much of the variability in the water quality parameters measured in this study is well represented by basic water quality measures, such as specific conductance, hardness, alkalinity, and total dissolved solids (Table 4); thus, more detailed measures, such as major ion analyses, could be done less frequently. Fish assemblages do respond to water quality, habitat, or some combination of the two. Thus, monitoring of fish assemblages offers a method for quickly evaluating a large number of sites for overall environmental quality. Continued monitoring during changes in patterns of water use would provide the opportunity to document effects on resident fishes, to better understand how they respond to water quality and physical habitat, and to formulate standards for the assessment of fish assemblages in the San Joaquin River drainage.

For questions or further information regarding the work described in this chapter, please contact:

**Larry R. Brown, Biologist**
U.S. Geological Survey, WRD
Placer Hall
6000 J Street
Sacramento, CA 95819-6129
e-mail: lrbrown@usgs.gov

## REFERENCES

1. Brown, L.R. 1996. Aquatic biology of the San Joaquin-Tulare basins: analysis of available data through 1992. Water-Supply Paper 2471, U.S. Geological Survey, Denver, CO.
2. San Joaquin Valley Drainage Program. 1990a. A management plan for agricultural subsurface drainage and related problems on the westside San Joaquin Valley. Final Report of the San Joaquin Valley Drainage Program, U.S. Department of the Interior and California Resources Agency, Sacramento, CA.
3. San Joaquin Valley Drainage Program. 1990b. Fish and wildlife resources and agricultural drainage in the San Joaquin Valley, California. Report of the San Joaquin Valley Drainage Program, U.S. Department of the Interior and California Resources Agency, Sacramento, CA.
4. Hill, B.H. and R.J. Gilliom. 1993. Streamflow, dissolved solids, suspended sediment, and trace elements, San Joaquin River, California, June 1985–September 1988. Water-Resources Investigations Report 93-4085, U.S. Geological Survey, Denver, CO.
5. Kahrl, W.L., W.A. Bowen, S. Brand, M.L. Shelton, D.L. Fuller and D.A. Ryan. 1978. The California Water Atlas. The Governor's Office of Planning and Research, Sacramento, CA.
6. Moyle, P.B. and R.D. Nichols. 1973. Ecology of some native and introduced fishes of the Sierra Nevada foothills in central California. *Copeia* 1973:478–490.
7. Moyle, P.B. and R.D. Nichols. 1974. Decline of the native fish fauna of the Sierra Nevada foothills, central California. *Am. Midl. Nat.* 92:72–83.
8. Saiki, M.K. 1984. Environmental conditions and fish faunas in low elevation rivers on the irrigated San Joaquin Valley floor, California. *Calif. Fish Game* 70:145–157.
9. Jennings, M.R. and M.K. Saiki. 1990. Establishment of red shiner, *Notropis lutrensis*, in the San Joaquin Valley, California. *Calif. Fish Game* 76:46–57.
10. Brown, L.R. and P.B. Moyle. 1993. Distribution, ecology, and status of the fishes of the San Joaquin River drainage, California. *Calif. Fish Game* 79:96–114.
11. Gilliom, R.J., W.M. Alley and M.E. Gurtz. 1995. Design of the National Water Quality Assessment Program: occurrence and distribution of water quality conditions. Circular 1112, U.S. Geological Survey, Denver, CO.
12. Shelton, L.R. 1994. Field guide for collecting and processing stream-water samples for the National Water-Quality Assessment Program. Open-File Report 94-455, U.S. Geological Survey, Denver, CO.

13. Crawford, J.K. and S.N. Luoma. 1993. Guidelines for studies of contaminants in biological tissues for the National Water-Quality Assessment Program. Open-File Report 92-494, U.S. Geological Survey, Denver, CO.
14. Shelton, L.R. and P.D. Capel. 1994. Guidelines for collecting and processing samples of stream bed sediment for analysis of trace elements and contaminants for the National Water-Quality Assessment Program. Open-File Report 94-458, U.S. Geological Survey, Denver, CO.
15. Meador, M.R., T.F. Cuffney and M.E. Gurtz. 1993. Methods for sampling fish communities as part of the National Water-Quality Assessment Program. Open-File Report 93-104, U.S. Geological Survey, Denver, CO.
16. Meador, M.R., C.R. Hupp, T.F. Cuffney and M.E. Gurtz. 1993. Methods for characterizing stream habitat as part of the National Water-Quality Assessment Program. Open-File Report 93-408, U.S. Geological Survey, Denver, CO.
17. Zaugg, S.D., M.W. Sandstrom, S.G. Smith and K.M. Fehlberg. 1995. Methods of analysis by the U.S. Geological Survey National Water Quality Laboratory — Determination of pesticides in water by C-18 solid-phase extraction and capillary-column gas chromatography/mass spectrometry with selected-ion monitoring. Open-File Report 95-181, U.S. Geological Survey, Denver, CO.
18. Werner, S.L., M.R. Burkhardt and S.N. DeRusseau. 1996. Methods of analysis by the U.S. Geological Survey National Water Quality Laboratory — Determination of pesticides in water by carbopak-b solid-phase extraction and high-performance liquid chromatography. Open-File Report 96-216, U.S. Geological Survey, Denver, CO.
19. Crepeau, K.L., J.L. Domagalski and K.M. Kuivila. 1994. Methods of analysis and quality-assurance practices of the U.S. Geological Survey Organic Laboratory, Sacramento, California — Determination of pesticides in water by solid-phase extraction and capillary-column gas chromatography/mass spectrometry. Open-File Report 94-362, U.S. Geological Survey, Denver, CO.
20. Fishman, M.J. and L.C. Friedman. 1970. Methods of determination of inorganic substances in water and fluvial sediments. Techniques of Water Resource Investigations, book 5, chapter 1, U.S. Geological Survey, Denver, CO.
21. Patton, C.J. and E.P. Truitt. 1992. Methods of analysis by the U.S. Geological Survey National Water Quality Laboratory — Determination of total phosphorus by a kjeldahl digestion method and an automated colorimetric finish that includes dialysis. Open-File Report 92-146, U.S. Geological Survey, Denver, CO.
22. Fishman, M.J. 1993. Methods of analysis by the U.S. Geological Survey National Water Quality Laboratory — Determination of inorganic and organic constituents in water and fluvial sediments. Open-File Report 93-125, U.S. Geological Survey, Denver, CO.
23. Leiker, T.J., J.E. Madsen, J.R. Deacon and W.T. Foreman. 1995. Methods of analysis by the U.S. Geological Survey National Water Quality Laboratory — Determination of chlorinated pesticides in aquatic tissue by capillary-column gas chromatography with electron-capture detection. Open-File Report 94-710, U.S. Geological Survey, Denver, CO.
24. Foreman, W.T., B.F. Connor, E.T. Furlong, D.G. Vaught and L.M. Merten. 1995. Methods of analysis by the U.S. Geological Survey National Water Quality Laboratory — Determination of organochlorine pesticides and biphenyls in bed sediment by dual capillary-column gas chromatography with electron-capture detection. Open-File Report 95-140, U.S. Geological Survey, Denver, CO.
25. California Department of Pesticide Regulation. 1994. Pesticide use data for 1993 (digital data). Sacramento, CA.

26. Hill, M.O. 1979a. TWINSPAN — A FORTRAN program for arranging multivariate data in an ordered two-way table by classification of the individuals and attributes. Ecology and Systematics, Cornell University, Ithaca, NY.
27. Hill, M.O. 1979b. DECORANA — A FORTRAN program for detrended correspondence analysis and reciprocal averaging. Ecology and Systematics, Cornell University, Ithaca, NY.
28. ter Braak, C.J.F. 1987. CANOCO: A FORTRAN program for canonical community ordination by (partial) (detrended) (canonical) correspondence analysis, principal components analysis and redundancy analysis. TNO Institute of Applied Computer Science, Wageningen, The Netherlands.
29. Jongman, R.H.G., C.J.F. ter Braak and O.F.R. van Tongeren. 1987. *Data Analysis in Community and Landscape Ecology*. Centre for Agricultural Publishing and Documentation (Pudoc), Wageningen, The Netherlands.
30. Hill, M.O. and H.G. Gauch. 1980. Detrended correspondence analysis, an improved ordination technique. *Vegetatio* 42:47–58.
31. Domagalski, J.L. 1995. Nonpoint sources of pesticides in the San Joaquin River, California: input from winter storms, 1992–1993. Open-File Report 95-165, U.S. Geological Survey, Denver, CO.
32. Eisler, R. 1989. Atrazine hazards to fish, wildlife, and invertebrates: a synoptic review. U.S. Fish and Wildlife Service, Biological Report 85 (1.18). U.S. Fish and Wildlife Service, Patuxent Wildlife Research Center, Laurel, MD.
33. Eisler, R. 1985. Carbofuran hazards to fish, wildlife, and invertebrates: a synoptic review. U.S. Fish and Wildlife Service, Biological Report 85 (1.3). U.S. Fish and Wildlife Service, Patuxent Wildlife Research Center, Laurel, MD.
34. Odenkirchen, E.W. and R. Eisler. 1988. Chlorpyrifos hazards to fish, wildlife, and invertebrates: a synoptic review. U.S. Fish and Wildlife Service, Biological Report 85 (1.3). U.S. Fish and Wildlife Service, Patuxent Wildlife Research Center, Laurel, MD.
35. Eisler, R. 1986. Diazinon hazards to fish, wildlife, and invertebrates: a synoptic review. U.S. Fish and Wildlife Service, Biological Report 85 (1.9). U.S. Fish and Wildlife Service, Patuxent Wildlife Research Center, Laurel, MD.
36. Eisler, R. 1990. Paraquat hazards to fish, wildlife, and invertebrates: a synoptic review. U.S. Fish and Wildlife Service, Biological Report 85 (1.22). U.S. Fish and Wildlife Service, Patuxent Wildlife Research Center, Laurel, MD.
37. Moyle, P.B. 1976. *Inland Fishes of California*. University of California Press, Berkeley, CA.
38. Foe, C. and V. Connor. 1991. San Joaquin watershed bioassay results, 1988–90. Central Valley Regional Water Quality Control Board, Sacramento, CA.
39. Kuivila, K.M. and C.G. Foe. 1995. Concentrations, transport and biological effects of dormant spray pesticides in the San Francisco estuary, California. *Environ. Toxicol. Chem.* 14:1141–1150.
40. Amato, J.R., D.I. Mount, E.J. Durhan, M.T. Lukasewycz, G.T. Ankley and E.D. Robert. 1992. An example of the identification of diazinon as a primary toxicant in effluent. *Environ. Toxicol. Chem.* 11:209–216.
41. Adams, W., L. Davis, J. Giddings, L. Hall, Jr., R. Smith, K. Solomon and D. Vogel. 1996. An ecological risk assessment of diazinon in the Sacramento and San Joaquin river basins. Ciba Crop Protection, Greensboro, NC.
42. Elder, J.F. 1989. Applicability of ambient toxicity testing to national or regional water-quality assessment. Open-File Report 89-55, U.S. Geological Survey, Denver, CO.

43. Gurtz, M.E. 1994. Design of biological components of the National Water-Quality Assessment (NAWQA) Program. In S.L. Loeb and A. Spacie, eds., *Biological Monitoring of Aquatic Systems*. Lewis Publishers, Boca Raton, FL, pp. 323–354.
44. Moyle, P.B. 1994. Biodiversity, biomonitoring and the structure of stream fish communities. In S.L. Loeb and A. Spacie, eds., *Biological Monitoring of Aquatic Systems*. Lewis Publishers, Boca Raton, FL, pp. 171–186.
45. Karr, J.R. 1981. Assessment of biotic integrity using fish communities. *Fisheries* 6:21–27.
46. Fausch, K.D., J.R. Karr and P.R. Yant. 1984. Regional application of an index of biotic integrity based on stream fish communities. *Trans. Am. Fish. Soc.* 113:39–55.
47. Karr, J.R., K.D. Fausch, P.L. Angermeier, P.R. Yant and I.J. Schlosser. 1986. Assessing biological integrity in running waters: a method and its rationale. Special Publication 5, Illinois Natural History Survey, Champaign, IL.
48. Leonard, P.M. and D.J. Orth. 1986. Application and testing of an index of biotic integrity in small, coolwater streams. *Trans. Am. Fish. Soc.* 115:401–414.
49. Oberdorff, T. and R.M. Hughes. 1992. Modification of an index of biotic integrity based on fish assemblages to characterize rivers of the Seine Basin, France. *Hydrobiologia* 228:117–130.
50. Miller, D.L., P.M. Leonard, R.M. Hughes, J.R. Karr, P.B. Moyle, L.H. Schrader, B.A. Thompson, R.A. Daniels, K.D. Fausch, G.A. Fizhugh, J.R. Gammon, D.B. Halliwell, P.L. Angermeier and D.J. Orth. 1988. Regional applications of an index of biotic integrity for use in water resource management. *Fisheries* 13:12–20.

# 4 Subsurface Contaminant Fate Determination Through Integrated Studies of Intrinsic Remediation

*Scott W. Hooper, Kevin J. Daye, Kevin A. Shuttle, and Richard A. Williams*

## CONTENTS

Abstract .................................................................................................................. 63
Introduction ........................................................................................................... 64
Decreasing Concentrations Over Time ................................................................. 66
Biodegradation Potential ....................................................................................... 67
Proof of *In Situ* Biodegradation .......................................................................... 68
References ............................................................................................................. 70

## ABSTRACT

Subsurface contaminant fate determination requires multiple integrated lines of evidence. No single methodology is sufficient to build a robust case. Furthermore, integrated investigations are crucial for designing engineered solutions to instances of unsatisfactory environmental fate.

While many groups and organizations have proposed protocols and methods for documenting intrinsic bioremediation, no single protocol has been adopted as a universal standard. In an attempt to clarify the situation, the National Research Council developed guidance for determining the biodegradative fate of subsurface contaminants.[4] This guidance is in the form of three criteria to be met for demonstrating intrinsic bioremediation. The criteria are documenting loss of the contaminants from the subsurface, demonstration of the potential for biodegradation at the site, and, finally, proof that subsurface contaminant degradation is occurring. The demonstration of these criteria can be complicated by site factors, including continuous recharge of contaminants from a subsurface reservoir, unknown mixtures of contaminants, and the general inaccessibility of the subsurface to thorough examination. These complications can require the implementation of alternate routes to satisfying a criterion.

Although each criterion could be satisfied with a single line of evidence, the number and diversity of integrated lines of evidence satisfying the criteria is an indication of the robustness of the argument supporting the intrinsic bioremediation assertion. Ideally, these multiple lines of evidence should come from differing study approaches. Through the integration of field, laboratory, and modeling efforts, the biodegradative environmental fate of contaminants can be reasonably determined.

**Keywords:** *biodegradation, bioremediation, contaminant fate, in situ, intrinsic bioremediation*

## INTRODUCTION

Ecosystem health is directly affected by the environmental fate of xenobiotic chemicals. This fate is governed by the physical and chemical properties of the compound (which dictate the transport and dispersion patterns of the chemicals) and also the biological transformations, mobilizations, and effects that result from introduction of the chemicals into the environment.

Once introduced, chemicals are dispersed by physical and biological mechanisms. This dispersal may increase or decrease the concentration of the compound at any particular point in the ecosystem. Removal of the compounds from the ecosystem occurs through routing to an environmental sink. These sinks may be physically driven (as in the cases of sedimentation or photolysis), chemically driven (as in the case of hydrolysis), or biologically driven (as in the case of biodegradation). Of these processes, *in situ* biodegradation is the most difficult to document because *in situ* biodegradation is an intrinsically complex process. Natural *in situ* biodegradation (i.e., without deliberate steps to enhance the degradation process) has been termed intrinsic bioremediation.

Intrinsic bioremediation has been less prominent than some other remediation alternatives for a variety of reasons, among which include increased requirements for documentation, lower profit prospects for implementation by commercial remediation concerns, fewer instances of regulatory acceptance resulting in a cycle of nonacceptance, and the need for expertise and work activities unfamiliar to many trained in traditional engineering disciplines. Indeed, minimal emphasis on intrinsic remediation as a treatment alternative is evident in several recent texts concerning bioremediation.[1,2] However, these perceptions are changing as arguments for intrinsic remediation are increasingly accepted by state and federal regulatory bodies. This increasing acceptance has been recognized by environmental engineering companies, prompting a closer look at their microbiological and biogeochemical analytical capabilities. This trend should accelerate as documentation of intrinsic remediation successes, in the form of compendia of intrinsic remediation studies, become available (e.g., Reference 3).

In spite of the increasing acceptance of intrinsic remediation, no single accepted standard for what constitutes documentation of intrinsic remediation exists. This is partly due to the economic and regulatory factors listed previously, but it is also a result of differing contaminant types and differing site geologies particular to the sites under consideration. As a result, step-by-step intrinsic remediation documentation protocols for one industry, site, and/or contaminant are not applicable to all situations.

In an attempt to provide overarching philosophical guidance to intrinsic remediation studies, The National Research Council has recommended three criteria for demonstrating intrinsic remediation.[4]

1. Documenting a decrease in contaminant concentrations at the site
2. Showing experimentally that microorganisms in site samples have the *potential* to transform the contaminants under expected site conditions
3. Developing one or more pieces of evidence showing that the biodegradation potential is *actually realized* in the field

These criteria, while providing excellent guidance to remediation managers and regulators and appearing to be easily quantified or qualified, are often not easily demonstrated in real-world situations. A discussion of each criterion, potential pitfalls for each, and a few examples of alternate means for demonstration of the criteria are given in the following. These examples demonstrate that multiple, integrated lines of evidence are needed to support a claim of intrinsic bioremediation.

The first two criteria, decreasing contaminant concentrations and the existence of biotransformation potential on site, are usually easier to demonstrate than *in situ* biodegradation. The evidence supporting decreasing contaminant concentrations and partial support for biotransformation potential may be developed as part of a RCRA facility investigation. During the investigation, the extent of the contamination is defined, and some monitoring over time is performed. This time frame, however, is often too short to see statistically meaningful changes in concentration. Data partially supporting the biotransformation potential criterion is usually gathered as a result of problem definition and delineation. These data are in the form of temperature, pH, and inorganic and organic chemical profiles. Further evidence of the potential for *in situ* biodegradation frequently comes from laboratory studies of site microorganisms.

Developing conclusive evidence that *in situ* remediation is occurring is difficult due to limited access to the subsurface and the potential for developing misleading evidence due to inadvertent laboratory and procedural artifacts. For these reasons, the third criterion is often satisfied with demonstration of the presence of contaminant metabolites and the correlation of various biologically influenced parameters with contaminant concentration.

Ideally, each criterion should be satisfied with multiple lines of evidence, each supporting the same conclusion. The number and strength of the lines of evidence provide an indication of the robustness of the argument supporting intrinsic remediation. Consideration of multiple lines of evidence, however, often runs afoul of biochemical and statistical interrelationships. As an example, the distribution of various nutrients, their metabolites, and other biologically influenced parameters will be controlled not only by the carbon source, but also by other parameters affecting biomass proliferation. For this reason, significant covariance often exists among many easily measured groundwater parameters.[5-7]

## DECREASING CONCENTRATIONS OVER TIME

Criterion 1, demonstration of a decrease in contaminant concentration over time, though sometimes easy to measure, can be complicated by continuous leaching from contaminant reservoir sources. As shown in Figure 1, loss of contaminants from a system is typically due to biodegradation (rate term b) and physical losses (rate term t). Even though transport and dispersion can be estimated, the true rate of release of contaminants from the reservoir is usually not known, resulting in the size of the biodegradation term also being unknown. Even if these rates were known, the typically nonhomogeneous distribution of contaminants in soil and groundwater can cause wide concentration variations between samples from the same point. The elevated scatter in the data can mask statistically significant rates of contaminant loss.

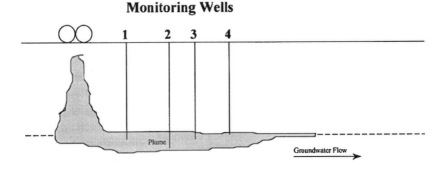

$$dC/dt = iC - (bC + tC)$$
$C$ = *Contaminant Mass in Plume*
$i$ = *Rate of Contaminant Introduction to Groundwater*
$b$ = *Rate of Biodegradation of Contaminant*
$t$ = *Rate of Transport and Dispersion*

*Apparent Rate when b Unknown:*
$$dC/dt = i_a C - t_a C$$

**FIGURE 1** Vadose zone contaminant reservoir leaching into groundwater.

The masking of true *in situ* biodegradation by nonhomogeneous contaminant distribution and by contaminant recharge from reservoirs results in the need for alternative indirect indicators of contaminant removal. Some examples are demonstration of increasing concentrations of a contaminant metabolite, indirect metabolic evidence such as depletion of electron acceptors and generation of metabolic byproducts, and decreases in concentration of less recalcitrant compounds relative to equivalently dispersed more recalcitrant ones along a flow path. Historically, inorganics such as chloride ion have been used as recalcitrant tracer molecules. Monitoring more recalcitrant tracers that are chemically and physically similar to the chemical of interest would be an improvement over using inorganic ions as the conserved tracer, due to diminished transport and partitioning differences between the molecules. An example would be monitoring degradation of nonhalogenated contaminants using halogenated co-contaminant analogs as tracers.[7] Figure 2 gives an example in which the ratio of the concentration of a lesser recalcitrant contaminant to the concentration of a more recalcitrant contaminant (i.e., tracer) is plotted over distance in a plume. If the two compounds were dispersed or dispersed and degraded at the same rate, a horizontal line would be expected. In Figure 2, the sloped line indicates that the less recalcitrant compound has been attenuated at a faster rate than the more recalcitrant tracer. Since the tracers are usually not completely recalcitrant, this method will give a conservative assessment of the degradation rate of the less recalcitrant compound.

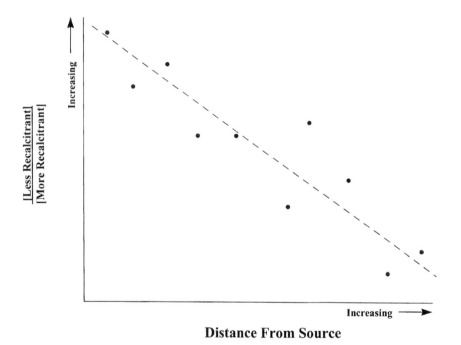

**FIGURE 2** Relationship between the concentrations of more and less recalcitrant compounds if biodegradation of the less recalcitrant compound is occurring at a faster rate than biodegradation of the more recalcitrant compound.

## BIODEGRADATION POTENTIAL

Criterion 2 is often easily achieved due to the metabolic diversity of most subsurface microorganisms. At first glance, it would appear that the single greatest challenge to satisfying Criterion 2 is defining and providing laboratory conditions similar to the subsurface. The criterion, however, specifies that one demonstrate the *potential* for intrinsic remediation. This means that the conditions do not necessarily need to exactly replicate the subsurface conditions. Furthermore, as has been pointed out by Hamer and Heitzer,[8] instrument measurements of the subsurface, which are by necessity macroscopic, do not necessarily reflect the microscopic conditions to which the microorganisms are subjected. For example, since most microorganisms are aquatic by nature, an increase in the water content of soil or sediment samples may have little apparent macroscopic effect on the environment of the microorganisms. It does, however, modify the partitioning and distribution of contaminants and nutrients within a laboratory microcosm. Because the criterion qualitatively queries biodegradation potential, partitioning, distribution, and water content are less critical to satisfying Criterion 2 than in studies intended to quantitatively measure contaminant degradation. Indeed, it is usually easier to obtain statistically meaningful data from aqueous or soil slurry experiments than from soil experiments because analytical recovery of contaminants is typically more efficient from aqueous samples (e.g., see Reference 9).

## PROOF OF *IN SITU* BIODEGRADATION

Criterion 3 is widely recognized as being the most difficult to satisfy. If one is fortunate, the presence of a chemical metabolite unique to the contaminant of concern will indicate *in situ* intrinsic remediation. In practice, however, uncertainty as to the exact composition and concentrations of the contaminants released to the subsurface often clouds the interpretation. For example, does the presence of chloride represent a metabolite of chlorobenzene, a contaminant of the spilled chlorobenzene, or was a brine also spilled? Is the detected cresol a metabolite of the toluene in the subsurface or leachate from an unknown railroad tie or telephone pole? Substantial difficulty can also be encountered in ensuring that metabolites detected in the laboratory studies of site materials are not artifacts resulting from contamination by surface microorganisms or from artificially stimulating subsurface microorganisms with surface conditions.

If no distinct metabolite is found, indirect methods of demonstrating *in situ* metabolism must be explored. A variety of methods have been described (see Reference 4). Several of these methods involve demonstration of electron acceptor and metabolic by product gradients consistent with *in situ* biodegradation. Some caution, however, is necessary when applying indirect methods of assaying *in situ* activity. This is to due to the interdependence of most subsurface biological and biologically influenced parameters. This interdependence can cause results opposite to those that would be expected from a site undergoing intrinsic remediation. An example of differing results for two demonstrated intrinsic remediation sites are given in Table 1. This table gives the correlation coefficients[10] for a number of parameters at two different sites. Site A has been demonstrated to have been undergoing intrinsic remediation,[7] while site B is also undergoing intrinsic remediation, but has known nutrient limitations (unpublished data). In the case of site A, multiple lines of evidence point to the conclusion that intrinsic remediation is occurring. No statistically significant unexpected or unexplained results were obtained. Site B, however, has several parameter correlations contradictory to those expected from a site undergoing intrinsic remediation. Looking solely at dissolved oxygen, sulfate concentration, ammonium ion concentration, and aerobic heterotrophs appear to provide promise that intrinsic remediation is occurring. Other parameters, however,

## TABLE 1
### Results for Two Demonstrated Intrinsic Remediation Sites

| | Site A<br>Correlation Coefficient<br>of Parameter<br>and VOC Concentration[a] | Site B<br>Correlation Coefficient<br>of Parameter<br>and VOC Concentration[a] |
|---|---|---|
| Dissolved oxygen concentration | Negatively correlated, as expected | Negatively correlated, as expected |
| Oxidation-reduction potential | Positively correlated, as expected | ND |
| Conductivity | Positively correlated, as expected | Negatively correlated, opposite of expected |
| Salinity | Positively correlated, as expected | Negatively correlated, opposite of expected |
| Nitrate concentration | Negatively correlated, as expected | NS |
| Ammonium ion concentration | NS | Positively correlated, as expected |
| Sulfate concentration | Negatively correlated, as expected | Negatively correlated, as expected |
| Aerobic heterotrophs (cfu/ml) | Negatively correlated, as expected | Negatively correlated, as expected |
| Aerotolerant anaerobes (cfu/ml) | Positively correlated, as expected | Negatively correlated, opposite of expected |
| Degraders of specific compounds of concern (cfu/ml) | Positively correlated, as expected | NS |

[a] Abbreviations are as follows: positively (or negatively) correlated, as expected means significant as defined by Colton[11] and the sign of the correlation (positively or negatively correlated) is as expected from hydrogeological and microbial ecological theory; positively (or negatively) correlated, opposite of expected means significant as defined by Colton[11] and the sign of the correlation (positively or negatively correlated) is opposite of that expected from hydrogeological and microbial ecological theory; ND = not determined; NS = not significant; cfu/ml = colony forming units per milliliter sample. All correlation coefficients were determined as the correlation between the parameter in the left column and VOC (volatile organic hydrocarbon) concentration.

have statistically significant correlation coefficients opposite in sign. Further investigations led to experimental demonstration that the site was undergoing intrinsic remediation, but had nutrient limitations. The summary data are presented as a note of caution: multiple lines of evidence must be pursued and integrated in order to provide a robust argument concerning the fate of environmental contaminants.

Strong arguments about the biological fate of chemicals released to the environment can be developed only through the concordance of multiple lines of evidence developed from a variety of methodologies. Ideally, these methodologies will involve field, laboratory, and modeling approaches that together determine the fate of environmental contaminants.

For questions or further information regarding the work described in this chapter, please contact:

**Scott Hooper**
Merck & Co., Inc.
2778 South East Side Hwy.
Elkton, VA 22827
e-mail: scott_hooper@merck.com

## REFERENCES

1. Alexander M. 1994. *Biodegradation and Bioremediation.* Academic Press, San Diego, CA.
2. Baker KH, Herson DS. 1994. *Bioremediation.* McGraw-Hill, New York.
3. Hinchee RE, Wilson JT, Downey DC. 1995. *Intrinsic Remediation.* Battelle Press, Columbus, OH.
4. National Research Council. 1993. *In Situ Bioremediation.* National Academy Press, Washington, D.C.
5. Borden RC, Gomez CA, Becker MT. 1995. Geochemical indicators of intrinsic bioremediation. *Ground Water* 33:180–189.
6. Chapelle FH, Bradley PM. 1997. Alteration of aquifer geochemistry by microorganisms. In: Hurst CJ, ed, *Manual of Environmental Microbiology.* ASM Press, Washington, D.C., pp. 558–564.
7. Williams RA, Shuttle KA, Kunkler JL, Madsen EL, Hooper SW. 1997. Intrinsic remediation in a solvent-contaminated alluvial groundwater. *J. Ind. Microbiol.* 18:177–188.
8. Hamer G, Heitzer A. 1991. Polluted heterogeneous environments: macro-scale fluxes, micro-scale mechanisms, and molecular control. In: Sayler GS, Fox R, Blackburn JW, eds, *Environmental Biotechnology for Waste Treatment.* Plenum Press, New York, pp. 233–248.
9. United States Environmental Protection Agency. 1986. *Test Methods for Evaluating Solid Waste.* U. S. Environmental Protection Agency, Office of Solid Waste and Emergency Response, Washington, D.C.
10. Sokal RF, Rohlf FJ. 1981. *Biometry,* 2nd ed. W. H. Freeman and Company, San Francisco, CA.
11. Colton T. 1974. *Statistics in Medicine.* Little, Brown Inc., New York.

# 5 The Cantara Spill: A Case Study — Pesticide Transport in a Riverine Environment

*Camilla M. Saviz, John F. DeGeorge, Gerald T. Orlob, and Ian P. King*

## CONTENTS

Abstract ............................................................................................................. 71
Introduction ...................................................................................................... 72
    Study Objectives ....................................................................................... 74
Methods ............................................................................................................ 74
    Geometry of the Upper Sacramento River .............................................. 75
    Hydrodynamic Modeling ......................................................................... 76
    Run-Riffle-Pool Modification .................................................................. 76
    Hydrodynamic Boundary Conditions ..................................................... 77
    Water Quality Modeling .......................................................................... 78
    Constituent Relations ............................................................................... 79
    Metam Sodium ......................................................................................... 80
    Methyl Isothiocyanate .............................................................................. 81
    Water Quality Boundary Conditions ....................................................... 83
    Model Calibration .................................................................................... 84
Results .............................................................................................................. 84
    Metam Spill Simulations ......................................................................... 85
Conclusions ..................................................................................................... 89
Acknowledgments ........................................................................................... 90
References ....................................................................................................... 91

## ABSTRACT

On July 14, 1991, an accidental spill of the pesticide metam sodium ("metam") occurred by derailment of a tank car of the Southern Pacific Railway at the Cantara Loop near Dunsmuir, CA. The tank car rupture resulted in the discharge of up to 27,000 kg of the pesticide into the Upper Sacramento River. The contaminant was subsequently transported from the spill site downstream to Shasta Lake, a distance of approximately 77.5 km. During transit, metam decomposed by hydrolysis and photolysis into methyl isothiocyanate (MITC) that was diminished in aqueous concentration by volatilization to the atmosphere along the river.

The spill event was simulated mathematically using the two-dimensional hydrodynamic model RMA-2, modified to represent the run-riffle-pool character of the steep gradient stream in the mountainous terrain, coupled with the companion finite element water quality model RMA-4Q, modified to represent the unique physical and chemical processes governing the fate and transport of metam and MITC.

The models were calibrated against field observations of the passage of MITC at downstream locations during several days following the accident. Sensitivity testing of alternative scenarios considering various combinations of hydrolysis, photolysis, and volatilization rates revealed the relative importance of these processes in determining the fates of metam and MITC in the stream and in the overlying atmosphere.

**Keywords:** *methyl isothiocyanate, metam sodium, water quality, hydrodynamics, modeling*

## INTRODUCTION

At about 9:50 p.m. on July 14, 1991, several cars of a Southern Pacific cargo train were derailed at the Cantara Loop crossing on the Upper Sacramento River, approximately 3.7 km downstream of Box Canyon Dam and Lake Siskiyou in the vicinity of Dunsmuir, CA, as shown in Figure 1. One car containing a 32.7% solution of the herbicide metam sodium (sodium *N*-methyl dithiocarbamate, hereinafter referred to as "metam") in aqueous solution[1] fell into the river and ruptured, discharging its contents into the river. Because of the steep canyon walls and limited accessibility at the spill site, the damaged tank car remained in the river for 34 h before it was removed by emergency response teams.

Metam is used in California principally as a preplanting soil fumigant to control fungi, bacteria, insects, unwanted seeds, plants, and nematodes.[2] It is stable in an enclosed environment, but, upon dilution with water and exposure to the atmosphere and soils, it rapidly decomposes into its major breakdown product, methyl isothiocyanate (MITC), with the production of hydrogen sulfide ($H_2S$) and other by-products.[3,4] Upon exposure to light, it decomposes rapidly by photolysis with yields of 15 to 60% of the volatile compound MITC.[5] Hydrolysis is also an important pathway for conversion of metam to MITC (G. Miller, University of Nevada, Reno, personal communication, 1994).

Estimates of the quantity of active metam reaching the Upper Sacramento River ranged from 19,000 to 27,000 kg.[6] This material, released over a short period (probably less than 34 h), was transported and dispersed along the river reach below the spill site and into the Sacramento River arm of Shasta Lake during a period of about three days. MITC concentrations measured at Doney Creek, 66.9 km downstream of the spill site, showed that the peak concentration of the dispersed plume

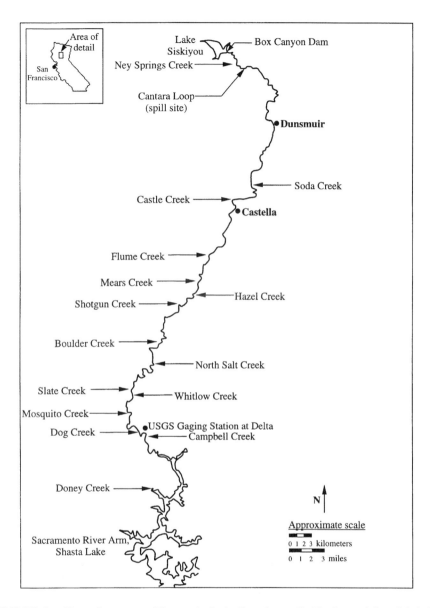

**FIGURE 1** Upper Sacramento River: principal tributaries applied in the model are labeled.

passed in approximately 55.5 h, i.e., the plume traveled at an average velocity of about 0.33 m/s. Upon entering the lake, the mixture of cold water and contaminants mixed with the warmed lake water and submerged to a level of neutral buoyancy. By mechanical mixing of contaminated water, the volatile nature of MITC was used advantageously to reduce concentrations below detectable levels in Shasta Lake. Exposure to the dilute mixture of metam, MITC, and decomposition products resulted in virtually total destruction of aquatic life in the river, including fish, crustaceans, algae, and other aquatic plants. Much of these biota were scoured from the bed and banks and transported into Shasta Lake.

Because of the volatile character of MITC and its relatively short half-life, in the range of 2 to 12 h,[7] the potential for human exposure resulted in temporary evacuation of inhabited areas along the river and considerable uncertainty concerning the potential impacts on vegetation in the riparian corridor. The greatest risks of exposure by people occurred at times and locations where maximum concentrations of volatile products evolved from the water surface to the atmosphere. Assessment of exposure consequences to humans and affected ecosystems required quantification of the dynamic processes involved and prediction of maximum concentrations of MITC volatilized to the atmosphere along the length of the river. Such assessments are facilitated with the aid of mathematical models, provided the relevant processes can be identified, cast in numerical form, and supported by field and laboratory observations. In the case of accidental spills in remote areas, it is seldom possible to obtain the desired direct observations in detail sufficient to support formulation and application of simulation models. The Cantara spill case may be an exception in that a detailed set of MITC concentrations were measured, allowing for quantitative description of the spill episode.

## STUDY OBJECTIVES

The objectives of the study were to develop and apply mathematical models to simulate the fate and transport of MITC and metam in the Upper Sacramento River for the period following the Southern Pacific Cantara Spill, July 14 to 17, 1991. The models were to be capable of representing quantitatively the important hydrodynamic and water quality characteristics of the spill episode. Due to the proximity of the river to populated areas, modeling efforts focused on simulation of maximum concentrations of the contaminants as they were transported downstream. Model results were utilized by an air quality assessment group in determining risks to public health associated with exposure to concentrations of MITC volatilized from the Upper Sacramento River.

## METHODS

Existing finite element models with compatible pre- and post-processing components were modified to simulate the transport and fate of the contaminant plume in the Upper Sacramento River. The models used included the hydrodynamic model, RMA-2, which is capable of simulating velocities and water depths, and the companion model, RMA-4Q, which was used to simulate water quality. Both models are capable of solving the respective flow or water quality equations applied to surface water systems represented solely as one- or two-dimensional, or in systems represented by a combination of one- and two-dimensional elements.

For many rivers, particularly those that are relatively shallow such as the Upper Sacramento River, approximations of vertical mixing are appropriate. A criterion defined by Fischer et al.[8] was used to approximate the distance downstream from the spill site at which the plume could be assumed to be fully mixed across the river. The "mixing length" relation, assuming a discharge at the center of the river, is given by

$$L = 0.1 u_{avg} W^2 / \varepsilon_t \qquad (1)$$

where  L = mixing length (m)
         $u_{avg}$ = average river velocity (0.33 m/s)
         W = average river width (19 m)
         $\varepsilon_t$ = transverse mixing coefficient (m²/s)
         $\varepsilon_t$ represents transverse or lateral turbulent mixing in a river as defined by Fischer et al.[8]

$$\varepsilon_t \cong 0.15 d u^* \qquad (2)$$

where  d = average depth (0.5 m)
         $u^*$ = shear velocity (m/s), defined by

$$u^* = \sqrt{gdS} \qquad (3)$$

where  g = acceleration due to gravity (9.81 m/s²)
         d = average depth (0.5 m)
         S = average slope of the river (estimated to be 0.008)

Using the average values defined previously, the following quantities are calculated: $u^* \cong 0.2$ m/s, $\varepsilon_t \cong 0.015$ m²/s, and $L \cong 0.8$ km.

Since the mixing length, L, was calculated to be less than 1 km and the actual river length was greater than 70 km, simplifying assumptions of full lateral and vertical mixing were appropriate, and the river could be represented as one dimensional, with variations occurring only along the principal axis, i.e., along the river length.

A one-dimensional network geometry representing the Upper Sacramento River and its tributaries was developed and utilized by both RMA-2 and RMA-4Q. RMA-2 was used to obtain velocities, water depths, and cross-sectional areas describing the flow field. RMA-4Q, capable of simulating an array of conservative and non-conservative water quality constituents, was modified to include constituent relations needed to simulate the fate of metam and MITC.

## GEOMETRY OF THE UPPER SACRAMENTO RIVER

The Upper Sacramento River extends from Box Canyon Dam at Lake Siskiyou to the Sacramento River arm of Shasta Lake as shown in Figure 1. At the upstream end, the river below the dam is at an elevation of about 914 m above mean sea level. From this point, the river flows down a steep gradient and through a narrow canyon at its headwaters, dropping at a gradually decreasing gradient until is enters the lake, which at the time of the spill stood at about 323 m above mean sea level. The river is characterized by randomly spaced pools, riffles, and cascades with highly irregular cross sections.

A finite element representation of the river geometry was produced using RMA-GEN, a pre-processor network generation program, utilizing geographic information digitized from 1:24000 U.S. Geological Survey (USGS) topographic maps to define the course of the river. The network was initially discretized into one-dimensional reaches, or elements, each 250 m in length. Refinement of the network in the vicinity of Cantara Loop and at tributary junctions resulted in a finite element network containing 546 elements ranging in length from 30 to 250 m and extending over a total distance of 77.5 km.

The river slope over each reach was estimated using 1:24000 USGS topographic maps. Because the river is shallow in its upper reaches, between 0.4 and 0.6 m deep, river slopes were interpolated between topographic contours. A Manning's roughness coefficient of 0.044 was used to represent the frictional resistance of the rough streambed consisting largely of cobbles and boulders. Cross sections and channel widths were estimated at 30m intervals from 1:3600 Geographic Information System (GIS) riparian habitat maps of the system.[9]

Flows from 14 small streams tributary to the Upper Sacramento River were introduced into the finite element network at locations shown in Figure 1. Historic records of flows were used to estimate contributions from each tributary, since flow data were not available for the period of the spill. Information from aerial photographs of the river taken on July 20, 1991 was also used in estimating relative contributions of tributary flows. Photographs showed inflows from Castle, Shotgun, Soda, Mears, and Campbell Creeks and smaller contributions from the remaining nine tributaries shown in Figure 1. Other tributaries, those with flows less than 0.05 m$^3$/s, were not included explicitly in the model, but were considered in the overall water balance.

## HYDRODYNAMIC MODELING

The finite element network geometry was input to a modified version of the hydrodynamic model, RMA-2, developed originally by Norton et al.[10] and since updated extensively by King.[11] Using the finite element method, RMA-2 solves the depth and cross-sectionally averaged shallow water equations for momentum and continuity to provide temporal and spatial descriptions of depths and velocities within a water body. RMA-2 has been applied extensively to one- and two-dimensional representations of rivers, reservoirs, lakes, and estuaries in both steady and unsteady modes.[12-14] The model was modified for this study to simulate flow in a mountainous stream with the run-riffle-pool characteristics which existed in the Upper Sacramento River during the spill episode.

## RUN-RIFFLE-POOL MODIFICATION

Flow behavior in the Upper Sacramento River is typical of streams in mountainous terrain: nonuniform in both geometry and gradient and characterized by sequences of pools bounded by steep gradient riffles and cascades. Explicit simulation of the detailed hydrodynamics of such so-called run-riffle-pool systems was beyond the scope of this study, yet the modified RMA-2 model provided acceptable simulation of the flows, velocities, and travel times of the pollutant plume through the river reach. As illustrated schematically in Figure 2, the model was constructed to represent a series of regions each with unique properties of subcritical flow (pools and runs) and steep drops (riffles) at element boundaries. A bed slope factor, $\beta$, was introduced into the governing momentum equation to accommodate the increased resistance to flow evident in the stream's behavior. The modified version of RMA-2 solved the following equation set for depth averaged flow.

Conservation of mass:

$$A\frac{\partial u}{\partial x} + u\frac{\partial A}{\partial x} + \frac{\partial A}{\partial h}\frac{\partial h}{\partial t} = 0 \quad (4)$$

Conservation of momentum:

$$\rho\left[A\frac{\partial u}{\partial t} + Au\frac{\partial u}{\partial x}\right] - \frac{\partial}{\partial x}\left(AE_{xx}\frac{\partial u}{\partial x}\right) + \rho gA\left[\beta\frac{\partial a}{\partial x} + \frac{\partial h}{\partial x}\right] - \frac{gAh}{2}\frac{\partial \rho}{\partial x} - A\Gamma_x = 0 \quad (5)$$

where  $A$ = cross-sectional area
  $\rho$ = density
  $u$ = depth- and cross-sectionally averaged velocity
  $E_{xx}$ = longitudinal eddy viscosity
  $a$ = bottom elevation
  $g$ = acceleration of gravity
  $h$ = water depth
  $\Gamma_x$ = body force
  $t$ = time
  $\beta$ = bed slope scaling factor; $0 < \beta < 1$

A singular modification of the original RMA-2 equation set for this study was the introduction of a bed slope scaling factor, $\beta$, for the run-riffle-pool representation. As $\beta$ is decreased, the effective bed slopes within elements are reduced, and the effective head losses due to riffles increase, resulting in deeper water over elements and slower average velocities. $\beta$ was used to calibrate the hydrodynamic model.

## Hydrodynamic Boundary Conditions

Flow discharged from Box Canyon Dam is regulated at the Lake Siskiyou Power Plant and provides the upstream boundary condition. Only minor fluctuations, less than 0.06 m³/s about a mean of 1.17 m³/s, were observed during the 72-h period of pesticide transport.[15] Because the spill occurred in July during the middle of an extended drought, tributary inflows were likely to be steady. Since upstream and tributary flows together with gaged flows at Delta about 60 km downstream were virtually steady during the period simulated,[16] it was decided to operate the model in a steady-state mode. The downstream boundary condition was defined by a stage-discharge relationship based on Manning's equation and assuming steady flow.

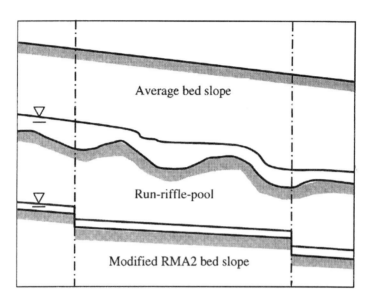

**FIGURE 2** Model representation of the run-riffle-pool sequence.

The difference of 4.49 m$^3$/s between the assumed steady flow at Box Canyon Dam (1.17 m$^3$/s) and the USGS gaging station at Delta (5.66 m$^3$/s), was distributed among the 14 larger tributaries to the Upper Sacramento River. Since actual tributary flow data were unavailable, the tributary contributions were scaled in proportion to the drainage areas, historical flow surveys performed by the California Department of Fish and Game, and estimates of relative flows from aerial photographs. Ungaged groundwater accretions along the length of the Upper Sacramento River were included with the tributary inflows in the overall hydrologic balance. Using the approximated steady boundary conditions at the upstream boundary and for the tributary inflows, the hydrodynamic model was calibrated by adjusting the frictional resistance and bed slope scaling factors, n and β, although Manning's n was modified minimally from its initially estimated value. Flow continuity was checked by comparison of model flows to those observed at both Delta and Castella. Once calibrated, the hydrodynamic simulation results provided estimated velocities and water depths throughout the system which were then used to run the water quality model.

## WATER QUALITY MODELING

The finite element water quality model, RMA-4Q,[17,18] was used to simulate transport of metam and MITC over the 72-h period following the spill. The water quality model utilized the same network geometry used in the hydrodynamic simulations, together with the velocities and depths generated by RMA-2. RMA-4Q simulated the transport and fate of multiple constituents by solving the one-dimensional, cross-sectionally averaged advection-dispersion equation for each constituent modeled:

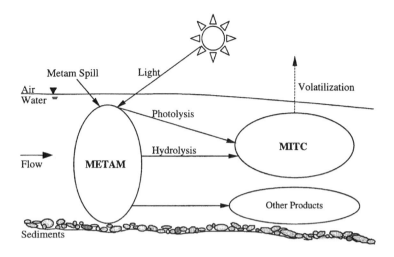

**FIGURE 3** Processes simulated for fate of metam and MITC in the water column.

$$A\left[\frac{\partial C}{\partial t} + u\frac{\partial C}{\partial x}\right] + C\left[\frac{\partial A}{\partial t} + \frac{\partial uA}{\partial x}\right] + \frac{\partial}{\partial x}AD_{xx}\frac{\partial C}{\partial x} - rC - S = 0 \qquad (6)$$

where  A = cross-sectional area
t = time
u = depth- and cross-sectionally averaged velocity
r = effective growth (or decay) rate
C = constituent concentration
S = mass source rate
$D_{xx}$ = longitudinal diffusion coefficient

The effective first order growth or decay rate, r, is constituent-dependent and may be a nonlinear function involving multiple constituents and parameters. Computation of the effective first order rate terms for metam and MITC, the two constituents simulated in this study, is described in the following sections. The conservative form of the advection-dispersion equation was used in RMA-4Q to ensure that mass conservation was enforced in the water quality simulations.[19]

## CONSTITUENT RELATIONS

When diluted with water and exposed to sediments, suspended matter, and light, metam rapidly decomposes into its primary breakdown product, MITC. To model the fate and transport of each compound in the river, constituent relations were written for the decay of metam and its decomposition to MITC due to photolysis and hydrolysis and for volatilization of MITC. The principal decomposition processes simulated for metam and MITC are shown schematically in Figure 3.

## METAM SODIUM

Metam sodium (metam) decomposition in water is influenced by the presence of light, suspended solids, dissolved oxygen content, water temperature, and some other environmental factors that are not readily quantifiable and are uncertain in their effects. Consequently, estimates of the degradation half-life of metam vary widely, from as little as 30 min to as much as 180 h.[4] The decay rate constant for metam was approximated using first order kinetics according to the following relation:

$$C_{MS} = C_o \exp(-k_{MS} t) \quad (7)$$

where  $C_{MS}$ = concentration of metam (mg/L)
$C_o$ = initial concentration of metam (mg/L)
$k_{MS}$ = decay rate constant for metam (1/s)
$t$ = time (s)

Using the half-life of metam, $t = (t_{1/2})_{MS}$, Equation 7 may be written as

$$\frac{C_o}{2} = C_o \exp\left(-k_{MS} (t_{1/2})_{MS}\right) \quad (8)$$

Solving for $k_{MS}$, the decay rate constant:

$$k_{MS} = \frac{\ln 2}{(t_{1/2})_{MS}} \quad (9)$$

$k_{MS}$ may be used to determine the rate of decay of metam, estimated by

$$\frac{dC_{MS}}{dt} = -k_{MS} C_{MS} \quad (10)$$

Metam decomposes in the presence of light (photolysis) and water (hydrolysis), each with different relative contributions to the breakdown based on environmental conditions. Concentrations of other degradation products present in low concentrations, e.g., hydrogen sulfide, were not included in this study. The degradation half-life of metam is shorter for photolysis than for hydrolysis, but hydrolysis reactions can produce higher concentrations (yields) of MITC. Some values recommended for the photolysis half-life of metam in aqueous solution are 30 min, 1.6 h, and 2.5 h.[3]

At the time this study was conducted, the behavior of metam in water had not been well described since application to soil was its principal purpose and its primary pathway into the environment. Laboratory work by Miller performed using dilute aqueous solutions showed hydrolysis to be an important reaction pathway for metam conversion to MITC and suggests that in the presence of sediments, hydrolysis half-life can drop to as low as a few hours with a yield of MITC up to 85% (G. Miller,

University of Nevada, Reno, personal communication, 1994). Values recommended for the hydrolysis half-life of metam varied between 4 and 12 h. Although sorption to sediments was not explicitly modeled, the estimates of hydrolysis yield were chosen to correspond to those values indicated for hydrolysis in the presence of sediments. Recently, researchers have proposed that oxidation may be a principal decomposition mechanism in water instead of hydrolysis.[20]

## METHYL ISOTHIOCYANATE

Methyl isothiocyanate (MITC) is a volatile compound produced by the breakdown of metam. Volatilization, the primary degradation pathway for MITC, generally occurs more rapidly than hydrolysis and is dependent on concentration, wind velocity, stream turbulence, stream velocity, temperature, and other environmental factors. Estimates of the half-life of MITC range from 20 days in still water to 7.7 to 16 h in a steady flow, although greater rates can be expected as a result of turbulence in rapidly flowing riffles of the Upper Sacramento River.[4] To estimate the decay rate of MITC, a first order decay relation was used.

$$C_{MITC} = C_o \exp(-k_{MITC} t) \quad (11)$$

where $C_{MITC}$ = concentration of MITC (mg/L)
$C_o$ = initial concentration of MITC (mg/L)
$k_{MITC}$ = decay rate constant for MITC (1/s)
$t$ = time (s)

Using the half-life of MITC, $t = (t_{1/2})_{MITC}$, Equation 11 may be written as

$$\frac{C_o}{2} = C_o \exp\left(-k_{MITC}(t_{1/2})_{MITC}\right) \quad (12)$$

Solving for $k_{MITC}$, the decay rate constant:

$$k_{MITC} = \frac{\ln 2}{(t_{1/2})_{MITC}} \quad (13)$$

The constituent relation for MITC used in the water quality model is given by

$$\frac{dC_{MITC}}{dt} = -k_{MITC} C_{MITC} + S_{MS} \quad (14)$$

where $S_{MS}$ is the source of MITC from degradation of metam (mg/L·s).

The first term in Equation 14 represents volatilization and is considered to be the only mechanism for loss of MITC. $k_{MITC}$ is evaluated using the expression given in Equation 13. The second term in Equation 14 represents MITC yield from metam, considered to be the only source. The combined production of MITC due to conversion from metam was calculated by

$$S_{MS} = \sum_{i=1}^{2} k_{MS,i} C_{MS} \left( \frac{MW_{MITC}}{MW_{MS}} \right) C_{y,i} \qquad (15)$$

where  $i$ = decomposition mechanism ($i = 1$ for decomposition by photolysis and $i = 2$ for decomposition by hydrolysis)
$k_{MS,i}$ = decay rates of metam for hydrolysis and photolysis, respectively
$MW_{MITC}$ = molecular weight of MITC = 73.1 g/mol
$MW_{MS}$ = molecular weight of metam = 129.2 g/mol
$C_{y,i}$ = conversion yield of MITC (amount of metam converted as a result of each decomposition mechanism)
$C_{y,1}$ = 20% for photolysis (reported values range from 10 to 60%)
$C_{y,2}$ = 85% for hydrolysis (reported values range from 60 to 90%)

The half-life of MITC is given by[3]

$$(t_{1/2})_{MITC} = \ln 2 \left( \frac{D}{K_L} \right) \qquad (16)$$

where  $D$ = average element water depth (m)
$K_L$ = mass transfer coefficient (m/s) defined by[3]

$$K_L = \frac{H \cdot k_g \cdot k_l}{(H \cdot k_g + k_l)} \qquad (17)$$

where  $H$ = Henry's Law Constant (air/water partition coefficient) = 0.007 for MITC at 20°C
$k_g$ = gas phase exchange coefficient (m/s)
$k_l$ = liquid phase exchange coefficient (m/s)

An important consideration in estimating volatilization of MITC from the river was the potential contribution of wind at the water surface and its effect on the volatilization rate.[21] $k_g$, the gas phase exchange coefficient, was estimated as a function of the stream velocity, water depth, and wind velocity from an empirical relation.[3]

$$k_g = 1137.5(V_w + V_c) \left( \frac{18}{MW_{MITC}} \right)^{1/2} \left( \frac{1}{3.6 \times 10^5} \right) \qquad (18)$$

where  $V_w$ = wind velocity (m/s)
  $V_c$ = average water velocity in the element (m/s)

$k_g$ varied for each element in the model as a function of the average water velocity ($V_c$) in the element. $k_l$, the liquid phase exchange coefficient, was estimated by the following relation in terms of the water depth (D) and the average water velocity ($V_c$) in each element:

$$k_1 = 23.51 \left( \frac{V_c^{0.969}}{D^{0.693}} \right) \left( \frac{32}{MW_{MITC}} \right)^{1/2} \left( \frac{1}{3.6 \times 10^5} \right) \tag{19}$$

The exponents and coefficients for the gas and liquid exchange coefficients were determined empirically.[3] By combining the expression for $(t_{1/2})_{MITC}$ from Equation 16 with $k_{MITC}$ from Equation 13, the decay rate of MITC could be calculated by RMA-4Q using other values predicted in the simulation.

$$k_{MITC} = \frac{K_L}{D} \tag{20}$$

It is noted in the California Environmental Protection Agency (CEPA)[3] report that Equation 20 may yield an MITC half-life that is too short when compared to MITC concentrations actually observed at Doney Creek.[6] The decay rate of MITC as predicted by Equation 20 is highly sensitive to several physical parameters including water velocity, water depth, and, in particular, wind velocity. In calculation of $k_g$, if the estimated wind velocity is higher than what actually occurs, rapid volatilization of MITC would result and less would remain in the lower reaches of the simulated system. Additionally, the estimated concentration of MITC in the river is not only a function of the MITC half-life, but also of the yield of MITC by decomposition of metam. Estimates of MITC yield from metam due to hydrolysis and photolysis vary over a wide range, possibly contributing to the difference between predicted and measured concentrations of MITC in the river.

## WATER QUALITY BOUNDARY CONDITIONS

Boundary conditions for the metam spill required estimating the quantity of active ingredient spilled and the duration of the spill into the river. For the purposes of this study, a maximum contaminant load of 27,000 kg was assumed to represent "worst case" conditions that would cause the highest concentrations of MITC in the river. In the simulations, the spill was estimated to have occurred as a single contaminant "slug" distributed uniformly over a 2-h period.

Additional information used by the water quality model to predict the fate and transport of the contaminant included experimentally determined estimates of the yield of MITC from metam based on the predominant decomposition mechanisms, i.e., hydrolysis and photolysis. The formulation used to estimate MITC concentrations required input of an estimated wind velocity, which for our case was assumed to be steady. The wind velocity was varied in calibration of the model to achieve a volatilization rate that would assure that simulated concentrations at Doney Creek, just upstream of Shasta Lake, matched observed data.

A "light" switch was included in the model to allow for photodecomposition of metam only during daylight hours. Daylight hours are computed within a meteorological subroutine in RMA-4Q which computes the times of sunrise and sunset based on a Julian day calendar. To compensate for the decreased exposure to sunlight due to shading by steep canyon sides, the daylight period was shortened by 1 h each for the times of sunrise and sunset during the simulation period.

## MODEL CALIBRATION

At the time of the spill, experimental procedures for the measurement of metam in water were generally considered less reliable than procedures for direct detection and measurement of MITC. Therefore, MITC data determined by direct field observation were used for calibration and verification of the hydrodynamic and water quality models. The most complete sampling analysis from hourly measurements showed a history of MITC concentrations as the plume passed the sampling location at Doney Creek from midnight to 10 a.m. on July 17, 1991.[6] At other locations between the spill site and Shasta Lake, the limited MITC concentration data available were insufficient for calibration of the model, but were used for checking of model results at intermediate locations.

Simulated flow results were calibrated against measured flows at Delta and Castella. Travel time was tested using the times of arrival of the leading edge of the plume and the peak of the contaminant distribution pattern at Doney Creek. Due to the lack of direct observations of stream velocities and depths, the best information available for calibration of the hydrodynamic model was the travel time of 55.5 h for the peak concentration of the plume to move from the spill site to Doney Creek. Calibration was achieved by adjusting the bed slope scaling factor within a range of 0.069 and 0.074 until the simulated peak occurred at a time equal to that observed. Sample calibration results are shown in Figure 4.

## RESULTS

Results of the hydrodynamic simulation are shown in Figure 5. Typical velocities in the system varied from 0.3 to 0.4 m/s and depths varied from 0.3 to 0.8 m. Effects of tributary inflows over different reaches of the river are seen as abrupt steps in the flow profile. Small oscillations in the flow profile are numerical artifacts of the finite element solution technique and indicate regions where numerical diffusion may prevent mass conservation. Variations within 5% of the predicted values are typically considered acceptable. The oscillations were minimized by refining the finite element network in regions of rapidly changing cross section.

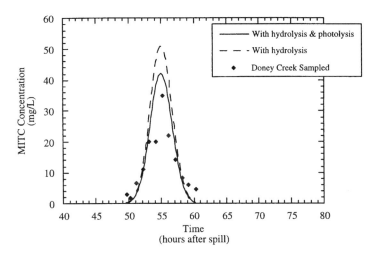

**FIGURE 4**  Calibration of travel time to MITC concentrations at Doney Creek.

## METAM SPILL SIMULATIONS

Ten alternative spill scenarios, summarized in Table 1, were developed using available information with approximations for most likely behavior, as well as for extreme or worst case conditions. All scenarios were based on the largest estimate of metam discharges from the ruptured railroad tank car, assumed to be 27,000 kg of active material spilled over a 2-h period. Scenarios considered conversion of metam to MITC due to either or both hydrolysis and photolysis. Conversion by photolysis was assumed to yield 20% MITC with a metam half-life of 2 h, active only during daylight hours. Conversion by hydrolysis was estimated to yield 85% MITC with half-lives of 4, 8, and 12 h.

Six model runs were made using the minimum volatilization rate achievable by setting the wind speed to zero. Runs 1 through 4 produced concentrations at Doney Creek which exceeded observed values. These scenarios were then repeated in Runs 7 through 10, increasing the volatilization rate by increasing the wind speed until the resulting concentration time histories matched observed data.

Simulation results for Runs 4 through 6 are shown in Figures 6 and 7, comparing model results to observed MITC concentrations at Doney Creek. Influences of both hydrolysis and photolysis are represented in Figure 6. The best approximation of the peak concentration observed is provided by an 8-h hydrolysis half-life. Results indicate accurate estimates of plume arrival time, but two data points near hour 60 indicate a long trailing edge of the plume which was not predicted by the model. The total mass of MITC, represented by the areas under the measured and simulated curves, is underpredicted by the model, but peak concentrations show reasonable agreement.

Several possible factors contribute to the discrepancy between the observed and simulated shapes of the contaminant concentration profile. First, the actual spill

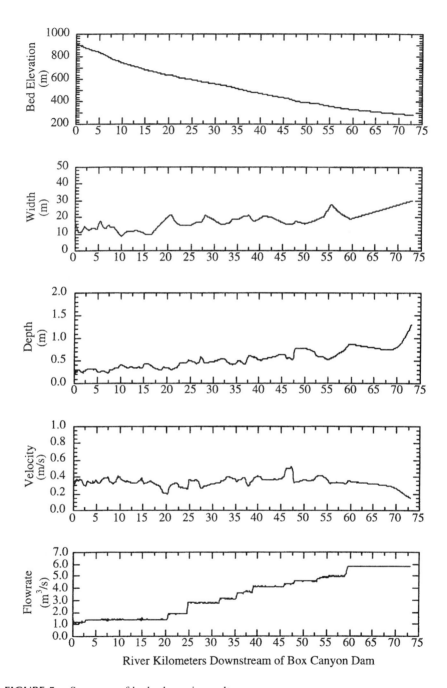

**FIGURE 5** Summary of hydrodynamic results.

configuration was probably different from the "worst-case" uniform 2-h spill scenario assumed. Second, it is likely that some of the contaminant may have been temporarily stored in pools, in local eddies, or in shallow water areas along banks,

### TABLE 1
### Alternative Spill Scenarios

| Run | Decomposition Mechanism | Hydrolysis Half-Life | Wind Speed |
|---|---|---|---|
| 1 | Hydrolysis only | 4 h | 0 m/s |
| 2 | Hydrolysis only | 8 h | 0 m/s |
| 3 | Hydrolysis only | 12 h | 0 m/s |
| 4 | Hydrolysis and photolysis | 4 h | 0 m/s |
| 5 | Hydrolysis and photolysis | 8 h | 0 m/s |
| 6 | Hydrolysis and photolysis | 12 h | 0 m/s |
| 7 | Hydrolysis only | 4 h | 0.15 m/s |
| 8 | Hydrolysis only | 8 h | 0.30 m/s |
| 9 | Hydrolysis only | 12 h | 0.45 m/s |
| 10 | Hydrolysis and photolysis | 4 h | 0.065 m/s |

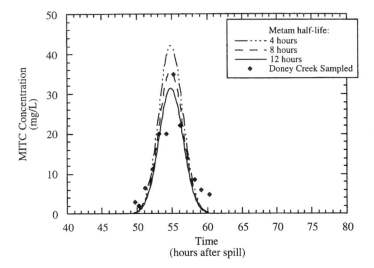

**FIGURE 6** MITC concentration vs. time at Doney Creek. Runs 4, 5, and 6 with metam decomposition by hydrolysis and photolysis.

then gradually released after the bulk of the contaminant had passed. RMA-4Q does not currently include capabilities for simulating temporary storage of water quality constituents. The lack of complete data at locations other than Doney Creek prevents assessment of the actual behavior of the plume at upstream locations.

As shown in Figure 7, volatilization of MITC was found to be very sensitive to wind velocity (represented by Equation 18). To simulate the concentrations observed at Doney Creek, the volatilization rate of MITC was adjusted by modifying the wind velocity within a range of 0.4 to 1.0 m/s, values of the average wind speed anticipated over the river. Actual meteorological and wind data were unavailable for the Upper

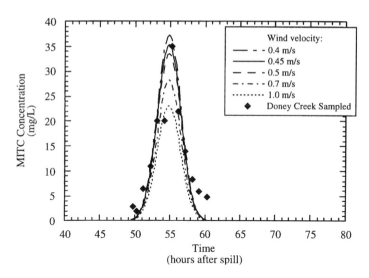

**FIGURE 7**  MITC concentration vs. time at Doney Creek. Comparison of wind effects (12-h metam hydrolysis half-life, hydrolysis only).

Sacramento River during the period of the spill. Among the values tested, a wind speed of 0.45 m/s produced the best model estimate of peak MITC concentration. This estimated wind speed was used in subsequent simulations to estimate the mass of MITC volatilized.

The water quality model was used to calculate the mass flux of MITC volatilized through the water surface at each node in the finite element network for each time step throughout the 72-h simulation. Integration over space and time provided an estimate of the total mass transferred to the atmosphere overlying the river surface. The spatial distribution of the mass transfer provided an indication of the locations along the river most likely to have experienced the greater atmospheric concentrations. As illustrated in Figure 8, considering only degradation of metam by hydrolysis and half-lives of 4, 8, and 12 h, a short half-life produced the maximum flux per 2-km reach and affected most severely the upper reaches of the river below the spill location. Longer half-lives caused volatilization to shift farther downstream, diminishing the flux in the upper reaches. A longer estimate of the hydrolysis half-life is likely to be more realistic for the actual spill episode since the headwater temperatures were relatively low during the spill period, decreasing the rate of hydrolysis.

Including photolysis in the model not only reduced the total MITC evolved to the atmosphere, but accentuated the effect of assuming a short (4-h) half-life, as shown in Figure 9. When photolysis was included, the hydrolysis half-life had little effect on volatilization downstream, but volatilization upstream was reduced. The volatilization rate is a function of the in-stream depth and velocity and the MITC concentration in the water column. The rapid decrease seen in volatilization immediately following river kilometer 20 was due primarily to dilution by inflows from Soda Creek and Castle Creek. For both scenarios, simulation results showed that elevated concentrations of volatilized MITC occurred along the river up to a distance

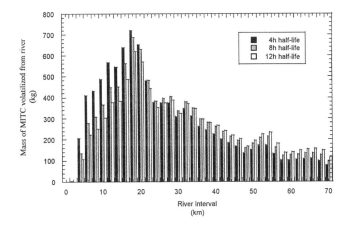

**FIGURE 8** Mass of MITC volatilized in 2-km river reaches downstream of Box Canyon Dam. Runs 7, 8, and 9 show metam decomposition by hydrolysis.

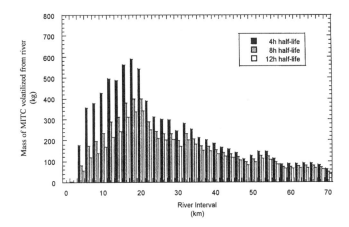

**FIGURE 9** Mass of MITC volatilized in 2-km river reaches downstream of Box Canyon Dam. Runs 5, 6, and 10 show decomposition by hydrolysis and photolysis.

of 25 km downstream of the spill site, including the river reach in the vicinity of the town of Dunsmuir.

## CONCLUSIONS

Coupled hydrodynamic and water quality models were used to simulate the transport and fate of a pesticide spill in the Upper Sacramento River. The hydrodynamic model, RMA-2, with a run-riffle-pool modification provided a simulated flow field for the water quality model, RMA-4Q. The water quality model, modified to include constituent relations for metam and its degradation product, MITC, was used to estimate the mass of MITC volatilized as the plume traveled downstream during the 3-day period following the Cantara spill. Results from several simulation scenarios were provided to an air quality modeling group to assist in evaluating atmospheric fate and transport of MITC and determining health risks associated with exposure to the contaminant.

For this study, a worst case condition was assumed for all simulations where the entire 27,000 kg of metam was released from the tank car at a uniform rate over a 2-h period, effectively causing the highest peak concentrations during transit downstream. Simulated concentrations were influenced by the flow field, spill boundary conditions, wind velocity, plume dispersion, and estimates of processes and decay rates for metam decomposition to MITC and volatilization of MITC. The assumed hydrolysis rate of metam strongly influenced the mass of MITC volatilized from the upstream reaches of the system. Due to the higher yield of MITC from metam by hydrolysis, it was found to have a greater effect than photolysis on the mass of MITC volatilized from the water surface. If photolysis was in fact an important mechanism, the total mass of MITC volatilized would be lower because of the relatively low yield of photolysis compared to hydrolysis.

Considering the uncertainty in estimating both the boundary conditions and model parameters required for these simulations, it is proper to view the model output as a best estimate of peak concentrations during the period following the actual spill. It is clear that there are alternative scenarios which can reasonably approximate the actual concentration time history observed at Doney Creek. However, given the assumptions made in developing these spill scenarios, it was possible to estimate the amount of metam that was volatilized as MITC over the entire river and the amount volatilized over the first 20 km downstream of the spill in the vicinity of Dunsmuir.

Among lessons learned from the Cantara Spill is the need for comprehensive monitoring of the contaminant in both water and air and those environmental conditions that influence its movement and concentration. Early deployment of monitoring teams at strategic locations along the river or in the reservoir downstream is critical. Continuous or frequent sampling is necessary to provide the data needed to understand the episode and its effects. In cases where a high potential exists for accidental spills, such as along transport routes located in the vicinity of sensitive aquatic systems, advance planning should include physical characterization of the system, *a priori* modeling of aquatic systems, identification of strategic monitoring sites, and organization of databases of useful information. The databases should contain hydrologic, meteorologic, and chemical data, as well as information on the aquatic ecosystem and habitat characteristics. The unique experience of modeling the Cantara Spill provides a foundation for improved assessment of future catastrophic episodes should they occur.

## ACKNOWLEDGMENTS

This study was supported by the California Department of Fish and Game in a project coordinated by Dr. Michael Fry at the University of California, Davis. Data, information, and assistance were generously provided by staff of the Central Valley Regional Water Quality Control Board and the California Department of Fish and Game. Dr. Glenn Miller of the University of Nevada, Reno and Dr. Robert Yamartino of Sigma Research Corporation provided additional input on MITC volatilization and metam decomposition processes.

For questions or further information regarding the work described in this chapter, please contact:

> **Dr. Gerald T. Orlab**
> Dept. of Civil and Environmental Engineering
> University of California
> One Shields Avenue
> Davis, CA 95616
> (530) 752-1424
> e-mail: gtorlob@ucdavic.edu

## REFERENCES

1. Alexeeff, G. V., D. J. Shusterman, R. A. Howd and R. J. Jackson. 1994. Dose-Response Assessment of Airborne Methyl Isothiocyanate (MITC) Following a Metam Sodium Spill. *Risk Analysis.* 14: 191–198.
2. Nihon Schering Agrochemical Division and Agrochemical and Animal Health Products Development Department, Shionogi & Company. 1990. Summary of Toxicity Data on Methyl Isothiocyanate. *J. Pesticide Research,* 15: 297–304.
3. California Environmental Protection Agency (CEPA), Office of Environmental and Health Hazard Assessment. 1992. Evaluation of the Health Risks Associated with the Metam Spill in the Upper Sacramento River. Hazard Identification and Risk Assessment Branch, Berkeley, CA.
4. California Department of Fish and Game (DFG). 1993. Natural Resource Damage Assessment Plan, Sacramento River: Cantara Spill, Shasta and Siskiyou Counties. Final Report. Redding, CA.
5. Draper, W. M. and D. E. Wakeham. 1993. Rate Constants for Metam-Sodium Cleavage and Photodecomposition in Water. *J. Agriculture Food Chemistry,* 41: 1129–1133.
6. Central Valley Region, Regional Water Quality Control Board (CVRWQCB). 1991. Final Water Sampling Report, Southern Pacific–Cantara Spill. Redding, CA.
7. Segawa, R. T., S. J. Marade, N. K. Miller and P. Y. Lee. 1991. Monitoring of the Metam Sodium Spill. Report no. EH 91-08. State of California Environmental Protection Agency, Department of Pesticide Regulation, Environmental Hazards Assessment Program. Sacramento, CA.
8. Fischer, H. B., E. J. List, R. C. Y. Koh, J. Imberger and N. H. Brooks. 1979. *Mixing in Inland and Coastal Waters.* Academic Press, San Diego, CA.
9. Center for Ecological Health Research (CEHR). 1994. Riparian Vegetation Maps of the Upper Sacramento River (1:3600 scale). University of California, Davis, CA.
10. Norton, W. R., I. P. King and G. T. Orlob. 1973. A Finite Element Model for Lower Granite Reservoir. Report prepared for the Walla Walla District, U.S. Army Corps of Engineers, Walla Walla, WA.

11. King, I. P. 1993. RMA-2 — A Two Dimensional Finite Element Model for Flow in Estuaries and Streams, v. 4.4. University of California, Davis, CA.
12. Deas, M. L., G. K. Meyer, C. L. Lowney, G. T. Orlob and I. P. King. 1997. Sacramento River Temperature Modeling Project. Final Report, no. 97-1. Center for Environmental and Water Resources Engineering, Department of Civil and Environmental Engineering, University of California, Davis, CA.
13. Shrestha, P. L. and G. T. Orlob. 1996. Multiphase Distribution of Cohesive Sediments and Heavy Metals in Estuarine Systems, *J. Environmental Engineering,* 122: 730–740.
14. Shrestha, P. L., C. M. Saviz, G. T. Orlob and I. P. King. 1993. Hydrodynamic Simulation of San Francisco Bay and Delta for Oil Spill Fate Studies. Final Report, no. 93-1. Center for Environmental and Water Resources Engineering, Department of Civil and Environmental Engineering, University of California, Davis, CA.
15. Lake Siskiyou Hydropower Plant. 1991. Ten-Minute Flow and Temperature Data Measured at Box Canyon Dam Outflow. Obtained from the California Department of Fish and Game, Redding, CA.
16. United States Geological Survey (USGS). 1991. Water Resources Data, California. Series title: U.S. Geological Survey Water-Data Report. Sacramento, CA: U.S. Department of the Interior, Geological Survey, Springfield, VA.
17. King, I. P. 1990. RMA-4 — A Two Dimensional Finite Element Water Quality Model, v. 3.1. University of California, Davis, CA.
18. King, I. P. and J. F. DeGeorge. 1994. RMA-4Q — A Two Dimensional Finite Element Model for Water Quality in Estuaries and Streams, v. 1.0. University of California, Davis, CA.
19. Saviz, C. M., J. F. DeGeorge, G. T. Orlob and I. P. King. 1994. Modeling the Fate of Metam Sodium and MITC in the Upper Sacramento River, the Cantara–Southern Pacific Spill. Report to the California Department of Fish and Game, Cantara Program, Redding, CA.
20. Wang, P. F., T. Mill, J. L. Martin and T. A. Wool. 1997. Fate and Transport of Metam Spill in Sacramento River. *J. Environmental Engineering,* 123: 704–712.
21. Schwarzenbach, R. P., P. M. Gschwend and D. M. Imboden. 1993. *Environmental Organic Chemistry.* John Wiley & Sons, New York.

# 6 Distribution and Transport of Air Pollutants to Vulnerable California Ecosystems

*Andrzej Bytnerowicz, John J. Carroll, Brent K. Takemoto, Paul R. Miller, Mark E. Fenn, and Robert C. Musselman*

## CONTENTS

Abstract ..................................................................................................................94
Introduction ............................................................................................................95
Air Pollutants .........................................................................................................95
Vulnerable California Ecosystems .........................................................................96
Transport of Pollutants to Vulnerable Ecosystems ................................................96
Methods Used for Field Air Pollution Monitoring ...............................................97
   Ozone Passive Samplers ....................................................................................98
   Gaseous Nitric Acid, Nitrous Acid, Sulfur Dioxide, Ammonia,
   Particulate Nitrate, and Ammonium ..................................................................98
      Annular Denuder Samplers .........................................................................
      Honeycomb Denuder Samplers ..................................................................
   Wet Precipitation ................................................................................................99
      Rain Collectors ............................................................................................
      Active Fog Collectors .................................................................................
      Passive Fog Collectors ................................................................................
   Wet Plus Dry Deposition Collectors ...............................................................100
   Dry Deposition to Plants .................................................................................101
      Estimates of Stomatal Uptake of Gaseous Pollutants by Plants ................
      Estimation of Surface Dry Deposition of Gases and Particles ..................
Results of Case Studies and Monitoring Networks ............................................103
   Ozone in the Mixed Conifer Forests of the Sierra Nevada and
   San Bernardino Mountains ..............................................................................103
   Wet Deposition of Nitrogen and Sulfur in the Sierra Nevada .......................103
   Gaseous and Particulate Nitrogenous Pollutants in the Sierra Nevada ..........107
   Dry Deposition of Nitrate and Ammonium to Trees in the
   Sierra Nevada ..................................................................................................108

Nitrogenous Pollutants on a Vertical Transect in a Mixed Conifer
Canopy of the San Bernardino Mountains......................................................... 108
Wet and Dry Deposition in the San Gorgonio Wilderness Area..................... 109
Nitrogen Deposition in the Coastal Sage Scrub Ecosystems
of Southern California .......................................................................................... 111
Development and Application of Deposition Models............................................ 111
Critical Doses/Loads Concept and Application .................................................... 112
    Ozone Critical Exposure Levels........................................................................ 112
    Nitrogen Critical Loads .................................................................................... 112
    Geographic Information Systems (GIS).......................................................... 113
Future Research Needs and Recommendations.................................................... 114
Acknowledgments........................................................................................................ 114
References ................................................................................................................... 114

## ABSTRACT

Exposures to ozone ($O_3$), nitrogenous (N), and sulfurous (S) pollutants may have serious effects on vulnerable ecosystems in California and elsewhere. Elevated $O_3$ concentrations have been suggested as a factor contributing to the decline of sensitive forest tree species in the San Bernardino and Sierra Nevada Mountains. Long-term exposure to elevated deposition of N pollutants may lead to N saturation of mountain ecosystems and may affect their health and sustainability. Direct toxic effects of ambient concentrations of some N air pollutants (e.g., gaseous nitric acid or peroxyacetyl nitrate) are also possible. The methodology used by the authors for measuring concentrations and deposition of air pollutants in remote locations of California is reviewed. Special emphasis is put on methodologies which can be applied in remote locations without access to electric power. Information on concentrations of $O_3$, gaseous and particulate N and S species, as well as nitrate, ammonium, and sulfate in wet precipitation for selected case studies and monitoring networks is presented. In general, summer season average 24-h concentrations of ozone were elevated on western slopes of the Sierra Nevada, approaching 75 ppb in the Mountain Home site. Wet deposition of N and S were the highest at elevations below 2100 m on the western slopes of the Sierra Nevada (approaching 2.83 kg N ha$^{-1}$ yr$^{-1}$ at Giant Forest) and similar to the values determined at Tanbark Flat in southern California. The sites above 2200 m in the eastern Sierra Nevada had the lowest values of N wet deposition values. Similar trends to wet N deposition were also determined for gaseous and particulate N pollutants as well as dry deposition to pines. Total levels of wet and dry N deposition in the Sierra Nevada sites are still well below saturation levels. In the Sierra Nevada, wet N deposition was about two- to threefold higher than wet deposition of sulfur. Results from the elevational gradient in the San Bernardino Mountains clearly indicated that deposition of $NO_3^-$, $SO_4^{2-}$, and $NH_4^+$ were proportional to the amount of rain precipitation. In the San Bernardino Mountains, the highest ozone concentrations were found at about 1500 m, with significantly reduced levels at higher elevations. This chapter also presents distribution of N pollutant concentrations and their deposition on a vertical gradient in the mixed conifer canopy, models for atmospheric deposition estimates, ecological consequences of elevated concentrations of ozone, N saturation in vulnerable ecosystems, and research recommendations.

**Keywords:** *remote sites, monitoring, ozone, N and S air pollution*

## INTRODUCTION

A good understanding of air mass transport patterns, spatial and temporal distribution of pollutants, as well as deposition rates of these potentially toxic compounds to plants and other landscape components are needed in order to estimate the potential threat to sensitive ecosystems. To accomplish this reliable methodology for monitoring air pollution concentrations and fluxes, models simulating deposition of pollutants at the plant/forest stand/ecosystem levels and methods for evaluation of critical doses and loads of pollutants are required.

This chapter presents selected methodologies which have been used or can be used for studying sensitive ecosystems in California, including national parks and other preservation areas.

Although part of the ozone and N and S wet deposition results were collected with standard methodologies utilizing the 110-V AC power equipment, such methods are not discussed in this chapter. Information on air pollution concentrations and fluxes in mixed conifer forests of the Sierra Nevada and San Bernardino Mountains and a few other ecosystems is presented from case studies and monitoring networks.

## AIR POLLUTANTS

In California, ozone is the main air pollutant which is phytotoxic at ambient levels. Classic examples of phytotoxic ozone effects on ponderosa and Jeffrey pines (*Pinus ponderosa* and *P. jeffrey*) dominated forests have occurred for more than 40 years in the San Bernardino Mountains[1,2] and in the last 20 years or so have also been seen on pines on the western slopes of the Sierra Nevada.[3]

However, other air pollutants may also affect sensitive terrestrial ecosystems in California. Probably the most important effects may be expected from inorganic and organic nitrogenous pollutants in dry (gaseous and particulate) and wet (rain, fog, snow) forms. Direct toxic effects caused by nitric oxide (NO), nitrogen dioxide ($NO_2$), ammonia ($NH_3$), or peroxyacetyl nitrate (PAN) are theoretically possible at very high ambient levels of those pollutants. However, they are rather unlikely to take place at the levels presently occurring in California. Occurrences of severe toxic effects caused by NO, $NO_2$, or $NH_3$ have been observed in selected locations in Europe or Canada near fertilizer plants or in the immediate vicinity of accidental spillage of liquid $NH_3$. In southern California in the 1960s and 1970s toxic effects of PAN were documented for sensitive agricultural species such as lettuce and tomatoes.[4] Toxic effects of elevated concentrations of gaseous nitric acid ($HNO_3$) have been shown on ponderosa pine and California black oak during short-term, elevated-level fumigations.[5] Toxic effects of $HNO_3$, however, have not yet been shown to occur under ambient field conditions. Potentially more important are indirect long-term effects of N dry and wet deposition which may lead to excess available N in forest stands. This, in turn, may impact forest sustainability and biodiversity, alter species composition, and lead to contamination of ground water and surface waters with excess nitrate.

## VULNERABLE CALIFORNIA ECOSYSTEMS

California is geographically diversified with a large variety of terrestrial ecosystems. More than 50 plant communities can be distinguished within the state.[6] In the Sierra Nevada, a strong biotic zonation influenced by elevation, latitude, rain-shadow, and slope effects is evident. Dramatic changes have recently taken place in the composition and distribution of some of the communities (e.g., redwood and red fir forests, chaparral, or dune scrub), due to increasing anthropogenic pressure.[6] Relatively little is known of how air pollution may be affecting various plant communities in California, with the exception of the mixed conifer forests of the San Bernardino and the Sierra Nevada Mountains. Portions of the mixed conifer forests have been seriously affected because of their proximity to the air pollution source areas and the presence of air pollution-sensitive species. Other types of ecosystems in the state have received much less attention. This could be because these ecosystems are not in the immediate zone of high air pollutant exposures, because the vegetation is insensitive to air pollutants, or because they are considered to be of a lesser ecological or economic value. Recently, some California ecosystems have received special attention because of concerns about their survival or irreversible changes. Coastal sage scrub and chaparral ecosystems in southern California have been seriously endangered by the growing anthropogenic pressure, including housing developments, uncontrolled fires, and possibly air pollution.[7] Sensitive components (lakes, fish, plants) of the subalpine and alpine ecosystems of the Sierra Nevada and the Transverse Ranges of southern California may also be threatened by elevated deposition of acidic precipitation, ozone, temperature changes or increasing levels of ultraviolet (UV-B) radiation.[8-10]

## TRANSPORT OF POLLUTANTS TO VULNERABLE ECOSYSTEMS

The influence of local topography on the movement of air masses across the state may be exemplified by a cross section of California at the latitude of Fresno. The moisture-laden air flows inland from the ocean, passes eastward over the Coastal Range, and descends into the Central Valley where temperature inversions are common. Air masses above the inversion layer rise over the Sierra Nevada and are adiabatically chilled as they reach higher elevations. The chilling causes precipitation on slopes facing the ocean. As the air descends into Owens Valley, it becomes heated, thus favoring high evaporation conditions typical for desert environments.[11] During summer the temperature inversion and low wind speed allow photochemical air pollutants to accumulate. At the Central Valley locations (e.g., Fresno), photochemical pollutant concentrations continue to rise to a midday peak, whereas on the western slopes of the Sierra Nevada the peak concentrations occur in the late

**FIGURE 1**  Summer wind flow patterns in California.[13]

afternoon and are caused by the upslope flow.[12] Flows of air masses within the state are driven by the topography and climate[13] and differ between the seasons. Summer wind flow patterns are characterized by movement of the polluted air from the valleys into the mountain ranges of the Sierra Nevada, Cascades, Coast Ranges, and the San Bernardino Mountains (Figure 1).

## METHODS USED FOR FIELD AIR POLLUTION MONITORING

Availability of electric power is probably the main factor deciding what types of air pollution monitoring can be carried out in remote locations. Other factors such as specific interests in particular types of data, quality control and assessment, availability of funding, difficulty to access remote sites with heavy equipment, possibility of vandalism, etc. also play an important role in deciding which monitoring techniques can be used.

At present, monitoring of ozone, nitrogen oxides, or sulfur dioxide has become relatively routine when electric power and adequate funding are available. Availability of data loggers, computers, appropriate software for data acquisition and statistical evaluation, as well as telephone communication capabilities allow for relatively trouble-free monitoring of these pollutants. This situation changes drastically when no electric power is easily available. Solar panel and battery-operated instruments or passive samplers not requiring any electric power have to be considered in such situations. Passive samplers depend on diffusion of the measured pollutant to a collecting or indication medium. In general, passive samplers provide the concentration of pollutants integrated over a relatively long (days, weeks) averaging time. The advantages of passive sampling devices are their simplicity of use, small size, and low costs. The disadvantages are relatively low sensitivity not allowing for determining short-term (minutes, hours) concentrations, their gradual loss of effectiveness during use or storage, and possible interference from other atmospheric constituents.

## OZONE PASSIVE SAMPLERS

The most widely used passive ozone sampler in the U.S. (distributed by Ogawa & Company, USA, Inc.) was designed for ambient[14] and indoor[15] measurements. The sampler contains two coated glass fiber filters saturated with nitrite, which in the presence of ozone is oxidized to nitrate which is later analyzed by ion chromatography.[14] In 1993, the Ogawa passive ozone samplers were compared with UV-photometric ozone analyzers at five sites operated by the National Park Service. Passive sampler measurements agreed well for each site and were within +10% accuracy for each measurement period (John Ray and Miguel Flores, unpublished). The samplers also have been used successfully in several studies in the U.S. and abroad (Paul Miller and Mark Fenn, personal communication; Sarah Brace and David Peterson, unpublished manuscript).[16]

## GASEOUS NITRIC ACID, NITROUS ACID, SULFUR DIOXIDE, AMMONIA, PARTICULATE NITRATE, AND AMMONIUM

### Annular Denuder Samplers

Concentrations of gaseous nitric acid, nitrous acid, sulfur dioxide, and ammonia as well as nitrate, sulfate, and ammonium in fine (<2.2 μm diameter) and coarse (>2.2 μm diameter) particles can be precisely measured with the annular denuder samplers.[17,18] Pumps operating on 12 V DC allow for collections in locations without electric power. Nitric acid is deposited on annular denuder tubes as nitrate ion, nitrous acid as nitrite, sulfur dioxide as sulfate, and ammonia as ammonium. Concentrations of nitrate, nitrite, and sulfate are determined with ion chromatography and concentrations of ammonium colorimetrically.

### Honeycomb Denuder Samplers

Although the currently available annular denuder samplers have been extremely useful, they are rather expensive and difficult to use in field conditions. As an alternative, a glass honeycomb denuder/filter pack system for measuring concentrations of nitric acid, nitrous acid, sulfur dioxide, ammonia and particulate nitrate, and sulfate and ammonium has been developed.[19] This relatively new system is easier to use in large-scale field monitoring and provides results which compare well with the classic annular denuder systems (Bytnerowicz, unpublished).

## WET PRECIPITATION

### Rain Collector

There is a great demand for reliable and inexpensive rain collectors for use in remote field locations where electric power is not available and where frequent visits of technical personnel are not possible. A reliable, light-weight, "flip-top" collector has been developed[20] which is tightly covered during dry periods and easily opened by the first drops of rain. This collector has been successfully used for collection of rain and throughfall samples during a three-year study in the San Bernardino Mountains.[21] Precipitation samples in the California Acid Deposition Monitoring Program (CADMP) have been collected on a weekly basis using a wet/dry collector (Model 301, Aerochem Metrics) in accordance with the protocols of the National Atmospheric Deposition Program (NADP). Rain volume is measured with an event-recording rain gauge (No. 5-780, Belfort Instruments). Concentrations of selected cations and anions are measured by ion chromatography.

### Active Fog Collectors

The originally developed, active, cloud water collector[22] excludes the collection of rain while facilitating collection of fog by actively pulling the fog droplets through a module of Teflon strands with a propeller-type cooling fan. The collector uses a 12-V fan, allowing the collector to be used where electric power is not available. The active fog collector has been modified by adding baffles to the inlet and outlet which prevent dry deposition from contaminating the fog collector strands during dry periods. The collector has also been automated so that the fan turns on only during fog events. The baffles remain closed until air drawn by the fan causes them to open. The system consists of the collector, a wetness sensor, a relative humidity sensor, and a data logger (Figure 2). Whenever the relative humidity exceeds 80% a small fan pulls air over the wetness sensor. After sufficient moisture accumulates on the wetness sensor, the fan propeller in the fog collector is activated to begin collecting fog water.

### Passive Fog Collectors

The passive collector described here is made with nylon strings and is for determination of the volume of fog or cloud water deposited per collector surface area. Fog water deposition fluxes determined with the passive collector can be used in conjunction with fog chemistry data to estimate ionic deposition from fog to the forest canopy if the canopy surface area is known. Using Teflon strings instead of nylon will make the passive collectors appropriate for measuring fog water chemistry as well as fog water deposition fluxes. The design is similar to the passive string collector described previously,[23] except that the materials for our collectors are readily available from a hardware store and general laboratory suppliers. The lines are connected on each end to the baseplate of a plastic Buchner funnel and are oriented parallel to intercept fog/cloud water droplets that pass through the collector (Figure 3). The fog/cloud water condensate runs down the lines, through a funnel,

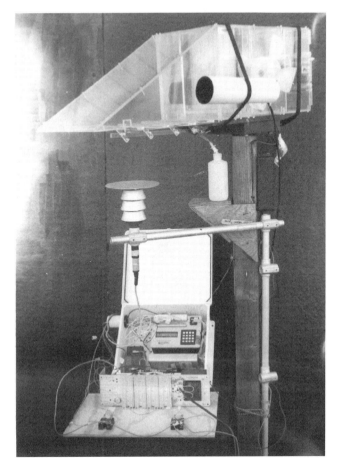

**FIGURE 2**  Active fog collector designed to collect fog or cloud water and to exclude rainfall.[22] Baffles have been added to the inlet and outlet of the collector to exclude contamination from dry deposition between fog events. The collector has also been modified to activate only during fog exposure.

and into a collection bottle attached at the bottom. In order to ensure that rain water is not collected, a plastic lid is attached to the top of the collector by a threaded rod.

## WET PLUS DRY DEPOSITION COLLECTORS

In remote sites, collecting dry deposition is problematic when wet deposition also occurs during the sampling period. In a recent study of air pollution inputs along an elevational gradient in the San Gorgonio Wilderness in the San Bernardino Mountains, collection devices for effective aerosol impaction, interception of dry deposition and fog, and rain collection were constructed. The collectors were a modified version of the "filter gauges"[24] consisting of a cylinder of polyethylene-coated wire mesh above a rain collector. The collector units consists of a 2-L

**FIGURE 3** Passive string fog or cloud water collector. Fog is intercepted by the parallel nylon or Teflon lines and collected in the funnel and bottle below. The collector top is to prevent rainfall collection.

collection bottle, a polyethylene funnel (21 cm diameter), and a mesh cylinder (7 cm diameter × 32 cm in length) nested in the funnel (Figure 4). The collectors were exchanged every two weeks, placed in plastic bags, and taken to the lab where the mesh and funnel were washed with deionized distilled water. Rinse water was combined with precipitation samples in the bottles, and concentrations of ions were determined by ion chromatography. These results were used for calculation of total ionic deposition during the 14-day exposure periods.

## DRY DEPOSITION TO PLANTS

### Estimates of Stomatal Uptake of Gaseous Pollutants by Plants

The amount of gases entering leaves through stomata can be calculated by multiplying ambient concentration of gases by the stomatal conductance value and by

**FIGURE 4** Bulk deposition collector for measuring wet and/or dry deposition in remote sites. The mesh screen provides a large surface area for efficient collection of dry deposition. The funnel allows for the capture of dry-deposited materials which may be rinsed from the screen during precipitation events.

the pollutant/water vapor correction factor for stomatal transport. Information on foliar surface area, spatial and temporal distribution of air pollutant concentrations, and stomatal conductances within a forest canopy are required to estimate the stomatal uptake of a gaseous pollutant by a forest stand.[25]

### Estimation of Surface Dry Deposition of Gases and Particles

The amount of gases and particles intercepted by surfaces of plant canopies are commonly estimated as throughfall deposition. The rain/throughfall "flip-top" collector[20] has been successfully used for determinations of dry deposited nitrogen, sulfur species, and metallic cations to mixed conifer canopies in the San Bernardino Mountains.[21] However, in arid areas of California rains occur very rarely limiting throughfall collection during the major "smog" season. Therefore, a substitute for throughfall, a branch rinsing technique[26,27] has been applied. Branch rinsing has been successfully used in various locations of California for determinations of dry deposition of nitrogen, sulfur and metal cations to several native plant species.[28,29]

# RESULTS OF CASE STUDIES AND MONITORING NETWORKS

## OZONE IN THE MIXED CONIFER FORESTS OF THE SIERRA NEVADA AND SAN BERNARDINO MOUNTAINS

Project FOREST (Forest Ozone Response Study) is a cooperative effort to relate foliar injury in sensitive pine species to ozone exposure in the California forests. Eleven Sierra Nevada sites and one site in the San Bernardino Mountains (Figure 5) were chosen at which ambient ozone measurements were made during the growing season. These sites ranged in elevation from about 1130 to 2040 m. The study sites were operated by the University of California at Davis (with the financial support of the California Air Resources Board) and the National Park Service. In all the sites, ozone concentrations were measured with Dasibi 1008 AH ozone monitors. The period of 1992 to 1994 was chosen for analysis since the environmental and tree data were most complete in this interval.[30]

Among all the sites, Mountain Home in the southern Sierra Nevada had the highest 24-h average ozone concentrations. Other sites in the Sierra Nevada, Jerseydale, Five-Mile, White Cloud, Giant Forest, and Giant Grove had very similar elevated concentrations of ozone. These concentrations were similar to the values recorded in the Barton Flats site in the San Bernardino Mountains. Concentrations of ozone in Camp Mather, Mount Lassen, and Wawona were consistently lower than in other sites. Average 24-h ozone concentrations in all the sites did not differ between the individual monitoring years (Table 1). The highest seasonal daily maximum 1-h concentrations for the Sierra Nevada sites were recorded at Mountain Home (99, 97, and 102 ppb in 1992, 1993, and 1994, respectively). At the Barton Flats site of the San Bernardino Mountains the values were higher (119, 120, and 124 ppb, respectively).

The data presented clearly indicates that elevated ozone concentrations occurred in most of the Sierra Nevada sites. These concentrations were similar to the values recorded in the Barton Flats site (located in the middle of the air pollution gradient in the San Bernardino Mountains). Discussion of the biological importance of these findings is presented in the "Critical Doses/Loads Concept and Application" section later in this chapter.

## WET DEPOSITION OF NITROGEN AND SULFUR IN THE SIERRA NEVADA

In the CADMP, wet deposition was monitored at 25 to 35 sites from July 1984 through June 1994 (Table 2).[31] Because the primary objective of the CADMP was to quantify inputs to urban areas, few sites were located in forested areas above 500-m elevation (Figure 6). Relative to annual wetfall patterns, most of California experiences a Mediterranean climate, in which 75 to 85% of annual precipitation is deposited during a 6-month wet season from October to March. Of these lower elevation sites, Montague exhibits a unique pattern of wetfall — only about 60% of

**FIGURE 5** Location of the FOREST ozone monitoring site in Sierra Nevada.

annual precipitation is deposited in the wet season. At these forested sites throughout the state, total wet N deposition (primarily rain and also some snow at the Soda Springs, Giant Forest, and Yosemite sites) ranges from 0.66 to 2.83 kg N ha$^{-1}$ yr$^{-1}$, with $NH_4^+$ and $NO_3^-$ providing near-equal quantities of N. Wet S deposition (from $SO_4^{2-}$ only) ranges from 0.26 to 0.98 kg S ha$^{-1}$ yr$^{-1}$.[32]

At elevations above 2200 m in the Sierra Nevada, where in subalpine/alpine ecosystems occur, snow is the principal form of precipitation. In 1990 through 1993, wet deposition was monitored at 11 sites in the Sierra Nevada as part of the Air Resources Board's Atmospheric Acidity Protection Program.[33] At these sites, wet-season (October through March) snowfall provided 80 to 95% of the annual

### TABLE 1
### Ozone 24-h Average Seasonal (June-September) in the Sierra Nevada and San Bernardino Mountains Locations (ppb)

| Site | Elevation (m) | 1992 | 1993 | 1994 |
|---|---|---|---|---|
| Manzanita Lake[a] | 1770 | 39.5 | 37.6 | 47.4 |
| White Cloud[b] | 1325 | 64.0 | 56.9 | 64.9 |
| Sly Park[b] | 1130 | 55.4 | 51.9 | 50.6 |
| Five-Mile Learning Center[b] | 1220 | 64.1 | 63.2 | 66.5 |
| Wawona[a] | 1220 | 42.4 | 40.9 | 37.4 |
| Camp Mather[a] | 1400 | 43.9 | 49.4 | 53.2 |
| Jerseydale[b] | 1140 | 65.8 | 64.4 | 63.1 |
| Shaver Lake[b] | 1830 | 56.2 | 52.7 | 55.7 |
| Giant Forest[a] | 1920 | 64.1 | 66.8 | 68.1 |
| Grant Grove[a] | 1980 | 62.0 | 61.1 | 68.1 |
| Mountain Home[b] | 1890 | 71.4 | 67.6 | 74.5 |
| Barton Flats[b] | 2040 | 63.0 | 63.9 | 66.6 |

*Note:* Locations are listed from N to S as presented in Figure 5.

[a] Sites operated by the Park Service
[b] Sites operated by the Forest Service

### TABLE 2
### Wet N and S Deposition at Selected CADMP Sites[a]

| Site[b] | Elevation (m) | Amount (cm) | Total N (kg N ha$^{-1}$ yr$^{-1}$) | NO$_3$–N (% of Total N) | SO$_4$–S (kg S ha$^{-1}$ yr$^{-1}$) |
|---|---|---|---|---|---|
| Montague | 815 | 30 | 0.66 | 47 | 0.26 |
| Quincy | 1060 | 83 | 1.20 | 55 | 0.62 |
| Soda Springs | 2065 | 138 | 1.97 | 55 | 0.98 |
| South Lake Tahoe | 1900 | 37 | 0.76 | 54 | 0.37 |
| Turtleback Dome | 1395 | 78 | 1.93 | 49 | 0.73 |
| Giant Forest | 1920 | 83 | 2.83 | 42 | 0.81 |
| Ash Mountain | 550 | 51 | 2.63 | 42 | 0.66 |
| Mount Wilson | 1740 | 64 | 1.57 | 56 | 0.83 |
| Tanbark Flat | 820 | 53 | 2.22 | 53 | 0.96 |

*Note:* Locations are listed from N to S as presented in Figure 6.

[a] CADMP: Wet Deposition Data Summaries — 1985 through 1994.
[b] Site elevations are 500 to 2100 m.

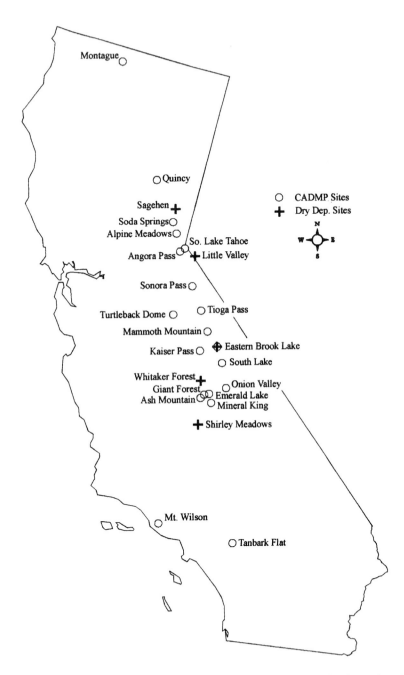

**FIGURE 6** Location of the CADMP wet N and S deposition sites and the dry N deposition sites in Sierra Nevada.

## TABLE 3
### Wet N and S deposition in the Sierra Nevada[a]

| Site[b] | Elevation (m) | Amount (cm) | Total N (kg N ha$^{-1}$ yr$^{-1}$) | NO$_3$–N (% of Total N) | SO$_4$–S (kg S ha$^{-1}$ yr$^{-1}$) |
|---|---|---|---|---|---|
| Alpine Meadows | 2164 | 125 | 1.29 | 52 | 0.53 |
| Angora Lake | 2286 | 156 | 1.58 | 48 | 0.72 |
| Sonora Pass | 2937 | 70 | 0.87 | 51 | 0.44 |
| Tioga Pass | 2993 | 120 | 1.25 | 49 | 0.58 |
| Mammoth Mountain | 2940 | 130 | 1.46 | 47 | 0.68 |
| Eastern Brook Lake | 3170 | 42 | 0.66 | 52 | 0.29 |
| Kaiser Pass | 2941 | 110 | 1.15 | 45 | 0.48 |
| South Lake | 3010 | 61 | 0.73 | 53 | 0.35 |
| Emerald Lake | 2824 | 129 | 1.48 | 44 | 0.65 |
| Onion Valley | 2800 | 59 | 0.89 | 49 | 0.39 |
| Mineral King | 2694 | 72 | 1.44 | 44 | 0.49 |

*Note:* Locations are listed from N to S as presented in Figure 6.

[a] Melack et al. 1995. Final Report, ARB Contract No. A932-081.
[b] Sampling site elevations are 2100 to 3100 m.

precipitation volume, and the remainder was deposited as rain in April through September (i.e., 4-year average rain inputs ranged from 7 to 19% per year). Total wet N deposition (i.e., primarily snow and some rain) was found to range from 0.66 to 1.58 kg N ha$^{-1}$ yr$^{-1}$ and wet S deposition from 0.29 to 0.72 kg S ha$^{-1}$ yr$^{-1}$ (Table 3). As in the lower elevation sites in the CADMP, wet-deposited N inputs from $NH_4^+$ and $NO_3^-$ were of similar magnitude. Although rain provides only a small portion of the total moisture input to these sites, it supplies a similar amount of N and S in the form of snow. Comparisons of analyte concentrations in snow and rain show that levels of N and S compounds in snow are 5 to 15 times lower than in rain.[33]

### GASEOUS AND PARTICULATE NITROGENOUS POLLUTANTS IN THE SIERRA NEVADA

Concentrations of gaseous and particulate pollutants have been measured with annular denuder systems in several sites in the Sierra Nevada (Figure 6, Table 4). These results indicated that concentrations of N trace pollutants in the Sierra Nevada were similar in various locations, but were much lower than concentrations recorded for the San Bernardino and San Gabriel Mountains of southern California.[27,34,35] Distribution of concentrations and deposition of N gases and particles in California wildlands have recently been reviewed by two authors of this chapter.[36]

## TABLE 4
## Average Concentrations of Gaseous and Particulate N Pollutants in the Sierra Nevada Locations

| Location | Elevation (m) | Concentration (mg m$^{-3}$) | | | | |
|---|---|---|---|---|---|---|
| | | HNO$_3$ | HNO$_2$ | NH$_3$ | NO$_3^-$ | NH$_4^+$ |
| Sagehen, August 1994 | 2100 | 0.64 (24 h) | 0.005 (24 h) | 0.35 (24 h) | 0.15 (24 h) | 0.18 (24 h) |
| Little Valley, August 1994 | 2010 | 2.02 (24 h) | 0.003 (24 h) | 1.14 (24 h) | 0.51 (24 h) | 0.14 (24 h) |
| Eastern Brook Lake, | 3170 | | | | | |
| Summer | | 0.36 (24 h) | | | | |
| Winter | | 0.14 (24 h) | | | | |
| Whitaker Forest, summers 1988–1990 | 1600 | 1.1–1.8 (day) | 0.08–0.30 (day) | 1.11–1.56 (day) | 0.30–1.40 (day) | 0.39–0.41 (day) |
| | | 0.3–0.7 (night) | 0.08–0.19 (night) | 1.92–2.48 (night) | 0.44–1.05 (night) | 0.11–0.33 (night) |
| Shirley Meadow, summers 1989–1990 | 1950 | 0.76–1.24 (day) | 0.17–0.76 (day) | 1.76–1.95 (day) | 0.44–0.46 (day) | 0.07–0.08 (day) |
| | | 1.89–2.28 (night) | 0.20–0.39 (night) | 2.06–2.31 (night) | 0.68–0.74 (night) | 0.10–0.26 (night) |

*Note:* Locations are listed from N to S as presented in Figure 6.

### DRY DEPOSITION OF NITRATE AND AMMONIUM TO TREES IN THE SIERRA NEVADA

Atmospheric deposition to branches of *Pinus contorta* and *P. albicaulis* was measured in a sub-alpine zone at Eastern Brook Lake in the eastern Sierra Nevada, and at Emerald Lake in the western Sierra Nevada. At Eastern Brook Lake, deposition fluxes of anions and cations were several times lower compared with the Emerald Lake site. There was a balance between cations and inorganic anions deposited to pine surfaces at the eastern Sierra Nevada site. In contrast, at the western Sierra Nevada site more cations than inorganic anions were deposited — this could be balanced by dry-deposited organic anions resulting from anthropogenic activities in the California Central Valley. At the eastern Sierra Nevada site a strong positive correlation between depositions of NO$_3^-$ and NH$_4^+$, as well as SO$_4^{2-}$ and Ca$^{2+}$, suggested that large portions of these ions might have originated from particulate NH$_4$NO$_3$ and CaSO$_4$ deposited on pine surfaces.[28,29]

### NITROGENOUS POLLUTANTS ON A VERTICAL TRANSECT IN A MIXED CONIFER CANOPY OF THE SAN BERNARDINO MOUNTAINS

In 1992 to 1994 an intensive study in a mixed conifer forest at Barton Flats, San Bernardino Mountains was performed. Concentrations of N gases and particles on a vertical gradient in the forest canopy were monitored. Dry deposition of N and other elements to mature ponderosa/Jeffrey pine and to seedlings of ponderosa pine, white fir, and California black oak on a vertical transect in the canopy were measured. In addition, deposition fluxes at the forest floor to seedlings and branches of mature trees were measured. This study provided data needed to estimate annual deposition of N and other elements to the mixed conifer stand and information for developing a deposition model for this type of forest. The study also revealed that concentrations of the measured nitrogen species were elevated. However, no differences along the vertical gradient were observed. In contrast, deposition fluxes of nitrate and ammonium changed with position within the canopy, with the highest fluxes at the canopy top. Deposition of N at the Barton Flats mixed conifer forest stand estimated from the results of branch rinsing and throughfall analysis was elevated (4 to 8 kg ha$^{-1}$ yr$^{-1}$), but several times lower than the estimated values (about 35 kg ha$^{-1}$ yr$^{-1}$) at the high end of the pollution gradient.[25,34]

## WET AND DRY DEPOSITION IN THE SAN GORGONIO WILDERNESS AREA

Wet plus dry deposition collectors were placed at six sites along an elevational gradient beginning about 300 m from Highway 38 in the San Bernardino Mountains and extending into the San Gorgonio Wilderness at elevations between 1192 and 2775 m. For comparison, samples were also collected at three high-elevation sites in southeastern Wyoming. The Ogawa passive ozone samplers were also placed at each of the California sites. All samplers were exposed for three consecutive 14-day exposures.

In the first exposure period at the California sites, rain increased from about 2 cm at the two lower sites (1277 and 1581 m) to 6 cm at 2493 m. At the highest California elevation (2797 m) rain decreased to about 1 cm (Figure 7a). Deposition of $NO_3^-$, $SO_4^{2-}$, and $NH_4^+$ followed a similar pattern, suggesting that deposition was largely precipitation driven. Deposition of $SO_4^{2-}$, and especially $NO_3^-$ were greatly reduced with lower rainfall — rain was slight in the second exposure and no rain occurred in the last exposure period. Ammonium deposition exhibited a more complex pattern, with very high $NH_4^+$ deposition at most of the higher elevations in the first two exposure periods, but only trace amounts in the third rain-free period. Surprisingly, in the first two exposures, $NH_4^+$ deposition was generally much greater than $NO_3^-$ deposition. Especially striking are the high $NH_4^+$ inputs in the second exposure period, when rainfall was negligible. The most likely explanation for the high $NH_4^+$ inputs is that considerable fog deposition occurred in the second exposure period. Ionic concentrations in fog and cloud water are well known to be highly elevated in areas with high air pollution.[37-39] However, if fog deposition was a major factor, $NO_3^-$ deposition should have been greater than measured in the second exposure period. Dry deposition of $NH_4^+$ to the collectors appears to be minimal, as shown by the data from the third rain-free exposure period (Figure 7c, d, and e).

These results suggest that the wet and dry collectors are especially useful for site to site comparisons for any given ion. However, caution must be used when comparing deposition of different ions, even at the same site, since the collection materials may have different affinities for different ions. The results of this study

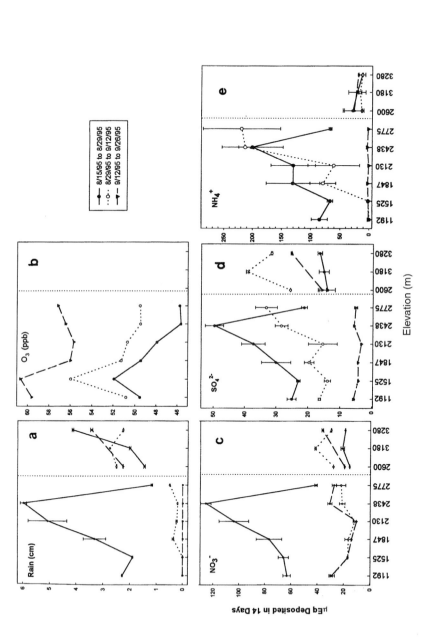

**FIGURE 7** (a) Rain precipitation, (b) cumulative average ozone concentrations, (c) nitrate, (d) sulfate, and (e) ammonium wet and dry deposition at sites along an elevational gradient in the San Gorgonio Wilderness (first six sites) and in southeastern Wyoming Rocky Mountains (three remaining sites).

illustrate the more constant rain and ionic deposition inputs during summer in the Rocky Mountain sites. Nitrate and sulfate deposition in the Rocky Mountain sites were similar, while $NH_4^+$ deposition was generally slightly lower than $NO_3^-$ and $SO_4^{2-}$ deposition.

Concentrations of ozone were the highest in the periods without rain and lowest in the period of high precipitation. Consistently, the highest ozone concentrations were recorded at an elevation of 1525 m. During two earlier periods, ozone concentrations were the lowest at the two highest elevation sites. In the last period, similar ozone concentrations were seen at the four highest elevation sites (Figure 7b). No measurements of ozone concentrations in the Wyoming sites were performed.

## NITROGEN DEPOSITION IN THE COASTAL SAGE SCRUB ECOSYSTEMS OF SOUTHERN CALIFORNIA

Coastal sage scrubland once covered large portions of the lower elevation topography of southern California. In recent years, urban development and other anthropogenic factors have led to severe losses of this ecosystem and poor condition of much of the remaining patches. Concerns regarding preservation and conservation of the remaining fragments adjacent to urban centers have led to questions about sustainability of this unique vegetation type, particularly as affected by the poor air quality of southern California. Results of a study in 1993 to 1995 indicated that concentrations of airborne $NO_3^-$ vary more than threefold along an N-S transect between Riverside and San Diego and that atmospheric $NO_3^-$ concentrations were two to three times higher than $NH_4^+$ concentrations. Also, significant increases in soil nitrogen in the more polluted sites were observed. The difference in soil nitrogen among study sites appears to be a combination of accumulation of directly deposited material during the dry summer period and modifications to the N cycle.[40]

## DEVELOPMENT AND APPLICATION OF DEPOSITION MODELS

Measurements of dry deposition of gases and particles to plant canopies and forest stands are difficult to perform on a regular basis. Therefore, simulation models for reliable estimates of deposition to vulnerable California ecosystems are needed. Models simulating dry deposition to plant canopies in relatively flat, horizontally homogenous terrains of the eastern U.S. (e.g., Big Leaf multiple resistance model[41]) did not perform satisfactorily in the San Bernardino Mountains.[25,42] This may be due mainly to the much higher complexity of the mountainous terrain and patchy vegetation cover in the western U.S. In addition, the existing models are not designed for calculating particulate deposition, which in the dry conditions of California mountains accounts for substantial input of airborne nutrients to forests. Predicting capabilities of the Big Leaf model are also limited when high concentrations of $NH_3$ are present,[41,43] a situation quite typical for the western Sierra Nevada[44] and other parts of the state.[36]

Based on three seasons of intensive field measurements of deposition,[25] a simple predictor was developed to estimate seasonal surface dry deposition of nitrogenous air pollution to ponderosa pine,[45] white fir, and California black oak[46] in the mixed conifer forest in the San Bernardino Mountains. The developed predictor explains 70% of the observed vertical canopy variation from N surface deposition as measured by foliar rinses. The estimator requires only foliar rinse measurements at the canopy top and foliar surface area estimates by species for the given forest site. The model assumes no vertical concentration gradient for the N pollutants and that $NH_4NO_3$ is the dominant form of deposited particulate nitrate and ammonium ions. The approach used for the described predictor may be widely used for estimates of deposition to vegetation of various ecosystems, especially for shrub systems.

## CRITICAL DOSES/LOADS CONCEPT AND APPLICATION

### Ozone Critical Exposure Levels

The most simple expression of ozone concentration is the 24-h average for monthly intervals or for the growing season. At Barton Flats in the San Bernardino mountains, the concentrations associated with the early growing season are sufficient to cause chronic injury to ponderosa and Jeffrey pines which is considered moderate.[21] The descriptive index (called the ozone injury index, OII)[47] stayed about the same during 1992, 1993, and 1994. Exposures during spring and early summer are more important than late summer because stomatal conductance and ozone flux to the interior of the leaf are highest. The injury to pines at Barton Flats did increase significantly between 1991 and 1992 because ozone flux to foliage was the higher in 1992 (the first year of adequate moisture following several years of drought).[48] Ponderosa and Jeffrey pines in the Sequoia National Park show levels of chronic injury only slightly less than at Barton Flats.

Ozone exposure indices, including Sum 0, Sum 60, and W126, are also useful for characterizing the exposure response of ponderosa and Jeffrey pines. These indices have been tested against the multiyear accretion of chlorotic mottle on exposed needles and the rate of defoliation of exposed needles of approximately 1600 pines in both the Sierra Nevada and San Bernardino Mountain plots.[49] The Weibul function was used to fit curves to the data, and the results indicated that Sum 0 had $r^2$ values of 0.57 for chlorotic mottle and 0.74 for needle loss. The $r^2$ values for Sum 60 and W126, which emphasize higher ozone concentrations, were only slightly lower than Sum 0.

### Nitrogen Critical Loads

There is a great amount of interest in setting critical loads for N deposition effects on forest ecosystems. The critical load may be defined as "the highest deposition of a compound that will not cause chemical changes leading to harmful effects on ecosystem structure and function."[50] For critical loads of N, both the acidification and eutrophication effects of N must be considered.[51,52] It is commonly accepted that the critical load for N will vary from ecosystem to ecosystem, and within an ecosystem it will depend on the environmental component to be protected.

While methods for determining critical N loads in forests are still being debated, their empirical estimates for different ecosystem types can be used as general guidelines of approximate critical loads. Empirical critical loads for changes in ground flora, N leaching, and nutrient imbalances for deciduous and coniferous forests ranged from 10 to 20 kg ha$^{-1}$ yr$^{-1}$.[53] From a survey of 65 forests in Europe, it was concluded that in forests with N deposition less than 10 kg ha$^{-1}$ yr$^{-1}$, little nitrate leaching occurred, while at intermediate levels of 10 to 25 kg ha$^{-1}$ yr$^{-1}$ leaching occurred at some sites.[54] At sites with N deposition greater than 25 kg ha$^{-1}$ yr$^{-1}$, significant nitrate leaching occurred at all sites. Studies on the N status of chaparral stands and mixed conifer forests in southern California agree with the broad categories reported for European ecosystems.[55,56] Stream water nitrate concentration data and N fertilizer experiments in the San Gabriel and San Bernardino Mountains of southern California suggest that when N deposition approaches 25 kg ha$^{-1}$ yr$^{-1}$ these forests are no longer N limited and export high concentrations of nitrate.[34,55,56] Forests with N deposition of 15 kg ha$^{-1}$ yr$^{-1}$ or less were N limited[55] with low nitrate concentrations in soil and in stream water (Mark Fenn, unpublished data).

Vector analysis and DRIS (Diagnosis and Recommendation Integrated System) are two techniques which can help evaluate to what extent N air pollution affects plant nutrient status. Vector analysis[55,57] is a graphical method of evaluating plant nutrient status based on foliar response to fertilization or other treatments. A vector indicating plant response to added N, for example indicates the effect of N on foliar biomass, N concentration, and total N content. Vector direction indicates N status (i.e., N deficiency, sufficiency, or toxicity), while vector length indicates the degree of the N status.

DRIS[58-60] is based on the concept of nutrient balance. DRIS indicates the most limiting nutrient and the order in which other nutrients are likely to limit yield. DRIS uses ratios of foliar nutrient concentrations to calculate indices that diagnose plant nutrient status. The value of DRIS lies in its potential to identify plant nutrient status with respect to many nutrients simultaneously. For example, with high N deposition and alleviation of N deficiency, DRIS can identify which other nutrient is limiting and which nutrients are approaching growth-limiting levels. However, in order to apply the DRIS system, DRIS norms have to be developed from a database of plant nutrient concentrations from stands spanning the range of forest productivity. Collecting the data to calculate DRIS norms is expensive because of the large number of samples and nutrient analyses required. Once the norms are available for a given species, however, DRIS provides a powerful tool for assessing tree nutrient status. Using DRIS to compare tree nutrient status across N deposition gradients can suggest possible trends of air pollution effects on tree nutrient status.

## GEOGRAPHIC INFORMATION SYSTEMS (GIS)

These systems provide opportunities for spatial display of information regarding topography, movement of air masses, distribution of pollution sources, vegetation cover, forest health monitoring stations, etc. The GIS techniques help in better visual presentation of air pollution concentrations/fluxes to plants for evaluating risk on a landscape level (see Chapter 13 by Miller et al.).

## FUTURE RESEARCH NEEDS AND RECOMMENDATIONS

Monitoring techniques and models described previously should be used on a much larger scale throughout California. Much more information on air pollution distribution and deposition, especially in the vulnerable ecosystems, is needed for adequate evaluation of the threat posed by increasing anthropogenic pressure. Because the available pollution transport models and pollution canopy deposition models cannot be applied in the western ecosystems, there is a need for development of the new ones. This will require more field data for model development and validation. The existing data sets are not adequate for this purpose.

There is a need for establishing critical load values for N in western ecosystems. Further studies such as the described vector analysis or DRIS techniques can serve this purpose. Much closer cooperation between researchers and land managers is needed. This is especially true for technology and data transfer, environmental risk assessment, and recommendations for abatement measures. Present and developing GIS techniques may be very useful in the area of implementation of research results into practice and for various types of ecosystem inventory, including information of distribution and transport of air pollutants to vulnerable ecosystems.

## ACKNOWLEDGMENTS

The authors thank Susan Schilling, Anthony Gomez, and David Salardino for their assistance in data processing and preparation of graphical materials.

For questions or further information regarding the work described in this chapter, please contact:

> **Andrzej Bytnerowicz**
> Pacific Southwest Research Station
> USDA Forest Service
> 4955 Canyon Crest Drive
> Riverside, CA 92507
> e-mail: andrzej@deltanet.com

## REFERENCES

1. Miller, P. R., J. R. Parmeter, Jr., O. C. Taylor and E. A. Cardiff. 1963. Ozone injury to the foliage of *Pinus ponderosa*. *Phytopathology*, 53:1072–1076.
2. Miller, P. R., J. R. Parmeter, Jr., B. H. Flick and C. W. Martinez. 1969. Ozone dosage response of ponderosa pine seedlings. *J. Air Pollut. Control Assoc.*, 19:435–438.

3. Durisco, D. M. and K. W. Stolte. 1989. Photochemical oxidant injury to ponderosa (*Pinus ponderosa* Laws.) and Jeffrey pine (*Pinus jeffreyi* Grev. and Balf.) in the national parks off the Sierra Nevada of California. In: R. K. Olson, and Lefohn, A. S., eds., *Effects of Air Pollution on Western Forests*, Air & Waste Management Association, Anaheim, CA, pp. 261–278.
4. Temple, P. J. and O. C. Taylor. 1983. World-wide ambient measurements of peroxyacetyl nitrate (PAN) and implications for plant injury. *Atmos. Environ.*, 17:1583–1587.
5. Krywult, M., J. Hom, A. Bytnerowicz and K. Percy. 1996. Deposition of gaseous nitric acid and its effects on foliage of ponderosa pine (*Pinus ponderosa*) seedlings. In: R. Cox, K. Percy, K. Jensen and C. Simpson, eds., *Air Pollution and Multiple Stresses*, 16th Int. Meeting for Specialists in Air Pollution Effects on Forest Ecosystems, IUFRO — Canadian Forest Service, Fredericton, Canada, pp. 45–51.
6. Barbour, M. G. and J. Major, 1977. *Terrestrial Vegetation of California*. John Wiley & Sons, New York.
7. Allen, E. B., P. E. Padgett, A. Bytnerowicz and R. Minnich. 1997. Nitrogen deposition effects on coastal sage vegetation of southern California. Proc. Symp. *The Effects of Air Pollution and Climate Change on Forest Ecosystems*, Riverside, CA, February 5–9, 1996. USDA Forest Service, Pacific Southwest Research Station, General Technical Report 164 <www.rfl.fs.fed.us/pubs/psw-gtr-164/index.html>.
8. Caldwell, M. M., A. H. Teramura and M. Tevini. 1989. The changing solar ultraviolet climate and the ecological consequences for higher plants. *Trends Ecol. Evol.*, 4:363–367.
9. Melack, J. M. and J. L. Stoddard. 1991. Sierra Nevada, California. In: D. F. Charles and S. Christie, eds., *Acidic Deposition and Aquatic Ecosystems. Regional Case Studies*. Springer-Verlag, New York, 503–530.
10. McColl, J. G. 1988. Hydrogen ion concentration and alkalinity of reservoir water in the Sierra Nevada, California, and correlations with air pollution. *J. Environ. Qual.*, 17:425–430.
11. Schoenherr, A. A. 1992. *A Natural History of California*. University of California Press, Berkeley, CA, 772 pp.
12. Miller, P. R., M. H. McCutchan and B. C. Ryan. 1972. Influence of climate and topography on oxidant air pollution concentrations that damage conifer forests in southern California. In: *Effects of Air Pollutants on Forest Trees*, VII Int. Symp. on Forest Fume Damage Experts, Forstliche Bundesversuchanstalt, Vienna, Austria, Vol. I, pp. 585–608.
13. Hayes, T. P., J. J. R. Kinney and N. J. M. Wheeler. 1984. *California Surface Wind Climatology*. State of California, Air Resources Division, Aerometric Data Division, Sacramento, CA, 73 pp + tables.
14. Koutrakis, P., J. M. Wolfson, A. Bunyaviroch, S. E. Froehlich, K. Hirano and J. D. Mulik. 1993. Measurement of ambient ozone using a nitrite-coated filter. *Anal. Chem.*, 65:210–214.
15. Liu, L.-J., M. P. Olson, III, G. A. Allen and P. Koutrakis. 1994. Evaluation of the Harvard ozone passive sampler on human subject indoors. *Environ. Sci. Technol.*, 28:915–923.
16. Bytnerowicz, A., R. Glaubig, M. Cerny, M. Michalec, R. Musselman and K. Zeller. 1995. Ozone concentrations in forested areas of the Brdy and Sumava Mountains, Czech Republic. Presented at the 88th *Annual Meeting & Exhibition of the Air & Waste Management Association*, San Antonio, TX, June 18–23, 1995.

17. Possanzini, M., A. Febo and A. Liberti. 1983. New design of a high-performance denuder for the sampling of atmospheric pollutants. *Atmos. Environ.*, 17:2605–2610.
18. Peake, E. and A. H. Legge. 1987. Evaluation of methods used to collect air quality data at remote and rural sites in Alberta, Canada. *EPA/APCA Symposium on Measurements of Toxic and Related Air Pollutants*, EPA/APCA, Research Triangle Park, NC, May 3–6, 1987.
19. Koutrakis, P., C. Sioutas, S. T. Ferguson and J. M. Wolfson. 1993. Development and evaluation of a glass denuder honeycomb denuder/filterpack system to collect atmospheric gases and particles. *Environ. Sci. Technol.*, 27:2497–2501.
20. Glaubig, R. and A. Gomez. 1994. A simple, inexpensive rain and canopy throughfall collector. *J. Environ. Qual.*, 23:1103–1107.
21. Fenn, M. E. and A. Bytnerowicz. 1997. Summer throughfall and winter deposition in the San Bernardino Mountains in southern California. *Atmos. Environ.*, 31:673–683.
22. Daube, B., Jr., K. D. Kimball, P. A. Lamar and K. C. Weathers. 1987. Two new ground-level cloud water sampler designs which reduce rain contamination. *Atmos. Environ.* 21:893–900.
23. Joslin, J. D., S. F. Mueller and M. H. Wolfe. 1990. Tests of models of cloud water deposition to forest canopies using artificial living collectors. *Atmos. Environ.*, 24A:3007–3019.
24. Miller, H. G. and J. D. Miller. 1980. Collection and retention of atmospheric pollutants by vegetation. In: D. Drablos and A. Tollan, eds., *Proceedings of the International Conference on the Ecological Impact of Acid Precipitation*, SNSF Project, Sandefjord, Norway, March 11–14, 1980, pp. 33–40.
25. Bytnerowicz, A., M. Fenn and M. Arbaugh. 1997. Concentrations and deposition of N air pollutants in a ponderosa/Jeffrey pine canopy. Proc. Symp. *The Effects of Air Pollution and Climate Change on Forest Ecosystems*, Riverside, CA, February 5–9, 1996. USDA Forest Service, Pacific Southwest Research Station, General Technical Report 164 <www.rfl.fs.fed.us/pubs/psw-gtr-164/index.html>.
26. Lindberg, S. E. and G. M. Lovett. 1985. Field measurements of particle dry deposition rates to foliage and inert surfaces in a forest canopy. *Environ. Sci. Technol.*, 19:238–244.
27. Bytnerowicz, A., P. R. Miller and D. M. Olszyk. 1987. Dry deposition of nitrate, ammonium and sulfate to a *Ceanothus crassifolius* canopy and surrogate surfaces. *Atmos. Environ.*, 21:1749–1757.
28. Bytnerowicz, A., P. J. Dawson, C. L. Morrison and M. P. Poe. 1991. Deposition of atmospheric ions to pine branches and surrogate surfaces in the vicinity of Emerald Lake Watershed, Sequoia National Park. *Atmos. Environ.*, 25A:2201–2210.
29. Bytnerowicz, A., P. J. Dawson, C. L. Morrison and M. P. Poe. 1992. Atmospheric dry deposition on pines in the Eastern Brook Lake Watershed, Sierra Nevada, California. *Atmos. Environ.*, 17:3195–3201.
30. Carroll, J. J. and A. J. Dixon. 1993. Sierra Cooperative Ozone Impact Assessment Study: Year 3. Final Report: Interagency Agreement #A132-188. Volume 1. California Air Reseources Board, Sacramento, CA, 74 pp.
31. Takemoto, B. K., B. T. Cahill, C. K. Buenviaje, R. Abangan and T. E. Houston. 1995. California Acid Deposition Monitoring Program: Wet Deposition Data Summary — July 1990 to June 1993. Research Division, California Air Resources Board, Sacramento, CA, 120 pp.
32. Blanchard, C. L. and H. Michaels. 1994. Regional Estimates of Acid Deposition Fluxes in California. Final Report, ARB Contract No. A132-149, Sacramento, CA, 96 pp.

33. Melack, J. M., J. O. Sickman, F. Setaro and D. Dawson. 1995. Monitoring of Wet Deposition in Alpine Areas in the Sierra Nevada. Final Report, ARB Contract No. A932-081, Sacramento, CA, 66 pp.
34. Fenn, M. E., M. A. Poth and D. W. Johnson. 1996. Evidence for nitrogen saturation in the San Bernardino Mountains in southern California. *For. Ecol. Manage.*, 82:211–230.
35. Grosjean, D. and A. Bytnerowicz. 1993. Nitrogenous air pollutants at a southern California mountain forest smog receptor site. *Atmos. Environ.*, 27A:483–492.
36. Bytnerowicz, A. and M. E. Fenn. 1996. Nitrogen deposition in California forests: a review. *Environ. Pollut.*, 92:127–146.
37. Collett, J. L., B. C. Daube, Jr. and M. R. Hoffmann. 1990. The chemical composition of intercepted cloud water in the Sierra Nevada. *Atmos. Environ.*, 24A:959–972.
38. Miller, E. K., J. A. Panek, A. J. Friedland, J. Kadlecek and V. A. Mohnen. 1993. Atmospheric deposition to a high-elevation forest at Whiteface Mountain, New York. *Tellus,* 45B:209–227.
39. Waldman, J. M., J. W. Munger, D. J. Jacob and M. R. Hoffman. 1985. Chemical characterization of stratus cloud water and its role as a vector for pollutant deposition in a Los Angeles pine forest. *Tellus,* 37B:91–108.
40. Padgett, P. E., E. B. Allen, A. Bytnerowicz and R. A. Minnich. 1999. Changes in soil inorganic nitrogen as related to atmospheric nitrogenous pollutants in southern California. *Atmos. Environ.*, 33: 769–781.
41. Hicks, B. B., D. D. Baldocchi, T. P. Meyers, R. P. Hosker and D. R. Matt. 1987. A preliminary multiple resistance routine for deriving dry deposition velocities from measured quantities. *Water Air Soil Pollut.*, 36:311–330.
42. Taylor, G. E., Jr., K. Beck, N. Beaulieu and M. Gustin. 1996. Modelling dry deposition of nitrogen, sulfur and ozone to the forest stand at Barton Flats. In: Miller, P. R., Chow, J. and Watson, J., eds., 1996. Assessment of Acidic Deposition and Ozone Effects on Conifer Forests in the San Bernardino Mountains. Final Report for the California Air Resources Board, Contract No. A032-180, A-1 through 8-21, Sacramento, CA.
43. Baldocchi, D. D., B. B. Hicks and P. Camara. 1987. A canopy stomatal resistance model for gaseous deposition to vegetated surfaces. *Atmos. Environ.*, 21:91–101.
44. Bytnerowicz, A. and G. Riechers. 1995. Ozone and nitrogenous air pollutants at the mixed coniferous stand of the western Sierra Nevada, California. *Atmos. Environ.*, 29:1369–1377.
45. Arbaugh, M., A. Bytnerowicz and M. Fenn. 1997. Predicting N flux along a vertical canopy gradient in a mixed conifer forest stand of the San Bernardino Mountains. Proc. of the Symp. *The Effects of Air Pollution and Climate Change on Forest Ecosystems*, Riverside, CA, February 5–9, 1996. USDA Forest Service, Pacific Southwest Research Station, General Technical Report 164 <www.rfl.fs.fed.us/pubs/psw-gtr-164/index.html>.
46. Arbaugh, M., A. Bytnerowicz and M. Fenn. 1999. A predictor of seasonal nitrogenous dry deposition in a mixed conifer stand in the San Bernardino Mountains. (This volume, Chapter 2).
47. Miller, P. R., K. W. Stolte, D. M. Duriscoe and J. Pronos (Technical Coordinators). 1996. Evaluationg Ozone Air Pollution Effects on Pines in the Western United States. Gen. Tech. Report 155, Albany, CA: Pacific Southwest Research Station, Forest Service, U.S. Department of Agriculture, 79 pp.

48. Miller, P. R., R. Guthrey and S. Schilling. 1997. Ozone exposure response of ponderosa and Jeffrey pines in the Sierra Nevada and southern California — the Forest Response Study. Proc. Symp. *The Effects of Air Pollution and Climate Change on Forest Ecosystems*, Riverside, CA, February 5–9, 1996. USDA Forest Service, Pacific Southwest Research Station, General Technical Report 164 <www.rfl.fs.fed.us/pubs/psw-gtr-164/index.html>.
49. Temple, P. J. and P. R. Miller. 1997. Seasonal influences on ozone uptake and foliar injury to ponderosa and Jeffrey pines at a southern California site. Proc. Symp. *The Effects of Air Pollution and Climate Change on Forest Ecosystems*, Riverside, CA, February 5–9, 1996. USDA Forest Service, Pacific Southwest Research Station, General Technical Report 164 <www.rfl.fs.fed.us/pubs/psw-gtr-164/index.html>.
50. Nilsson, J. and P. Grennfelt. 1988. Critical Loads for Sulfur and Nitrogen. Report from a workshop held at Skokloster, March 19–24, 1988. Nordic Council of Ministers, Miljorapport 1988:15, Copenhagen.
51. Bull, K. R., J. J. Brown, H. Dyke, B. C. Eversham, R. M. Fuller, M. Hornung, D. C. Howard, J. Rodwell and D. B. Roy. 1995. Critical loads for nitrogen deposition for Great Britain. *Water Air Soil Pollut.*, 85:2527–2532.
52. Sverdrup, H., P. Warfvinge and K. Rosén. 1994. Critical loads of acidity and nitrogen for Swedish forest ecosystems, and the relationship to soil weathering. In: H. Raitio and T. Kilponen, eds., Proc. of the Finnish-Swedish Environmental Conference on Critical Loads and Critical Limit Values, Vaasa, Finland, October 27 and 28, 1994, pp. 109–138.
53. Bobbink, R. and G. M. Roelofs. 1995. Nitrogen critical loads for natural and semi-natural ecosystems: the empirical approach. *Water Air Soil Pollut.*, 85:2413–2418.
54. Dise, N. B. and R. F. Wright. 1995. Nitrogen leaching from European forests in relation to nitrogen deposition. *For. Ecol. Manage.*, 71:153–161.
55. Kiefer, J. W. and M. E. Fenn. 1997. Using vector analysis to assess nitrogen status of ponderosa and Jeffrey pine along deposition gradients in forests of southern California. *For. Ecol. Manage.*, 94:47–59.
56. Riggan, P. J., R. N. Lockwood and E. N. Lopez. 1985. Deposition and processing of airborne nitrogen pollutants in Mediterranean-type ecosystems of southern California. *Environ. Sci. Technol.*, 19:781–789.
57. Haase, D. L. and R. Rose. 1995. Vector analysis and its use for interpreting plant nutrient shifts in response to silvicultural treatments. *For. Sci.*, 41:54–66.
58. Timmer, V. R. and E. L. Stone. 1978. Comparative foliar analysis of young balsam fir fertilized with nitrogen, phosphorus, potassium, and lime. *Soil Sci. Soc. Am. J.*, 42:125–130.
59. Hockman, J. N. and H. L. Allen. 1990. Nutritional diagnoses in loblolly pine stands using a DRIS approach. In: S.P. Gessel, D.S. Lacate, G.F. Weetman and R.F. Powers, eds., *Seventh North American Forest Soils Conference, Sustained Productivity of Forest Soils*, University of British Columbia, Vancouver, B.C., Canada, July 1988, pp. 500–514.
60. Needham, T. D., J. A. Burger and R. G. Oderwald. 1990. Relationship between diagnosis and recommendation integrated system (DRIS) optima and foliar nutrient critical levels. *Soil Soc. Am. J.*, 54:883–886.

# 7 The History of Human Impacts in the Clear Lake Watershed (California) as Deduced from Lake Sediment Cores*

*Peter J. Richerson, Thomas H. Suchanek, Jesse C. Becker, Alan C. Heyvaert, Darell G. Slotton, Jae G. Kim, Xiaoping Li, Laurent M. Meillier, Douglas C. Nelson, and Charles E. Vaughn*

## CONTENTS

Abstract .................................................................................................................... 120
Introduction ............................................................................................................. 120
Methods ................................................................................................................... 124
Results ..................................................................................................................... 127
    Dating .............................................................................................................. 127
    Physical and Chemical Changes Downcore .................................................. 129
    Pollen Profile .................................................................................................. 133
    Diatom Profile ................................................................................................ 133
Discussion ............................................................................................................... 137
Conclusions ............................................................................................................. 141
Acknowledgments ................................................................................................... 142
References ............................................................................................................... 143

---

* Although the information in this chapter results in part from funding by the U.S. Environmental Protection Agency, it does not necessarily reflect the views of the agency and no official endorsement should be inferred.

## ABSTRACT

We have raised sediment cores to investigate multiple stresses on Clear Lake, CA over the past 250 years. Earlier work suggested the hypothesis that the use of heavy earth-moving equipment was responsible for erosion, mercury, and habitat loss stresses. Such stresses would have first become significant about 1925 to 1930. The cores are about 2.5 m long and span 200 to 300 years of the lake's history. We present the results for our, as yet, most thoroughly analyzed core. $^{210}$Pb dating yields an estimated 1.2 cm yr$^{-1}$ average sedimentation rate for this core. Total (primarily inorganic) mercury and a number of other parameters were measured at 5-cm intervals down the core. Nearly all parameters show major changes at depths of 75 to 80 cm, corresponding to an estimated date of 1927. Organic matter, total carbon, water content, and total nitrogen all show significant decreases above this depth. A peak in inorganic deposition rate and minimum values for percent water is present at a depth corresponding to about 1971. Inorganic mercury concentrations show major increases in concentration (roughly tenfold) above the 1927 horizon. There is also a smaller uptick in total mercury at 145 to 150 cm deep in the core. This horizon is beyond $^{210}$Pb dating capabilities, but most likely represents the early episodes of mercury mining which started in 1873 at the Sulphur Bank Mercury Mine located on the lake's eastern shore. Peak total mercury levels occur at an estimated date of 1961 (last mining was in 1957), and a modest decline has occurred since. Interestingly, the first 75 years of European settlement in the Clear Lake basin (including the most productive years of Sulphur Bank Mercury Mine) appeared to have had barely detectable effects on core properties despite considerable presence after the 1870s. Changes since 1925 are much more dramatic. The large increase in mercury beginning about 1927 corresponds to the use of heavy equipment to exploit the ore deposit at Sulphur Bank Mine with open pit methods. The increase in inorganic sediment load during the last 75 years is substantial in this core, but is not replicated in other cores. Increases in sulfate and/or acidity loading from the mine may be responsible for the dramatic changes seen in the upper 75 to 80 cm of the core.

**Keywords:** *human impacts, mercury, watersheds, sediment cores, mining*

## INTRODUCTION

Clear Lake is a large (177 km²), shallow, eutrophic, polymictic lake in the Central Coast Range of Northern California. It sits at about 404 m elevation and has a typical Mediterranean climate with cool, rainy winters and hot, dry summers. The topography of the drainage basin is mostly steep, with typical elevations along the rim running 1000 m. About 195 km² of the total drainage basin (1219 km²) are level enough for agriculture and urban development. Mean annual high temperature is about 22°C, and rainfall averages about 700 mm yr$^{-1}$. The lake lies in a tectonic basin blocked by a landslide at one end and a volcanic dam at the other. Since 1914, the top 3 m of water depth has been controlled by a dam. Clear Lake is thought to be the oldest lake in North America, with continuously recorded lake sediment cores dating to approximately 500,000 ybp (years before present).[1] The lake is divided

# History of Human Impacts in the Clear Lake Watershed

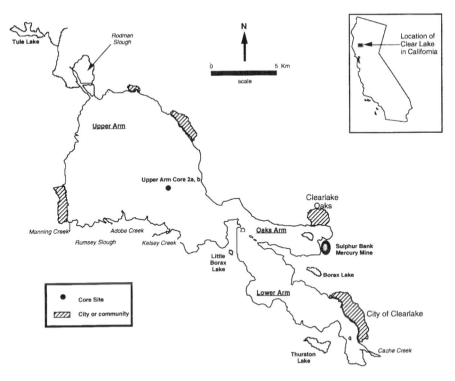

**FIGURE 1** Map of the Clear Lake area showing core site Upper Arm 2 and the location of Sulphur Bank Mercury Mine and other features mentioned in the text.

into three main basins or arms (see Figure 1), the largest of which is the Upper Arm. Nearly all of the winter inflow of water into Clear Lake is into the Upper Arm (approximately 90%, using the estimates in Reference 2). Forty-five percent of the inflow is from the Scotts Creek and Middle Creek watersheds[2] which enter the lake through Rodman Slough. There are some very small, seasonal inflows into the two other arms. The outflow is through Cache Creek at the southern end of the Lower Arm. Aside from the natural stresses of fire, floods, drought, etc., there are many anthropogenic stresses to the system. The most important management problems at the lake are the frequent occurrence of blooms of scum forming cyanobacteria (bluegreen algae) and the contamination of the aquatic ecosystem by mercury and, in the 1950s, by the pesticide DDD.

The objective of our study is to estimate the degree of stress imposed on the Clear Lake watershed by the successive stages of population growth and economic development, especially European growth and development. Were any of the stages of European population and economy "sustainable"? The history of human settlement in the Clear Lake basin is typical of rural California. Native American settlement reaches back to the end of the Pleistocene. One of the first "early man" (Paleo-Indian) sites in North America was discovered at Borax Lake,[3] a small lake on the peninsula between the Lower and Oaks Arms of Clear Lake. Because of an abundance of acorns, fish, and other resources, the Native American populations at the time of European contact were relatively dense; approximately 3000 people lived in some 30 villages mostly near the lake.[4] The acorn-intensive subsistence strategy developed about 2500 years before present and continued more or less unchanged until European contact. No significant changes in anthropogenic effects during the last few hundred years of the Native American period[5] are expected from the archaeological record, and none were observed in our cores. The history of European settlement in the basin is described by Simoons.[6] Mexican land grants in the 1840s initiated European settlement, but only a handful of ranchers resided in the Clear Lake watershed. Agricultural settlement began in 1854, and the rate of settlement increased after 1866. Early censuses showed nearly 1000 people in 1860, nearly 3000 in 1870, and just over 7000 in 1890 in Lake County, most residing in the Clear Lake watershed. Farming, ranching, and farm services were the most important occupations, followed by mining for sulfur, borax, and cinnabar and a small lumber industry. Tourism, centered on hot spring spas, began in 1852. It was well developed in the late 19th century, with several resorts having capacities for hundreds of guests. Early farming focused on wheat, barley, beef, and wool. Sheep numbers peaked sharply in the late 1870s at around 50,000 head and declined thereafter due to deteriorating pastures. Goats and hogs peaked at the turn of the 20th century at around 10,000 head each. Cattle numbers were much more stable, with around 10,000 head from 1861 through the 1980s. Mining for sulfur began at the Sulphur Bank Mine at the east end of the Oaks Arm in 1865, and cinnabar was soon discovered at the same site. Large amounts of mercury (Hg) were mined from 1873 to 1957, with the largest episodes of mining from 1873 to 1883 (3000 metric tons [MT]) and 1927 to 1944 (1100 MT), with a total of about 4700 MT of mercury removed.[7,8] Thus, 19th century European settlement in the watershed was extensive, and the economy was substantially based on the exploitation of natural resources, leading to the potential for significant anthropogenic stresses before 1900.

Presently, agriculture (especially orchard crops) and tourism are the most important industries in the basin. Orchard crops (pears, walnuts, and grapes) have replaced annual crops over almost all of the tillable acreage. Most level acreage near the lake is irrigated. There is recreational development around much of the shoreline (this development followed the construction of a state highway in the 1920s). The two incorporated cities of Clearlake (about 13,000 residents) and Lakeport (about 5000 residents) lie on the lake shore, as do several other smaller communities. The total population of the basin in 1990 was about 50,000.

European activity has generated a number of stresses in the Clear Lake watershed. During the 19th century, clearance for farmland, road building, livestock grazing, logging, and firewood cutting could have generated substantial erosion and eutrophication, as similar settlements did in the eastern U.S.[9] Mercury mining could have resulted in the contamination of lake sediments. Several activities are likely to have contributed to increased stress in the 20th century. Our investigation of the causes of algal blooms in Clear Lake[2] concluded that nutrient loading to the lake increased substantially between 1925 and 1938 due to the beginning of the use of heavy earth-moving equipment. Two early scientific reports describe the lake as having abundant rooted aquatic vegetation, as do elderly residents' recollections of conditions before 1930, recorded in the 1960s.[2,10,11] Histories taken in the 19th century describe the lake's water quality in good terms, though one of them is quite candid about problems with noxious swarms of the Clear Lake Gnat (*Chaoborus* sp.).[12,13] By 1938, when a series of secchi disk transparency measurements were made, the lake had become too turbid for rooted aquatic vegetation to flourish, and cyanobacterial scums had become a perennial problem.[14,15] Since heavy equipment was used in the Clear Lake watershed for many kinds of earth-moving projects, we expected to find multiple impacts of its use. The cores were designed to reach well below the advent of European settlement to put the heavy equipment hypothesis into the context of earlier and later impacts of European activities. Until 1987, gravel mining in streambeds was common[16] (S. Zalusky, personal communication). The Sulphur Bank Mercury Mine was operated using open pit methods after 1927, and approximately $10^6$ MT overburden and waste rock were moved. Some excavated material was pushed directly into the lake, and eventually 400 m of lake shore were covered with waste rock and tailings piles about 10 m high at the angle of repose. Subsequent erosion from these piles into the lake was significant.[17-19] Road building and lake filling activities also increased after 1920. Rough estimates[20] suggest that current erosion rates are roughly twice pre-European rates, with stream channel disturbance and road cuts/fills being the most important causes. In 1928, a 2000-acre reclamation project, also constructed using heavy equipment, eliminated most of wetland at the northwest end of the lake. The project created Rodman Slough (Figure 1) to direct flood flows directly into the lake, thus destroying a major trap for nutrients and sediment.[2] Thus, several lines of evidence suggest that the advent of heavy earth-moving equipment was the ultimate source of the most important anthropogenic stresses on the Clear Lake watershed and lake ecosystems. Natural processes such as drought could have impacts on the lake and its watershed of the same magnitude as anthropogenic stresses.

Sediment cores provide many quantitative indicators of past conditions in lakes and their drainage basins. The extension of historical data beyond the usually short timeframe of monitoring data often taps natural "experiments" that can provide more critical tests of hypothesis than recent data alone. Brush[9,21,22] and others have used core data successfully to reconstruct the extent of human impacts on aquatic ecosystems. Here, we describe results from our ongoing Clear Lake coring project based on Upper Arm Core 2a and a replicate core taken at the same site, Upper Arm Core 2b (Figure 1). We measured parameters designed to estimate inorganic sedimentation rate and the degree of contamination from the Sulphur Bank Mercury Mine. Inorganic mercury was measured to estimate the load of this heavy metal to the aquatic ecosystem. Changes in organic matter content, nitrogen, phosphorus, and diatoms were measured to detect a change in trophic status. Pollen grains were counted to determine if anthropogenic disturbances due to logging, land clearance, grazing, or changes in fire frequency were likely to have been significant. This reasonably broad suite of indicators of physical, chemical, and biological conditions in the lake should detect changes in many of the most important anthropogenic and natural stresses that we believe could plausibly have occurred in the last few centuries.

The cores document a suite of major changes in the Clear Lake system dating from about 1927, much as suggested by the heavy equipment hypothesis. However, the evidence for an increase in inorganic sedimentation rate in recent decades is ambiguous. An increase occurred in the Upper Arm 2 cores, but preliminary data from four other cores from other locations in the lake suggest that sedimentation rates may have fallen in the last 75 years. A plausible alternative hypothesis is that the sulfate and/or acid load, generated by the operation of the Sulphur Bank Mine using open pit methods, substantially altered the microbiology and chemistry of the lake's sediments, leading to eutrophication as well as to a greatly increased level of mercury contamination.

## METHODS

The cores were taken from a 22-ft research vessel using a push-rod operated piston corer. The core sleeve was constructed from 2 in. i.d. × 10 ft Schedule 80 PVC conduit. The piston was a modified 2 in. pipe test plug with a nylon rope attached to the boat to keep the piston at the sediment–water interface during the insertion of the core. The push-rod was 1.5 in. galvanized steel pipe in 11-ft sections. The corer was lowered to just above the sediment–water interface, and the piston rope was tied off. The core was driven in one steady push, as smoothly as possible, and retrieved using a winch on the boat. Just before the core was removed from the water, the bottom was capped to prevent any loss of material. Once on board, the core was split into three sections, approximately 1 m long, by sawing around the core sleeve and then cutting through the core with a knife. The sections were then capped top and bottom and sealed using silicone sealant to prevent moisture loss during the short storage period. Core sections were transported and stored vertically. Any storage was done at approximately 4°C in the dark. The core was extruded in 5-cm sections within 48 h of retrieval. Approximately 30 cm at the top of extrusion had to be done vertically, as the sediments were very soft and easily disturbed. Below this level, cores were extruded horizontally onto an aluminum foil sheet, marked in 5-cm intervals, and sectioned. All sections were placed in preweighed 4-oz specimen containers, weighed, homogenized, and then stored at 4°C under no-light conditions.

Once all sections had been weighed and homogenized, subsamples were taken for total mercury, carbon, nitrogen, and phosphorous analyses. Subsamples for the carbon and nitrogen (C/N) and $^{210}$Pb dating were dried at 60°C and crushed by mortar and pestle to pass through a #60 mesh (250 µm) sieve. For use in pollen analysis, the remaining sample was stored in darkness at 4°C. Percent loss on ignition (equivalent to organic matter) and percent water were determined at the UC Davis Clear Lake Environmental Research Center. Loss-on-ignition subsamples were dried at 105°C and then combusted at 500 to 550°C for 2 h. The remaining sample after combustion is an estimate of the inorganic matter in the sample. The percent water presented is the average between the percent water values for the C/N and loss-on-ignition subsamples. The inorganic accumulation rate was calculated assuming that inorganic matter has a specific gravity of 2, that the midpoint of 80- to 85-cm core slice represents 1927, and the 145- to 150-cm slice core represents 1873 (see the following discussion of date estimates). Total phosphorus (P) concentrations were determined at the University of California

Hopland Research and Extension Center using the methods of Sherman.[23] C/N concentrations were determined at the DANR Analytical Lab at UC Davis using a Carlo-Erba 1500 series Nitrogen Carbon Analyzer.

Total mercury was determined at the UC Davis Environmental Mercury Laboratory on dry, powdered subsamples using a modified cold vapor atomic absorption method.[24]

An accurate reconstruction of the depositional record is dependent upon establishing a reliable sediment chronology. One method for establishing this geochronology in sediments of recent origin is with $^{210}$Pb. The usefulness of this natural radionuclide is limited to sediments of the last 100 to 130 years because its half-life is about 22 years. An alternative sediment dating method matches pollutant profiles, such as mercury, to periods of known production and input. Application of multiple methods generally improves the accuracy and confidence in age estimates. Concentrations of sediment $^{210}$Pb were determined in the laboratory of D.N. Edgington (Center for Great Lakes Studies, University of Wisconsin, Milwaukee) by measuring the activity of $^{210}$Po, a decay product of $^{210}$Pb. Sample size was approximately 0.5 g of dried sediment prepared for counting by the methods specified by Robbins and Edgington,[25] with $^{209}$Po added as an internal yield tracer. Alpha activity was measured on an argon purged, low background counter. Eight sets of replicate extractions yielded an average coefficient of variation less than 12%. Most of these replicate sets were taken from sediment sections below 100 cm and, therefore, contained low excess activity. The deposition dates were computed from $^{210}$Pb concentrations using a constant flux, constant sedimentation rate model described in Robbins.[26] We carefully inspected the data (Figure 2) and determined that a more complex model, assuming heterogeneous sedimentation rates, was not warranted. The appropriate model for excess $^{210}$Pb activity at depth z ($A_z$) is therefore

$$A_z = A_o * \exp(-\lambda * m/r) + A_s$$

where
- $A_o$ = "unsupported" $^{210}$Pb activity due to atmospheric input
- $\lambda$ = the radioactive decay constant for $^{210}$Pb
- m = the cumulative dry mass at depth z
- r = the mass sedimentation rate
- $A_s$ = the "supported" $^{210}$Pb background activity generated by within-sediment production of $^{210}$Pb from the *in situ* decay of $^{226}$Ra

Cumulative dry mass sedimentation was estimated using the dry weight (105°C) of each core section. The unknown parameters ($A_s$, $A_o$, and r) were determined by nonlinear fit of this model to the $^{210}$Pb data.

Pollen counts were done by modifying the method of Faegri and Iverson.[27] Each 1-g, fresh weight sediment subsample was boiled 10 min in 10% KOH to remove organic materials and washed with distilled and deionized water. Ten percent HCl and 95% ethanol were added to remove calcium carbonate and boiled in water bath. HF was added, and samples were boiled in water bath for 5 min to remove siliceous matter. The ethanol/HCl treatment was repeated. To remove cellulose on pollen

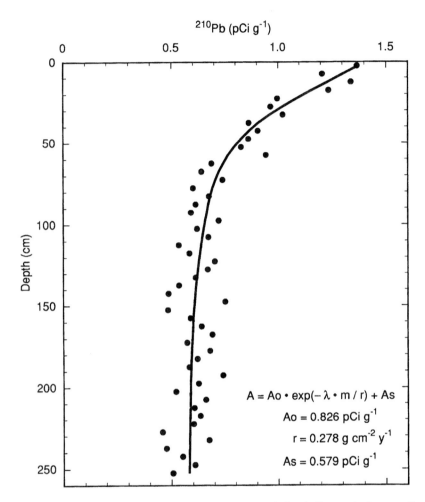

**FIGURE 2**  $^{210}$Pb profile for core Upper Arm 2a. Smooth line indicates the best nonlinear fit to the data.

surface, a 3-min acetolysis with glacial acetic acid was performed in a water bath. After three washings with distilled, deionized water, glycerol was added. One drop of this solution was mounted on slide glass, and a minimum of 300 pollen grains was counted. The remaining solution was stored in a small glass vial. All data analyses were based on pollen percentage.

Samples of 100 to 150 mg of dried sediment were prepared for diatom enumeration by oxidizing with 30% hydrogen peroxide and washing with deionized water. Homogenized sediments in 2-ml aliquots were placed on a Battarbee plate (evaporation plate fitted with four cover slip depressions).[30] Cover slips were mounted onto slides with Cargile mounting media. For each slide, 300 diatom valves were counted using a Zeiss phase contrast microscope at 1000X. Diatom taxonomy was based on standard references.

## RESULTS

### DATING

The $^{210}$Pb data for the Upper Arm 2a core is presented in Figure 2. The best nonlinear fit to this data was derived with simplex and quasi-Newton minimization algorithms. The data are described by an exponential decay equation, as given earlier, with the following parameters and their asymptotic standard errors: r = 0.272 (±0.032) g cm$^{-2}$ y$^{-1}$, A$_o$ = 0.829 (±0.048) pCi g$^{-1}$, and A$_s$ = 0.580 (±0.018) pCi g$^{-1}$. The residuals are random, homogenous, and monotonic. Table 1 shows the deposition dates corresponding to midpoints for each sediment section. Any error in the estimated sedimentation rate would introduce an error in assignment of the deposition date which increases cumulatively with depth. This is evident in the upper and lower 95% confidence limits for each date, as calculated from the error associated with our estimate of sedimentation rate (r). Although dates are extended to the bottom of this core, the deeper dates are extrapolations beyond information actually contained in the signal of excess $^{210}$Pb, which typically decays below the limits of detection within approximately five half-lives (112 years). This extrapolation assumes a constant sedimentation rate below the excess $^{210}$Pb profile which is equal to the sedimentation rate calculated from the upper core profile. Similarly, confidence intervals were extrapolated to core bottom from the asymptotic standard error in our estimate of sedimentation rate. We should note that these confidence intervals were calculated from the standard error for an estimated parameter of a nonlinear function, and, thus, are not exact because they are based upon linearizing assumptions that tend to underestimate the true uncertainty.

The atmospheric flux of $^{210}$Pb can be estimated from these data if we assume that $^{210}$Pb has a short residence time in the water column and that its loss to outflow is negligible. These are reasonable assumptions given that Clear Lake is a shallow, eutrophic system with a hydraulic retention time of 4.6 years.[2] Further assuming that $^{210}$Pb is uniformly distributed across the lake surface and then evenly deposited to the underlying sediments, the atmospheric flux is calculated to be 0.23 pCi cm$^{-2}$ y$^{-1}$. This is equivalent to the lower range of values (approximately 0.2 to 1.1 pCi cm$^{-2}$ y$^{-1}$) calculated by Turekian[29] for the atmospheric flux to hemispheric latitudes between 15 and 55°N. At Clear Lake the deposition of $^{210}$Pb is low because it is close to the Pacific Ocean, where the prevailing maritime winds are depleted in $^{210}$Pb relative to continental sources.

Since the atmospheric flux of $^{210}$Pb at Clear Lake is relatively low and the mass sedimentation rate is relatively high, the specific activity of excess $^{210}$Pb in surface sediments is proportionately reduced. For this reason, the signal of excess $^{210}$Pb in the Clear Lake core has decayed nearly to the level of "supported" background activity by about 80 to 85 cm, which is equivalent to about 1929 from our estimated sedimentation rate for this core. With excess activity near the limits of detection at this depth, we cannot use $^{210}$Pb methods to directly test the hypothesis that sedimentation rates increased beginning at about 1929. Given the scatter in the $^{210}$Pb data and the low excess activity in these sediments, the most reliable approach in this case is an estimate of the average sedimentation rate from a

## TABLE 1
### Estimated Deposition Date of the Midpoint of Each Sediment Slice Derived from $^{210}$Pb Estimates

| Midsection Depth (cm) | Total $^{210}$Pb Activity (pCi g$^{-1}$) | Cumulative Section Mass (g cm$^{-2}$) | Midsection Deposition (date) | Confidence Intervals -95% | +95% |
|---|---|---|---|---|---|
| 2.5 | 1.37 | 0.4 | 1995 | 1994 | 1995 |
| 7.5 | 1.21 | 1.1 | 1992 | 1991 | 1993 |
| 12.5 | 1.34 | 2.1 | 1989 | 1986 | 1990 |
| 17.5 | 1.24 | 3.2 | 1985 | 1981 | 1987 |
| 22.5 | 1.00 | 4.3 | 1981 | 1975 | 1984 |
| 27.5 | 0.97 | 5.6 | 1976 | 1969 | 1980 |
| 32.5 | 1.02 | 6.9 | 1971 | 1963 | 1976 |
| 37.5 | 0.86 | 8.2 | 1966 | 1957 | 1972 |
| 42.5 | 0.91 | 9.5 | 1962 | 1951 | 1969 |
| 47.5 | 0.86 | 10.8 | 1957 | 1944 | 1965 |
| 52.5 | 0.83 | 12.1 | 1953 | 1938 | 1961 |
| 57.5 | 0.94 | 13.3 | 1948 | 1932 | 1958 |
| 62.5 | 0.69 | 14.5 | 1944 | 1927 | 1954 |
| 67.5 | 0.64 | 15.7 | 1940 | 1921 | 1951 |
| 72.5 | 0.74 | 16.8 | 1936 | 1916 | 1948 |
| 77.5 | 0.60 | 17.8 | 1932 | 1911 | 1945 |
| 82.5 | 0.68 | 18.7 | 1929 | 1907 | 1942 |
| 87.5 | 0.61 | 19.5 | 1926 | 1903 | 1940 |
| 92.5 | 0.59 | 20.2 | 1923 | 1899 | 1938 |
| 97.5 | 0.72 | 21.0 | 1920 | 1895 | 1935 |
| 102.5 | 0.62 | 21.9 | 1917 | 1892 | 1933 |
| 107.5 | 0.67 | 22.7 | 1915 | 1888 | 1931 |
| 112.5 | 0.53 | 23.5 | 1912 | 1884 | 1928 |
| 117.5 | 0.58 | 24.4 | 1908 | 1880 | 1926 |
| 122.5 | 0.70 | 25.2 | 1905 | 1875 | 1923 |
| 127.5 | 0.67 | 26.2 | 1902 | 1871 | 1921 |
| 132.5 | 0.61 | 27.1 | 1899 | 1866 | 1918 |
| 137.5 | 0.54 | 28.0 | 1895 | 1862 | 1915 |
| 142.5 | 0.49 | 28.9 | 1892 | 1858 | 1913 |
| 147.5 | 0.75 | 29.9 | 1888 | 1853 | 1910 |
| 152.5 | 0.48 | 30.9 | 1885 | 1848 | 1907 |
| 157.5 | 0.59 | 31.9 | 1881 | 1843 | 1904 |
| 162.5 | 0.64 | 32.9 | 1878 | 1839 | 1901 |
| 167.5 | 0.69 | 33.8 | 1874 | 1834 | 1899 |
| 172.5 | 0.57 | 34.8 | 1871 | 1830 | 1896 |
| 177.5 | 0.68 | 35.7 | 1868 | 1825 | 1893 |
| 182.5 | 0.62 | 36.7 | 1864 | 1820 | 1890 |
| 187.5 | 0.58 | 37.8 | 1860 | 1815 | 1887 |
| 192.5 | 0.74 | 38.8 | 1857 | 1810 | 1884 |
| 197.5 | 0.63 | 39.8 | 1853 | 1806 | 1881 |
| 202.5 | 0.52 | 40.8 | 1849 | 1801 | 1879 |

## TABLE 1 (CONTINUED)
### Estimated Deposition Date of the Midpoint of Each Sediment Slice Derived from $^{210}$Pb Estimates

| Midsection Depth (cm) | Total $^{210}$Pb Activity (pCi g$^{-1}$) | Cumulative Section Mass (g cm$^{-2}$) | Midsection Deposition (date) | Confidence Intervals -95% | +95% |
|---|---|---|---|---|---|
| 207.5 | 0.66 | 41.8 | 1846 | 1796 | 1876 |
| 212.5 | 0.61 | 42.9 | 1842 | 1791 | 1873 |
| 217.5 | 0.63 | 43.9 | 1838 | 1786 | 1870 |
| 222.5 | 0.60 | 45.0 | 1834 | 1781 | 1866 |
| 227.5 | 0.46 | 46.1 | 1830 | 1776 | 1863 |
| 232.5 | 0.67 | 47.2 | 1826 | 1770 | 1860 |
| 237.5 | 0.48 | 48.3 | 1822 | 1765 | 1857 |
| 242.5 | 0.55 | 49.4 | 1818 | 1760 | 1854 |
| 247.5 | 0.61 | 50.5 | 1814 | 1754 | 1850 |
| 252.5 | 0.51 | 51.6 | 1810 | 1749 | 1847 |

*Note:* Dates and the confidence interval for sediments deeper than 82.5 cm are extrapolations from the upper part of the sediments to which the decay model was fit.

nonlinear fit to the profile, as explained previously. Averaged over the length of this core above 82.5 cm, where the $^{210}$Pb dating is reasonably reliable, the calculated linear sedimentation rate is 1.2 cm y$^{-1}$. Any flattening of the $^{210}$Pb profile commonly associated with active bioturbation appears to be limited to the top 10 to 15 cm in this core. This would place a lower limit on the temporal resolution of specific events at approximately 10 years. Other features of the core, especially mercury concentrations, permit further inferences about the sediment chronology, as discussed in the next section.

### Physical and Chemical Changes Downcore

The most striking patterns in the record are the coincident changes of many parameters in the upper 80 cm of the core (see Figure 3 for mercury data, Figure 4 for physical data, and Figure 5 for chemical data). Above this horizon, sediments become markedly drier, have lower organic and nitrogen content, and contain about ten times more total (essentially inorganic) than the pre-European background. The $^{210}$Pb estimate for age of the last section before the sharp increase in mercury begins is 1929. Open pit operations began at Sulphur Bank Mine in 1927, and we have taken 1927 as the midpoint date for the 80- to 85-cm core slice. There is a doubling of mercury concentrations beginning at 145 to 150 cm in the Upper Arm 2a core, with an extrapolated $^{210}$Pb date of 1888. This increase probably represents the opening of the mine. Hg production began in 1873, but sulfur mining operations encountered mercury-contaminated material a few years before (sulfur mining began in 1865). The lower 95% confidence interval for the 1888 extrapolated $^{210}$Pb date is 1853

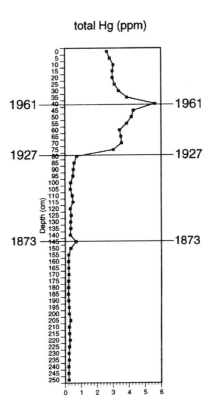

**FIGURE 3**  Mercury in the Upper Arm 2a core. The dates for the lines indicating estimated dates of 1961 and 1927 are derived from $^{210}$Pb estimates. The 1873 line is based on the assumption that the spike of mercury followed by the persistent doubling of mercury corresponds to the beginning of mercury production at Sulphur Bank Mine. Depths indicated are for the top of each 5-cm slice of the core.

(Table 1). Whether mercury first reached the lake in 1873 or a little before or after is impossible to tell, given the need to extrapolate dates below the level of excess $^{210}$Pb and an inexact history of mine operations. We have taken 1873 as the midpoint date for the 145- to 150-cm core slice.

If our hypothesis that sedimentation rates increased significantly above 80 to 85 cm is correct, then the $^{210}$Pb extrapolated age of the 145- to 150-cm section where the first mercury increase appears would be an underestimate. In support of the interpretation that the sedimentation rate changed after about 1927, water content in both cores taken at the this site shows a dramatic decrease above 80 to 85 cm as if the deposition rate of inorganic sediment increased sharply (Figure 4). Normally, water content would increase monotonically toward the sediment surface, as it does below 80 to 85 cm. Observations of sediment consistency, recorded during the core sectioning process, noted a dramatic change at this horizon as well. As with water content, nearly all of the analyzed constituents show a major change at the 80- to

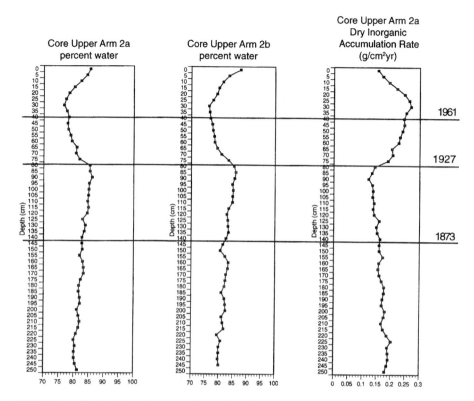

**FIGURE 4** Physical data for cores Upper Arm 2a and b. Date estimates as in Figure 3.

85-cm (~1927) horizon. Total C and N and loss-on-ignition (organic matter) percentages show a large and rapid decrease above this depth (Figure 5). Total P declines slightly, but rises again at the surface. The total P content of creek sediments is about 1000 ppm,[2] similar to values in the cores. The N-to-P ratio, an important index of the lake's nutrient status because of the role of N-fixing cyanobacteria, declines sharply after 1927.

If the date of the first mercury increase in the sediments is taken to be 1873, and the second larger increase is 1927, the rate of dry inorganic deposition changes substantially, from 0.14 gm cm$^{-2}$ yr$^{-1}$ 1873–1927 to 0.22 gm cm$^{-2}$ yr$^{-1}$ post-1927. The total sediment accumulation rate for 1873 to 1927 is almost exactly the same as the post-1927 rate (1.20 and 1.18 cm yr$^{-1}$, respectively), and the considerably higher content of inorganic material in the upper part of the core accounts for the higher inorganic deposition rate. A replicate core at Upper Arm 2 agrees, using the almost identical mercury profile for dating, almost exactly with these estimates, as indicated by the very similar water profiles shown in Figure 4. Cores from one other location in the Upper Arm, two locations in the Lower Arm, and one location in the Oaks Arm all have rather thinner sections of drier, mercury-rich sediments than the Upper Arm 2 core. $^{210}$Pb dates for the last section before the beginning of the large increase in mercury from a third Upper Arm core at a site distant from Station 2 is 1916,

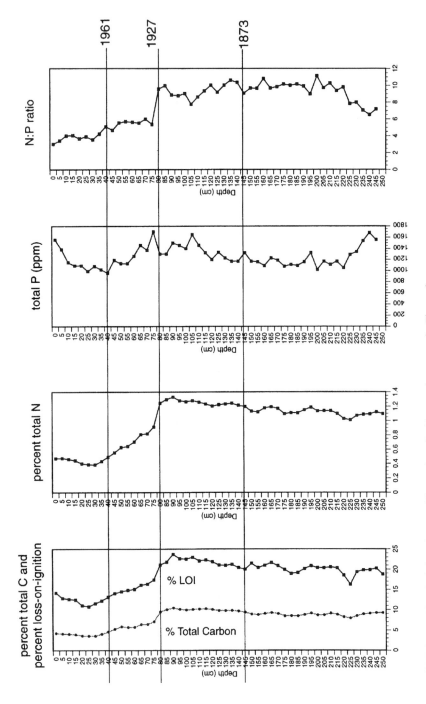

**FIGURE 5** Chemical data for core Upper Arm 2a. Date estimates as in Figure 3.

and the estimate from the Oaks arm core is 1937. The mean of the three dates is 1927.3 years, and 1927 is well within the 95% confidence interval for all three estimates. The changes in other chemical, physical, and biological parameters is also similar to Upper Arm 2 in the other cores, as far as these have been analyzed to date. Thus, so far we have detected only ambiguous evidence that inorganic sedimentation rates have increased basin wide since 1927, despite the large changes in sediment physical and chemical composition suggesting just such an increase. Indeed, the Upper Arm 2 cores are the only ones that do not imply a *decrease* in inorganic sedimentation rate since 1927.

For several parameters, the upper few centimeters of sediment show a partial return to pre-1927 values. This return is considerable for percent water, which rises to near the values one would expect by extrapolating the slowly rising pre-1927 values to the surface. Loss-on-ignition (organic matter) N content show the smallest returns to earlier concentrations. The decline in inorganic mercury is considerable.

## Pollen Profile

Pollen diagrams (Figures 6A and 6B) show a very different pattern from the physical and chemical constituents discussed previously. Major taxa in the pollen record show little or no evidence of changes over the whole period represented in the core. The dominant pine and oak pollens show no detectable pattern of change at all. TCT (Taxodiaceae, Cupressaceae, Taxaceae) group pollen shows a small decrease over the past 100 years as does Compositae. *Salix*, *Acer*, *Artemisia*, and *Chrysolepis* increase somewhat in recent times. Only *Juglans* (walnut) records the agricultural activities of the European settlers. There is a hint that *Potamogeton* was absent after around 1925 and reappeared recently.

## Diatom Profile

Figure 7 shows the diatom profile in the core. The dominant diatom taxa encountered in Clear Lake are various species of *Aulocoseiria* (formerly *Melosira*) and *Stephanodiscus*. The most dramatic change in the profile is the appearance and rise to subdominant status of *Aulocoseiria distans* (Ehrenberg) Simonsen, while *A. ambigua* (Ehrenberg) Simonsen declined. This change correlates exactly with the many changes in physical and chemical variables dating to about 1927 described earlier. As with some other variables, there is a reversal of this change at the very top of the core. Less abundant species also suggest changes post-1927, including the appearance of *Cyclostephanos costalimbus* (Kobayashi & Kobayashi) Stoermer, Håkansson & Theriot and *S. vestibulis* Håkansson, Theriot & Stoermer. Diatom diversity and equitability also increased after 1927, and diatom productivity declined, but irregularly, after 1927 (L. Meillier, unpublished data). As we remarked earlier, there is historical evidence that cyanobacteria became more abundant after 1927. In support of the sparse direct evidence, the N-to-P ratio in the sediments drops sharply after 1927 (Figure 5), suggesting that N became more limiting relative to P, a condition which favors N-fixing cyanobacteria. The decline in diatom abundance and species shifts evident in the core may reflect either the direct effects of altered

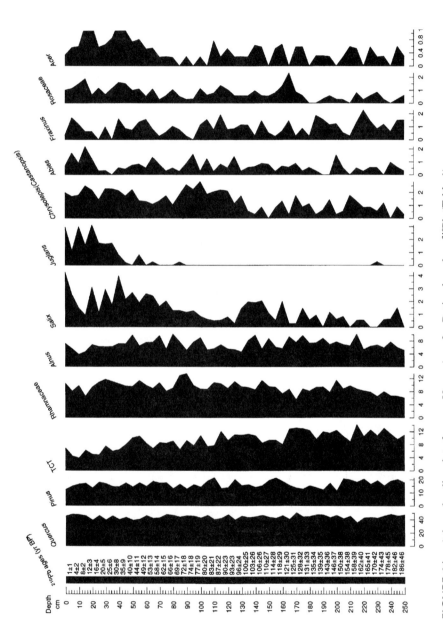

**FIGURE 6A** Arboreal pollen data for core Upper Arm 2a. Dating based on $^{210}$Pb (Table 1).

# History of Human Impacts in the Clear Lake Watershed

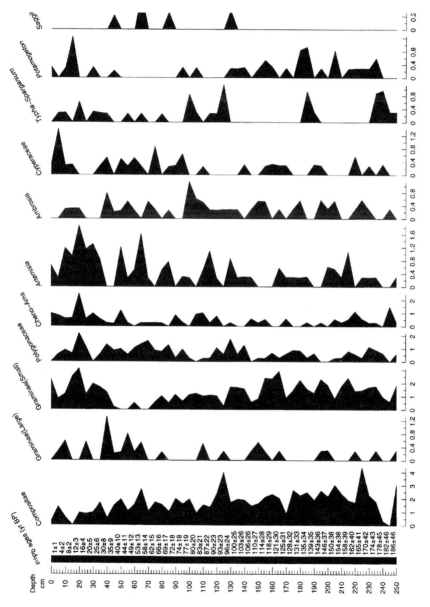

**FIGURE 6B** Aquatic and herbaceous pollen data for core Upper Arm 2a.

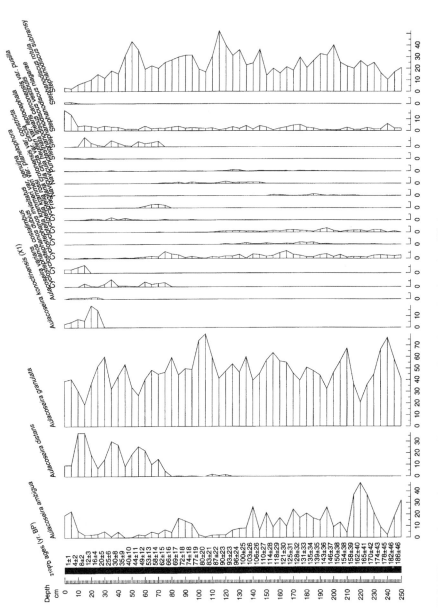

**FIGURE 7** Diatom abundances in Upper Arm Core 2a. Dating based on $^{210}$Pb.

nutrient status or indirect effects mediated through competition with cyanobacteria for nutrients and light.

## DISCUSSION

The early, relatively small increase in total mercury at 145 to 150 cm in Upper Arm Core 2 is almost certainly a result of the early mining at the Sulphur Bank Mercury Mine (Figure 3). The much larger amounts of Hg in the sediments after 1927 are almost certainly due to the use of open pit mining methods. The disturbance of large volumes of overburden and rock contaminated with metal sulfides, and the production of large volumes of tailings likely increased the loading of mercury to the lake ecosystem. The waste piles contain amounts of mercury ranging from nondetectable to 2400 mg kg$^{-1}$, with typical values of around 100 mg kg$^{-1}$.[18] About 25% of the surface of the regraded piles contains sulfide minerals in sufficient concentrations to prevent the growth of annual vegetation (authors' observations). Mercury and related contaminants have entered the lake by three routes.[7] First, mine waste rock containing cinnabar and elemental mercury was pushed directly into the lake as long as the mine was active. Much fine material was perhaps carried away from the immediate vicinity of the mine. Second, sheetwash erosion and mass wastage from wave undercutting of the waste piles transported mercury directly into the lake until the U.S. Environmental Protection Agency reduced the slope and rip-rapped the toe of the piles in 1992. Third, the mine currently discharges acid mine drainage into the lake, which we believe to be the only remaining significant source of mercury.

Since the toe of the waste rock pile rests in the lake, flow from the mine site into the lake is almost entirely subsurface and not easily measured. A rough estimate is possible. In the spring of 1995, a coherent layer of acid mine drainage precipitate covered about 1 km$^2$ of lake bottom adjacent to the mine about 10 cm deep, containing about 120 kg of mercury. Since the precipitate is very finely divided when it first forms, most of it was probably carried away from the vicinity of the mine before it could be incorporated into the distinctive layer near the mine. A plausible calculation suggests that the acid drainage source could account for all of the approximately 400 kg yr$^{-1}$ of mercury necessary to contaminate sediments to the levels recorded in our cores (Chapter 7 in Reference 19). We have also measured depressed pH and elevated sulfate in lake water near the mine (authors' unpublished data). Thus, acid mine drainage may have been the main, if not the only, source of mercury contamination of the lake, with the immediate area of the mine excepted. The modest declines in total mercury from post-1927 peaks (estimated turning point 1961) appear to have begun too long ago to be explained by the recent drought or the erosion remediation work at the mine site in 1992. We have done microstratigraphic analyses on the upper few centimeters of some short cores which indicate a considerable recent drop in inorganic mercury loading, probably due to the remediation (Chapter 5b in Reference 19).

The other coincident changes in physical and chemical characteristics of the core above the 75 to 80 cm level are consistent with the heavy equipment hypothesis, except for our failure to replicate an increase in deposition of inorganic matter per unit time in other cores. Drier sediments with less organic matter and N are consistent with a greater ratio of inorganic to organic deposition. This hypothesis was previously proposed, based on historical records[2] and limited erosion surveying.[20] In addition to the internal data from the cores, historical evidence suggests a significant increase in watershed disturbance around 1927. At this time, the application of powered earth-moving technology to streambed gravel mining, road construction, wetland reclamation projects, and mercury mining began. This type of machinery was developed just before and after World War I, and its use became widespread in the late 1920s.[30] By the early 1930s, Caterpillar had a line of earth-moving machinery much like contemporary types. The end of active mercury mining in 1957 and the end of gravel mining in most streams by 1987[16] should have resulted in a reduction in sediment loading. Consistent with this expectation, there is a trend toward moister, more organic sediments above 30 to 35 cm after the peak of inorganic deposition (~1971). Beginning in the winter of 1971 to 1972 and persisting until the present, water quality monitoring data[2] show less winter turbidity, which is normally due to inorganic material derived from streams. Unfortunately, water quality data are only available back to 1968. Since the pattern of post-1927 increase in inorganic deposition is not present in other cores, it is quite possible that the Upper Arm 2 cores reflect disturbances to the nearest stream, Kelsey Creek (Figure 1). Kelsey Creek began a major episode of downcutting in its lower reaches when its delta was channelized to construct a boat harbor in the mid-1960s. The peak of inorganic deposition in the core dates to about 1971.

Sediment deposition directly from mining operations cannot account for changes in bulk sediment characteristics as distant from the mine as the Upper Arm 2 station. Inorganic mercury in surficial sediments declines exponentially with distance from the mine.[7] Oaks Arm surficial sediments near the mine have total mercury concentrations of about 100 to 300 mg kg$^{-1}$ (dry weight basis). In contrast, concentrations are only 3 mg kg$^{-1}$ at Upper Arm 2, indicating that erosion products from the mine are disproportionately deposited near the mine. The fine precipitate produced when acid mine drainage is neutralized by lake water is likely to be disproportionately represented in the mercury deposited at the Upper Arm 2 site.[7,19]

The 1987 to 1992 drought, during which time there was very little sediment (and sulfate) input to the lake, may have had some influence on the upper few centimeters of the core. Bioturbation undoubtedly integrated the sediments deposited during these years with the four more normal years since the end of the drought. Nevertheless, recent changes appear to have started well before the drought. One exception is the surface peak of total P. We know[2] that increasingly large amounts of base extractable (iron and aluminum bound) P cycled from the water column (late summer) to sediments (winter) during the drought years. The large mass of P that appeared during the drought has resisted burial in the subsequent years, presumably due to its mobility and consequent upward diffusion during summer sediment anoxia.

The long 1986 to 1991 drought reduced the delivery of erosion products to the lake. Thus, it is a natural experimental test of the heavy equipment hypothesis. In several respects, the drought did induce conditions that were characteristic of the pre-1927 period. Historical records also suggest that rooted aquatic vegetation was abundant before 1925. Turbid water after 1927 inhibited the growth of *Potamogeton* and other submersed macrophytes for many decades. In the last few years, severe iron limitation,[31] the causes of which we are continuing to investigate, have produced clear water conditions again, and *Potamogeton* and other rooted aquatic species thrived, consistent with the record of *Potamogeton* pollen. The changes in the diatom flora are not dramatic, but certainly indicate changes in nutrient or light status. The decline in the N-to-P ratio to one third of predisturbance values is consistent with N becoming more limiting relative to P (Figure 5), which would have favored N-fixing cyanobacteria, consistent with the limited historical data indicating that cyanobacterial scums were not a serious problem before the late 1920s or early 1930s. Work by Horne[32] also suggested that winter light limitation due to inorganic turbidity favored winter buoyant *Aphanizomenon* populations and limited diatom growth. Clear water in winter favors diatoms that do not use buoyancy to stay near the surface. The significance of the shifts in the diatom community is obscure. It is interesting that the rise in *A. ambigua* and decline in *A. distans* right at the top of the core coincides with the clear-water conditions accompanying the recent drought event (Figure 7). However, diatom productivity has not increased in recent years. Thus, lake water quality in the period from 1991 to the present in some ways resembles the conditions reported by 19th and early 20th century observers,[10-13] and this resemblance is reflected in some core measurements.

The increases of *Salix* and *Acer* pollen may represent the colonization of disturbed stream channels by willow and box elder, whereas intact riparian vegetation tends to be dominated by oaks (*Quercus lobata*). The essentially constant proportions of *Pinus* and *Quercus* pollen, the dominant genera in upland communities, rule out wholesale disturbance of terrestrial ecosystems as a cause of changes seen in the core. If heavy equipment activity is the cause of the post-1927 changes, it must be due to increased erosion from relatively small areas such as road cuts and fills and disturbance to riparian areas. Whatever changes occurred in the upland areas of the basin with the advent of European settlement left very modest traces in the palynological record. As Simoons[6] noted in his history of the impacts of European settlement, there is very little evidence of gross anthropogenic impacts on upland watersheds.

Notwithstanding the considerable circumstantial evidence in favor of the heavy equipment hypothesis, support from cores collected to date is ambiguous. It is possible that with only two cores from the Upper Arm, and only five independent stations in the whole lake, we have failed to detect an increase in sedimentation rate. What are the alternative hypotheses to explain the coincident changes that began about 1927? It is conceivable that other anthropogenic changes have counterbalanced the use of heavy equipment in the basin. Most conspicuously, sheep and goat grazing and pig herding was significant in the late 19th century and highly detrimental to upland pastures, but all these stresses declined to insignificance well before 1927.[6] If high animal stocking was responsible for increased erosion, it left no signature in the pre-1927 sediments we have been able to detect. Unlike the watershed of the Chesapeake,[9,21] only a relatively small proportion of the Clear Lake watershed has been converted from grassland, chaparral, and oak and pine forests to intensive uses. Most intensive farming and human settlement is on level alluvial soils with little erosion potential. Distinguishing natural from anthropogenic causes of erosion is difficult, and the steep uplands in the headwaters of the basin will have high natural erosion rates. The only direct study of erosion in the basin was very cursory and could well have erred in its estimation of anthropogenic influences on erosion rates.[20] Alternatively, uncertainties in the $^{210}$Pb dates are appreciable, especially in the levels older than about 1927 where excess $^{210}$Pb activity is minimal. The 1873 date based on the first appearance of mercury is reasonably well corroborated by our $^{210}$Pb date of 1888. It is possible that we have misdated the core sufficiently to lead to significant errors in interpretation of sedimentation rates and other processes. It is also possible that sedimentation rates in the basin have increased, but that human activities have resulted in the storage of the sediment in the watershed rather than in the lake (J.F. Mount, personal communication). Perhaps the Upper Arm 2 core reflects the fact that significant erosion occurred on the lower reaches of Kelsey Creek and that sediment mobilized higher up in other drainages has yet to reach the lake in quantity.

We are currently focusing our main effort on testing the hypothesis that increases in sulfate and/or acidity loading are wholly or partly responsible for the changes observed in the core. The Sulphur Bank Mercury Mine potentially became a major source of sulfate and acidity loading after it began to be operated by open pit methods in 1927. Approximately $10^6$ m$^3$ of sulfide-rich overburden and waste rock were excavated in the course of open pit operations, leaving a flooded impoundment of 860 $10^3$ m$^3$ (data here and for the following calculations from the U.S. Environmental Protection Agency[18]). Sulfate concentrations in the impoundment waters, groundwater downgradient toward the lake from the impoundment, and surface runoff from waste piles are about 30 m$M$, with pHs around 2.5. The standing mass of sulfate dissolved in impoundment water alone is about 2500 MT. The standing mass in the lake is around 17,000 MT in nondrought years. The hydraulic residence time of the lake is about 4.6 years, so a sulfate load on the order of 2500 MT yr$^{-1}$ could be an important term in the sulfate budget of the whole lake.

How might sulfate and/or acidity loading cause dramatic changes in sediment physical and chemical properties? Independent lines of evidence do strongly suggest that major changes occurred about 1927, roughly contemporaneous with the beginning of open pit mining. There is little doubt that water quality deteriorated after Coleman's visits in 1925, as discussed earlier. The severe drought of the late 1980s and early 1990s again provides an interesting natural experiment. Sulfate loading from the mine and the basin was presumably sharply curtailed. The proximal reason for the return of clear water in 1991 was the result of severe iron limitation of plankton primary productivity and of N fixation by cyanobacteria.[31] At the same time, the amount of P cycling out of the sediments rose from a predrought average of 150 MT to a peak of 650 MT in 1990, falling back to 200 MT in 1995 and 1996. The P cycles almost entirely from the iron-bound pool in the sediments, so it is puzzling that iron fell during the drought[2] (authors' unpublished data). A peak of total iron in deep water occurred during the late summer peak of P release from the sediments.[31] This peak, however, was accompanied by a minimum in dissolved iron.

Geochemical modeling[31] suggests that iron was being bound as ferrous sulfide as rapidly as it was mobilized from ferric phosphate. Sulfate reduction is an important process in Clear Lake sediments.[33] During the drought, sulfate concentrations in the lake fell from about 125 to 60 $\mu M$ (California Department of Water Resources, unpublished data), reflecting the failure of sulfate loading to balance active net sulfate reduction. Gross sulfate reduction is much higher than net sulfide storage in the sediments.[33] Presumably, much sulfide is reoxidized at the sediment surface or in the overlying water. Sulfur thus acts as a conveyor of relatively high-energy-potential oxidizing power into the sediments. When sulfate is abundant, sulfide isotherms are, all else being equal, likely to lie deeper in the sediments. Since the sediment surface is less likely to be anoxic, ferric phosphate reduction is less intense, reducing the amount of P recycled from the sediments. However, by confining sulfide to greater depths in the sediment, ferrous iron is scavenged less efficiently, leading to more available iron in the water column. If this line of reasoning is correct, increased sulfate loading will tend to relieve the iron limitation of N fixation by cyanobacteria, leading to eutrophication of the lake. The effect of the sulfur conveyor will also be a more thorough mineralization of organic matter in the sediments, hypothetically leading to low organic matter and drier and more nitrogen-poor sediments, as we observed.

Kilham[34] discusses the case of a soft water lake in a basin with significant carbonate rock becoming alkalized by acid rain. After the acidity of the rain is neutralized by $CaCO_3$ in the watershed, sulfate reduction and denitrification by microbes in the lake's sediments generate alkalinity, ending with the apparent paradox of acid loading generating net alkalinity. In general, phytoplankton production rises with alkalinity. Clear Lake is presently a fairly alkaline system (pH 8, alkalinity 125 ppm). Presumably, the acidity from the acid mine discharge has been neutralized by sediment $CaCO_3$, and net alkalinity is generated by sulfate reduction, by analogy with Kilham's system.

## CONCLUSIONS

The Clear Lake ecosystem has experienced a major, continuing stress, or series of temporally coincident stresses, beginning about 1927 and continuing to the present. Several of the changes recorded in our sediment cores are consistent with increased inorganic sedimentation, such as lower water, nitrogen, and organic matter content after 1927. These data, together with some historical data and a limited erosion survey, suggest that the advent of heavy earth-moving equipment in the late 1920s and early 1930s was responsible both for the increased mercury in the sediments and the apparent increase in sedimentation rate. However, evidence of increased sedimentation rates is mixed. Four out of five cores so far raised suggest a *lower* sedimentation rate after 1927. Either these cores are misleading in some way or some other process caused the simultaneous change of several properties of the sediments beginning about 1927. The tight correlation of these changes with the increase in mercury in the sediments suggests that some direct link with mine operations is possible. An alternative to the heavy equipment hypothesis is that acidity and/or sulfate loading from the mine caused these changes. Our current research is devoted to distinguishing between the heavy equipment and acid/sulfate hypotheses.

The absence of major impacts from the early period of European settlement is striking. The transformation of the grasslands to grain fields and the replacement of native pasture grasses by Mediterranean weeds was a major impact which is unfortunately difficult to detect in the pollen record.[35] Nevertheless, the overall stability of the vegetation and the lack of a conspicuous increase in sediment yield are remarkable. Judging from impacts so far uncovered in our cores, grazing, wood cutting and lumbering, agricultural clearing, and the development of small towns and recreation facilities, as conducted from 1854 to 1927, were relatively low-impact activities from a watershed mass-balance perspective. It is encouraging to think that one of California's rugged, semi-arid environments apparently supported a fairly large, active human population in a reasonably sustainable fashion. A narrow range of activities would appear to have been responsible for the stresses visible in our cores.

## ACKNOWLEDGMENTS

We appreciate the advice and assistance of Lauri Mullen, Linnie Brister, Bradley Lamphere, and Cat Woodmansee. Erin Mack and Jeff Mount have given us very valuable counsel on the complexities of sediment biogeochemistry. Two anonymous referees made several helpful suggestions. A perceptive comment by Mike Johnson saved us from making a major error. Grace Brush was a source of encouragement and much sound advice, as was Reds Wolman. Steve Zalusky's knowledge of the relevant aspects of Lake County mining and earth-moving activities was invaluable. Steve Hill, Steve Why, Tom Smythe, and Sue Arterburn smoothed collaboration with Lake County on the field work for this project. Bob Bettinger gave us sound advice on issues of prehistory. Many thanks to David Edgington for help with the $^{210}$Pb analysis and to Shaun Ayers for mercury analyses. Support from the U.S. EPA Center for Ecological Health Research (R819658) and EPA Region 9 Superfund (68-S2-9005) made this project possible. Thanks to Carolyn d'Almeida and Jeri Simmons of EPA Region 9-Superfund and to Karen Moorehouse of EPA-ORD for assistance in connection with this funding. Dennis Rolston and Cheryl Smith made many things easier.

For questions or further information regarding the work described in this chapter, please contact:

> **Peter Richerson**
> Dept. of Environmental Science and Policy
> One Shields Avenue
> University of California
> Davis, CA 95818
> e-mail: pjricherson@ucdavic.edu

## REFERENCES

1. Sims, J.D. ed. 1988. *Late Quaternary Climate, Tectonism, and Sedimentation in Clear Lake, Northern California Coast Ranges.* Special Paper 214, Geological Society of America, Boulder, CO.
2. Richerson, P.J., T.H. Suchanek, and S.J. Why. 1994, The causes and control of algal blooms in Clear Lake, Clean Lakes diagnostic/feasibility study for Clear Lake, California. Report prepared for Lake County Flood Control and Water Conservation District, California Department of Water Resources, and United States Environmental Protection Agency. Division of Environmental Studies, University of California, Davis, CA.
3. Heizer, R.F. 1963. The West Coast of North America. In J.D. Jennings and E. Norbeck eds., *Prehistoric Man in the New World, A.D.* University of Chicago Press, Chicago, pp. 117–148.
4. Baumhoff, M.A. 1963. Ecological determinants of aboriginal California populations. *University of California Publications in American Archaeology and Ethnology* 49:153–236.
5. Basgall, M.E. 1987. Resource intensification among hunter-gatherers: acorn economies in prehistoric California. *Research in Economic Anthropology* 9:21–52.
6. Simoons, F.J. 1952. The settlement of the Clear Lake Upland of California. MA Thesis, Geography. University of California-Berkeley.
7. Suchanek, T.H., L.H. Mullen, B.A. Lamphere, P.J. Richerson, C.E. Woodmansee, D.G. Slotton, E.J. Harner, and L.A. Woodward. 1998. Redistribution of mercury from contaminated lake sediments of Clear Lake, California. *Water, Air, and Soil Pollution* 104: 77–102.
8. Suchanek, T.H., P.J. Richerson, L.J. Holts, B.A. Lamphere, C.E. Woodmansee, D.G. Slotton, E.J. Harner, and L.A. Woodward. 1995. Impacts of mercury on benthic invertebrate populations and communities within the aquatic ecosystem of Clear Lake, California. *Water, Air and Soil Pollution* 80:951–960.
9. Brush, G.S. and F.W. Davis. 1984. Stratigraphic evidence of human disturbance in an estuary. *Quaternary Research* 22:91–108.
10. Stone, L. 1974. XX. Report on operations in California in 1873. A. Clear Lake. 1. Field work in the winter of 1872–3. Report of the Commissioner for 1872 and 1873, Part II. United States Commission of Fish and Fisheries, Washington, D.C., pp. 377–381.
11. Coleman, G.A. 1930. A biological survey of Clear Lake, Lake County. *California Fish and Game* 16:221–227.
12. Menefee, C.A. 1873. *Historical and Descriptive Sketch Book of Napa, Sonoma, Lake and Mendocino.* Reporter Publishing House, Napa, CA.
13. Palmer, L.L. 1881. *History of Lake and Napa Counties, California.* Slocum, Bowen & Co., San Francisco, CA.
14. Lindquist, A.W. and C.C. Deonier. 1943. Seasonal abundance and distribution of larvae of the Clear Lake gnat. *Journal of the Kansas Entomological Society* 16:143–149.
15. Murphy, G.I. 1951. The fishery of Clear Lake, Lake County, California. *California Fish and Game* 37:439–484.
16. Zalusky, S. 1992. Lake County aggregate resource management plan. An element of the Lake County General Plan. Lake County Planning Department, Resource Management Division, Lakeport, CA.

17. Chamberlin, C.E., R. Chaney, B. Finney, M. Hood, P. Lehman, M. McKee, and R. Willis. 1990. Abatement and control study: Sulphur Bank Mine and Clear Lake. Report prepared for California Regional Water Quality Control Board. Environmental Resources Engineering Department, Humboldt State University, Arcata, CA.
18. United States Environmental Protection Agency. 1994. Remedial investigation/feasibility study. Sulphur Bank Mercury Mine Superfund Site Clearlake Oaks, California. United States Environmental Protection Agency Region 9, San Francisco, CA.
19. Suchanek, T.H., P.J. Richerson, L.H. Mullen, L.L. Brister, J.C. Becker, A. Maxon, and D.G. Slotton. 1997. Interim final report: the role of the Sulphur Bank Mercury Mine site (and associated hydrogeological processes) in the dynamics of mercury transport and bioaccumulation within the Clear Lake Aquatic Ecosystem. Report to the United States Environmental Protection Agency Superfund Program, Region IX. Department of Environmental Science and Policy, University of California, Davis, CA.
20. Goldstein, J. and T. Tolsdorf. 1994. An economic analysis of potential water quality improvement in Clear Lake: benefits and costs of sediment control, including a geological assessment of potential sediment control levels. United States Department of Agriculture Soil Conservation Service, Davis and Lakeport, CA.
21. Brush, G.S. 1989. Rates and patterns of estuarine sediment accumulation. *Limnology and Oceanography* 34(7):1235–1246.
22. Brush, G.S. 1991. Stratigraphic history helps us understand today's Chesapeake Bay. *Geotimes*, December 1991:21–23.
23. Sherman, M.S. 1942. Colorimetric determination of phosphorous in soils. *Industrial and Engineering Chemistry Analytical Edition* 14:182–185.
24. Slotton, D.G., J.E. Reuter, and C.R. Goldman. 1995. Mercury uptake patterns of biota in a seasonally anoxic northern California reservoir. *Water, Air, and Soil Pollution* 80:841–850.
25. Robbins, J. A. and D. N. Edgington. 1975. Determination of recent sedimentation rates in Lake Michigan using Pb-210 and Cs-137. *Geochimica et Cosmochimica Acta* 39:285–304.
26. Robbins, J. A. 1978. Geochemical and geophysical applications of radioactive lead. In J.O. Nriagu, ed., *Biogeochemistry of Lead in the Environment*. Elsevier, Holland, pp. 285–393.
27. Faegri, K. and J. Iverson. 1989. *Textbook of Pollen Analysis, 4th Edition*. Wiley, New York.
28. Battarbee, R.W. 1973. A new method for the estimation of absolute microfossil numbers, with reference especially to diatoms. *Limnology and Oceanography* 18:647–653.
29. Turekian, K.K., Y. Nozaki, and L.K. Benninger 1977. Geochemistry of atmospheric radon and radon products. *Annual Review of Earth and Planetary Sciences* 5:227–255.
30. Caterpillar Incorporated. 1984. *The Caterpillar Story*. Peoria, IL.
31. Li, X. 1998. Iron cycling and its effect on algal growth in Clear Lake, California. Ph.D. Thesis. University of California, Davis, CA.
32. Horne, A.J. 1975. The ecology of the Clear Lake phytoplankton. Clear Lake Algal Research Unit, Lakeport CA.
33. Mack, E.E. 1998. Sulfate reduction and mercury methylation potential in the sediments of Clear Lake (CA). Ph.D. Thesis. University of California, Davis, CA.
34. Kilham, P. 1982. Acid precipitation: its role in the alkalization of a lake in Michigan. *Limnology and Oceanography* 27:856–867.

35. West, G.J. 1989. Early historic vegetation change in Alta California: the fossil evidence. In Thomas, D.H., ed., *Colombian Consequences. Volume 1: Archaeological and Historical Perspectives on the Spanish Borderlands West.* Smithsonian Institution Press, Washington, D.C., pp. 333–348.

# Section II

## Effects

# 8 The Development of Cumulative Effects Assessment Tools Using Fish Populations

*Kelly R. Munkittrick, Mark E. McMaster, Cam Portt, Wade N. Gibbons, Andrea Farwell, Lisa Ruemper, Mark R. Servos, John Nickle, and Glen J. Van Der Kraak*

## CONTENTS

Abstract .................................................................................................................. 149
Introduction ........................................................................................................... 150
Philosophical Approach ........................................................................................ 152
Objectives .............................................................................................................. 155
Geographical Description ..................................................................................... 155
Study Design ......................................................................................................... 159
Phase I: Latitudinal Comparison of Reference Sites ............................................ 162
    Phase I Preliminary Results ........................................................................... 163
Phase II: Longitudinal Comparison of Undeveloped Sites .................................. 165
Phase III: Longitudinal Comparison at Developed Sites ..................................... 166
    Preliminary Phase III Results ........................................................................ 167
Phase IV: Model Development ............................................................................. 170
Phase V: Evaluation of Cause and Effect Relationships ..................................... 170
Conclusions ........................................................................................................... 171
Acknowledgments ................................................................................................. 171
References ............................................................................................................. 172

## ABSTRACT

Until recently, environmental assessments were conducted without regard to either upstream or downstream developments, such that the relative contributions of individual developments could not be separated. The environmental legislative agenda in Canada has been proceeding at a pace where there is an expectation to address the cumulative effects of existing developments prior to the development of methodology for conducting these assessments. We initiated a long-term study in 1991 which is attempting to address the cumulative aquatic effects of hydroelectric facilities and pulp mill discharges on fish populations within the Moose River basin in Northern Ontario. The focus is placed on the top-down interpretation of the impact of alterations in habitat and fish energetics associated with damage at the fish population level and not on detailed initial characterization of stressors. This approach allows follow-up studies to focus on the critical areas associated with responses of surviving fish within the watersheds in an attempt to identify limiting environmental conditions and the responsible stressors. The objectives of the project are to provide an interpretation framework using traditional fisheries information to assess the integrated effects of all stressors on the populations at risk and to isolate key areas and critical stages in life history associated with those changes in fish populations.

**Keywords:** *cumulative effects, fish populations, Moose River basin, pulp mills, hydroelectric dams*

## INTRODUCTION

The Canadian government has included a requirement for addressing cumulative environmental effects under the Canadian Environmental Assessment Act. The regulation requires that development proponents identify any potential cumulative environmental effects, analyze them, determine their significance, and identify possible mitigation measures. Although this requirement has been in place since 1995, there are no established or widely accepted scientific methods for systematically analyzing and evaluating cumulative environmental changes. While developmental assessment requires evaluation of the sociological impact, as well as terrestrial and aquatic environmental impacts, this chapter will deal only with attempting to define cumulative environmental effects within the aquatic environment.

The requirement for a cumulative effects assessment should provide a real step forward in the development of methods for addressing assimilative capacities of receiving waters for industrial, agricultural, and municipal wastes, as well as developing methods to deal more effectively with the long-term protection and sustainability of environmental integrity. Sustainability, as it is used here, is defined as a situation where the exploitation of resources and technological development occur within a framework which considers the environmental, economic, and social aspects of both future and present needs.[1] The expansion of methodologies for existing environmental assessments in order to deal with cumulative assessments requires that a number of issues be carefully considered, including (1) how do we define what will be considered an acceptable reference condition (background levels of performance), (2) what level of change is biologically significant (as opposed to statistically significant), (3) how do we attribute responsibility for existing changes within a system responding to multiple stressors, and (4) how do we predict the impacts of additional stressors when the existing performance is already adversely affected by preexisting multiple stressors?

To initiate some of these discussions, we examined a system affected by multiple stressors. Examination of fish near discharges from pulp mills indicated responses in terms of fish growth and reproductive development downstream of a number of Ontario mills.[2] Some of the pulp mills initially examined were on the Moose River system in Northern Ontario, and this system seemed an ideal candidate for looking at multiple stressors. Some background work had been collated previously to examine the potential of the Moose River basin as a model for addressing the issues required to understand cumulative effects assessment.[3,4] This region contains a number of existing and proposed industrial developments, including[4]

1. Ten existing hydroelectric developments with a capacity over 950 MW, 10 nonutility generating stations with a capacity of 120 MW, and 234 additional sites with hydroelectric potential. This river system has been the focus of previous, extensive studies investigating the potential effects of individual hydroelectric developments. There are also six co-generating facilities (i.e., nonhydraulic) with a capacity of 175 MW.
2. Three pulp mills, including a 310,000 tonne/year mill at Iroquois Falls which was recently converted to a thermomechanical process, a 170,000 tonne/year bleached kraft mill at Smooth Rock Falls (where $175 M has been spent on renovations since 1990[4]), and a 380,000 tonne/year thermomechanical mill at Kapuskasing which has recently changed mill process type. All three mills recently installed new waste treatment systems, and all are undergoing process changes.
3. An oriented strand board mill in Timmins, sawmills in Hearst and Timmins, and a plywood and veneer mill in Cochrane. Several of these sites are expanding and/or modernizing their facilities.
4. Eight operating mines and one refining facility, extracting a variety of commodities which include copper, zinc, lead, silver, and talc. There are also 8 advanced exploration sites in the basin (7 gold, 1 nickle) and over 100 additional exploration activities during 1994 alone (reviewed in Reference 4) Additional potential exists within the basin for mining of phosphates, aluminum, lignite, gypsum, diamonds, gold, copper, zinc, lead, silver, nickle, cadmium, indium, kaolin, and peat.
5. Portions of 23 forest management units which are harvested for timber. Previously, there has been extensive log driving operations on several of the rivers.
6. There are extensive tourism, angling, hunting, and recreational activities within the basin.

In spite of the scope and number of developments within the system, much of the drainage basin is still a relatively isolated, undisturbed system. The Moose River drainage basin is more than 100,000 km$^2$, and human population is sparse. Boundaries were required to limit the scope of these Moose River studies. The scope was limited to aquatic environmental effects in the middle reaches of the rivers and concentrated on the potential impacts of hydroelectric and pulp mill facilities. They occur in close proximity within the basin. Although some preliminary collections were conducted in 1991, the cumulative assessment nature of this project was initiated in 1993 and is continuing at this time.

The 1991 studies were designed to evaluate the responses of fish to pulp mill effluents and showed that female white sucker near the pulp mills at Kapuskasing and Smooth Rock Falls had smaller gonads and faster growth rates.[2] The cumulative effects project was designed to determine whether there were factors other than effluent discharge contributing to these responses in fish. This overview will outline the philosophical basis for the project, objectives for the field studies, a geographical description of the study site, and a summary of the study design and preliminary results.

## PHILOSOPHICAL APPROACH

The main requirement for cumulative effects assessment is a framework which will determine whether there are cumulative effects within a system. It is impossible to identify and prioritize existing stressors or to predict the impact of additional stressors without understanding the present status of the system. Once the status has been identified, it is necessary to identify a series of steps which will assist in determining the main stressors associated with the responses when they are present. The ultimate objective of our project is to try and develop a predictive model to determine what impact an additional stressor may have on an existing system. In the development of a predictive model, additional stressors would be evaluated for their potential to interact with the factors affecting or limiting current productivity.

Since fish represent the upper trophic level of the aquatic ecosystem, the performance characteristics of the populations present are a reflection of the flow of energy through the ecosystem. Any changes in the habitat which do not result in a change in energy flow at the level of the fish population will not change performance characteristics and will be assumed to reflect inconsequential impact.

There are various factors to consider when trying to establish which fish species should be monitored within a system.[5] It is assumed that monitoring will focus on key species which are sensitive to habitat and community-level changes. Once those species are established, a monitoring level must be selected. Approaches to monitoring fish performance range from community analysis to biochemical competency.[6] There has been considerable discussion with respect to the parameters which best indicate the individual performance potential of fish.[7,8] While community changes may be more reflective of long-term damage, community changes are a reflection of extinctions and changes in population abundance. Determination of the stressors responsible for such changes requires closer examination of individual, surviving populations. At the other extreme, changes in biochemical parameters in individual fish requires extrapolation to determine relevance. Closer examination of the performance of the adult fish is required to estimate relevance of the changes to some indicators of more permanent damage. In both cases, the follow-up studies require information on individual organisms to establish relevance or determine causes, essentially a "middle-out" approach.[6]

Our project is focused on changes in performance indicators of adult fish. The measurement of the whole organism characteristics of fish captured within a system requires no extrapolation and can be used to develop hypotheses about the origins of the stress on the system.[5,9-11] Any changes in energy utilization will be reflected in growth, reproduction, or energy storage, and changes in survival should be reflected in the age distribution of the individuals remaining in the population.

We modified a framework which attempted to discriminate the types of population response patterns that are expressed in adult fish.[9,10] Response pattern separation provides an indication of the stressors limiting performance of the fish population. Based on the responses of the fish, the framework should identify whether exposed fish are exhibiting altered characteristics relative to reference populations (or historical data) and the specific component of the adult population (if any) that has been affected. If the sample is representative of the population, fish populations respond in a predictable manner to specific stressors. The original version of the framework provided five patterns describing the responses of populations to exploitation (adult mortality), recruitment failure (larval mortality), multiple stressors, food limitation, and a niche shift.[9,10] Subsequent studies have expanded the framework to include additional possible responses, including responses to metabolic disruption, and to allow for intermediate patterns not previously identified.[5,11,12] This framework is still evolving, but can be used as a basis for describing the factors limiting fish performance. It is proposed that once the limiting factors are identified, the factors responsible for producing those limitations can be isolated and related to the stressors operating within the system.

Defining the response pattern requires information from the sampled fish population describing a variety of characteristics (the latest published version of the framework is outlined in Reference 11). It is possible for ease of presentation to reduce the description of characteristics to the basic groups which describe age structure, energy expenditure, and energy storage, but there are a variety of performance indicators which can be used for each category. Research is still ongoing to attempt to determine the most cost-efficient indicators. Regardless of the indicators chosen, a population showing no change in characteristics (age structure, energetic expenditure, or energy storage) would be represented by a null response (000); a population initially responding to severe egg or larval mortality would show an older age distribution (+00). There is still some debate about the sample size of fish required to establish these measurements, but our studies have shown that preliminary indications can be achieved with sample sizes of less than 20 adult individuals of each sex for the species we commonly use.[5] The success and accuracy of the sampling effort also depends on the reference sites selected. If an inappropriate reference site is selected, the response pattern identified will focus efforts on the main factors limiting performance.

Each of the response patterns and pathways can be presented as part of a theoretical cycle of response and adaptation to stressors. The established patterns which have been identified in the literature are highlighted in Figure 1 by shaded boxes. While all of the patterns have not been found in nature, the framework attempts to integrate the common relationships among fish responses. Some of the new patterns described are probably transitional, and the possibility of their detection in wild fish populations would depend on the time scale, responsiveness of the species being monitored, and the severity of impact. A more complete description

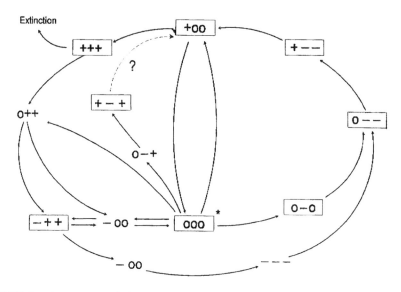

**FIGURE 1** A summary of the responses of fish populations to stressors (adapted from Reference 11). The responses are abbreviated in terms of the relative change (− less than, 0 equivalent to, + more than) in age structure, energy expenditure, and energy storage when compared with the reference sites or historical data. The asterisk signifies the starting point of the progression (000); response patterns in shaded boxes represent patterns documented in the literature.

of the patterns and their possible utility can be found in Reference 11. The main point for this summary is that a fish population showing a marked difference in energy storage and use is being limited by energy availability or utilization and this area would represent a key area for investigation. A population showing a dramatic skewing of the age distribution would require that mortality or reproduction be focused on to determine the stressors limiting (or altering) performance.

While the framework is incomplete and is undergoing further development, it has proved to be a useful tool in preliminary trials and should provide a focal point for developing cumulative effects studies using fish populations. This approach has several advantages.

1. It provides a framework to utilize traditional fisheries information, requiring initial demonstration of damage/changes prior to detailed sampling. The approach does not attempt to separate effects from the various developments, but emphasizes interpretation of the integrated effects of all stressors on the populations at risk, therefore ensuring that studies are directed in a top-down, effects-oriented manner.
2. It allows initial sampling to focus and direct follow-up studies to areas which are clearly impacted. Since initial sampling is relatively inexpensive, this approach allows funding to be conserved and subsequently focused on areas requiring detailed studies.

3. The focus is placed on interpretation of habitat changes associated with damage at the population level and not on detailed initial characterization of stressors, allowing follow-up studies to focus on the habitat changes which are impacting the populations.
4. It prevents expenditure of funds on follow-up studies at sites not showing changes and, therefore, prevents collection of data sets on physiological changes and habitat alterations which are not translated to damage at the population level.

## OBJECTIVES

The major objectives of the Moose River cumulative effects project relate to the development of techniques for the assessment of the cumulative effects of developments and instream changes on fish populations. The long-term objectives are to provide the following:

1. A characterization of performance parameters in fish inhabiting undeveloped and developed reaches of the Moose River basin
2. Key indicators which reflect the performance of fish populations within the Moose River watershed
3. The identity of habitat changes and environmental stressors associated with performance differences between sites
4. A common methodology and an interpretation framework for assessing cumulative effects as it applies to the performance characteristics of fish within a watershed affected by a variety of developmental scenarios
5. A model to make predictions of the effect of proposed industrial developments on fish populations

## GEOGRAPHICAL DESCRIPTION

The Moose River basin is an area of approximately 109,000 km$^2$ which drains northward into James Bay (Figure 2). The Moose River itself is a seventh order stream with a mean discharge of 780 m$^3$/sec. Major sub-basins within the watershed, moving from west to east, include the Missinaibi (22,500 km$^2$), the Kapuskasing (8600 km$^2$), the Groundhog (12,500 km$^2$), the Mattagami (41,700 km$^2$), and the Abitibi (34,000 km$^2$).[13]

The upper (southern) portion of the watershed is underlain by Precambrian Canadian Shield, which is covered by glacial, glaciofluvial, and glacolacustrine deposits. In most areas these deposits are thin and heterogeneous, but in the southeastern portion of the watershed there is a large glacolacustrine clay deposit, and the rivers which flow through this area are characterized by high turbidity. Within the Canadian Shield, numerous bedrock outcrops create falls and rapids, many of which are barriers to fish migration. Sedimentary deposits of the Mesozoic and Paleozoic lie beneath the northern part of the drainage basin,[14] a region known as the Hudson bay lowlands. Here, the bedrock is covered with marine and glacial

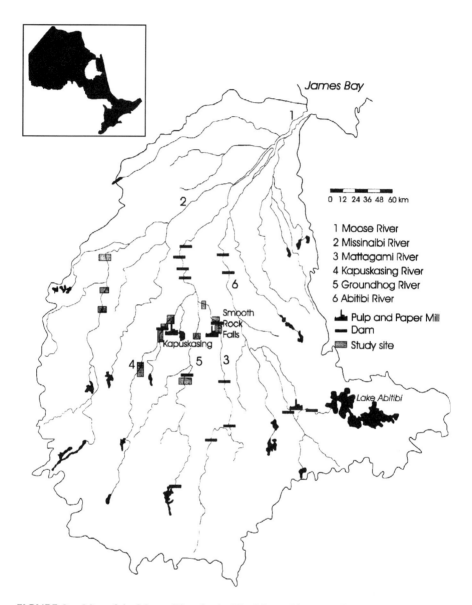

**FIGURE 2** Map of the Moose River basin. The Moose River empties into James Bay, and the major western tributaries are the focus of the current studies. (Inset map: Province of Ontario, Canada.)

deposits, typically overlain by peat deposits. The change in elevation from the headwaters to James Bay is in the order of 325 m, and reaches of particularly high gradient occur on the Missinaibi and Mattagami Rivers where they descend from the Canadian Shield to the lowlands (Figure 3).

Mean annual temperature ranges from 1.1°C in the south to –1.1°C in the north. Mean annual precipitation is between 700 and 900 mm/year, with 30 to 40% of this

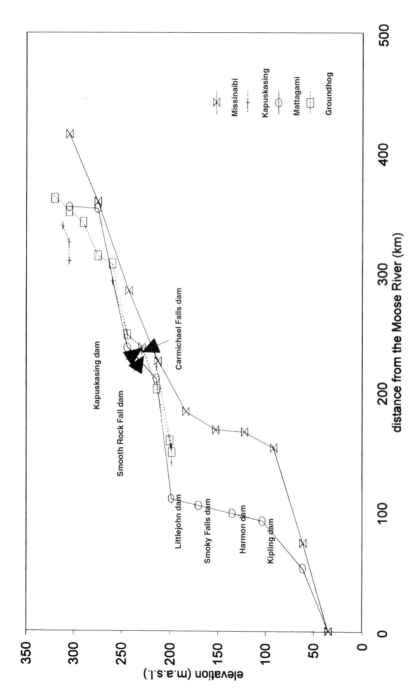

**FIGURE 3** Gradient profiles of the four major tributaries to the Moose River examined in this study (modified from Ontario Ministry of Natural Resources information).

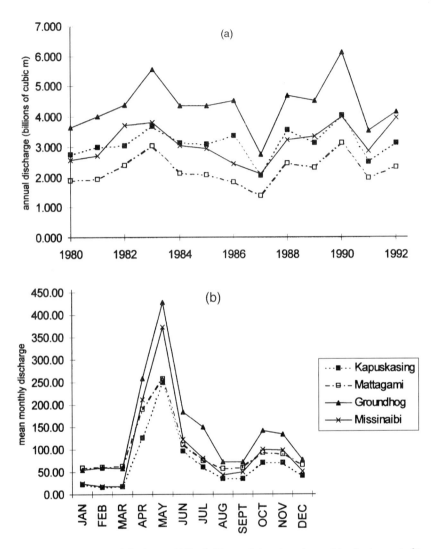

**FIGURE 4** Total annual discharge ($10^9$ m$^3$; Figure 4a) and total monthly discharge (m$^3$/sec; Figure 4b) for the study rivers.

falling as snow.[15,16] River discharge peaks in May and is lowest during the winter (Figure 4).

The present studies are restricted to the middle reaches of the western tributaries (Missinaibi, Kapuskasing, Groundhog, and Mattagami Rivers). All of the sampling sites lie within the Precambrian Shield, and surficial materials are dominated by glacial tills or shallow drift. Overall gradients are moderate, as the streams have not yet begun their descent over the edge of the Precambrian Shield. At this point in the watersheds, the discharge of the largest of the rivers (Groundhog) is roughly twice that of the smallest (Kapuskasing) (Figure 4). Seasonal discharge patterns and patterns of year-to-year variability in discharge are similar between rivers (Figure 4).

The Missinaibi represents a large, undeveloped river and has been designated as a Heritage River. As such, the river is protected from land use activities within the park boundary which extends from 120 to 200 m from the water's edge for much of the river's length. Virtually all of the other river systems within the Moose River Basin have undergone development to some extent, including hydroelectric dams and pulp and paper mills. The Kapuskasing River has hydroelectric development and a large pulp mill located at Kapuskasing. The Groundhog River has a hydroelectric facility at Carmichael Falls, which began operation in October 1991 (and we have historical samples from the preconstruction period). The Mattagami River has extensive hydroelectric development and a pulp mill at Smooth Rock Falls. Headwater mining developments on the upper Mattagami River near Timmins were not considered in the study design because of their distance upstream. The Abitibi River has a pulp mill and numerous hydroelectric developments.

## STUDY DESIGN

The dominant stressors in the river reaches where growth and reproduction of fish were affected in preliminary studies were hydroelectric dams and discharges of pulp mill effluent. The white sucker sampled downstream of the mills showed increases in fish size, concomitant with decreases in reproduction.[2] This response could be caused by several situations, including (1) increased energy available for growth due to the decreased reproductive investments associated with pulp mill contaminants, (2) increased energy flow associated with increased productivity due to impoundment of the river, or (3) increased energy flow associated with increased nutrients associated with effluent discharge. Both growth and reproductive performance are indications of energy expenditure, and it is necessary to identify trophic relationships in the various rivers in order to identify the types of stressors associated with changes.

Evaluation of stable isotopes of carbon and nitrogen trace nutrient flow through the system[17,18] and will be used to describe trophic relationships. Food web descriptions will include data on vegetation, dissolved organic carbon, algal and periphyton communities, benthic communities, and the fish communities.

Performance data on individual fish species will be restricted to two major species. The Moose River basin is home to a wide variety of fish species (Table 1), but, because it is a northern system with relatively low productivity, the density of predatory fish is relatively low. The study selected two species based on availability and density at all sites: the white sucker (*Catostomus commersoni*) and the trout-perch (*Percopsis omiscomaycus*). White sucker are a benthic feeding species, are known to be sensitive to pulp mill effluents,[12,19-22] and are widely distributed in the Moose River basin. Trout-perch are a forage species, are also abundant in most areas, and will provide evidence of productivity at another level of the fish community.

The performance parameters investigated for both species have included density, fish size (length and weight), age, gonad weights, fecundity (egg counts in females), liver weights, and abnormalities or lesions on 20 males and 20 females from each site. Samples were also collected for plasma levels of sex steroids; pituitary glands for determination of gonadotropin hormone content; liver samples for mixed function oxygenase activity; water, muscle, liver, or bile contaminant

## TABLE 1
## Fish Species Reported from the Moose River and Four of Its Tributaries

| Common Name | Scientific Name | Moose | Missinaibi | Mattagami | Groundhog | Kapuskasing |
|---|---|---|---|---|---|---|
| Lake sturgeon | *Acipenser fulvescens* | X | X | X | X | X |
| Lake whitefish | *Coreogonus clupeaformis* | X | X | X | X | X[a] |
| Northern pike | *Esox lucius* | X | X | X | X | X |
| Goldeye | *Hiodon alosoides* | | | X | | |
| Mooneye | *H. tergisus* | | X | | | |
| Longnose sucker | *Catostomus catostomus* | X | X | X | X | X[a] |
| White sucker | *C. commersoni* | X | X | X | X | X |
| Redhorse suckers | *Moxostoma sp.* | | X | | | |
| Lake chub | *Couesius plumbeus* | | | | X | |
| Golden shiner | *Notemigonus crysoleucas* | | | X | | X |
| Emerald shiner | *Notropis atherinoides* | | | X | X | X |
| Common shiner | *Luxilus cornutus* | | X | | X | X |
| Blacknose shiner | *N. heterolepis* | X | X | | X | |
| Spotfin shiner | *Cyprinella spiloptera* | | X | X | X | X |
| Fathead minnow | *Pimephales promelas* | | | X | X | |
| Longnose dace | *Rhinichthys cataractae* | X | X | X | X | X |

| Common Name | Scientific Name | Moose | Missinaibi | Mattagami | Groundhog | Kapuskasing |
|---|---|---|---|---|---|---|
| Fall fish | *Semotilus corporalis* | X | X | X | | |
| Pearl dace | *Margariscus margarita* | | | X | X | |
| Burbot | *Lota lota* | | | X | X | X |
| Threespine stickleback | *Gasterosteus aculeatus* | X | | | | |
| Brook stickleback | *Culaea inconstans* | | | X | | |
| Trout-perch | *Percopsis omiscomaycus* | X | X | X | X[a] | X |
| Pumkinseed | *Lepomis gibbosus* | | | X | | |
| Smallmouth bass | *Micropterus dolomieui* | | X | | | |
| Yellow perch | *Perca flavescens* | X | X | X | X | X |
| Walleye | *Stizostedium vitreum* | X | X | X | X | X |
| Iowa darter | *Etheostoma exile* | | | | X | X |
| Johnny darter | *Etheostomus nigrum* | X | X | X | X | X |
| Logperch | *Percina caproides* | X | X | X | X | X |
| Mottled sculpin | *Cottus bairdi* | X | X | X | X | X |
| Slimy sculpin | *C. cognatus* | X | | | | X |

*Note:* Only species thought to be present in the rivers are includes. Table modified from Reference 13.

[a] Not reported by Reference 13, but captured by C. Portt and Associates during 1994.

levels (mercury and chlorinated dioxins and furans and other pulp mill contaminants) and lipid levels; eggs for hormone production capability; and stomach contents (to verify feeding relationships).

The main questions to be addressed were

1. Is there similar performance of fish populations in upstream reaches of all rivers which are similar in habitat characteristics?
2. Is there similar performance of fish populations in reaches of the Missinaibi (reference river) which are similar in habitat characteristics (undeveloped reference location)?
3. Is fish performance at sites along developed rivers, where impacts are suspected, similar to their respective headwater reaches?
4. Is fish performance at sites along developed rivers, where impacts are suspected, similar to comparable sites (latitude) on the Missinaibi (reference river).

The objectives were divided into three main phases:

- Phase I Latitudinal Comparison of Reference Sites
- Phase II Longitudinal Comparison of Undeveloped Sites
- Phase III Longitudinal Comparison at Developed Sites

The final two phases of the project represent model development (Phase IV) and evaluation of cause and effect relationships (Phase V). These phases will be briefly described, preliminary data will be outlined, and future studies will be discussed.

## PHASE I: LATITUDINAL COMPARISON OF REFERENCE SITES

It is important to establish baseline conditions or expected conditions for performance criteria. The initial phase of the study was designed to compare fish performance across tributaries, upstream of the hydroelectric dams and pulp mill discharges. In the original study design, 2 years were allocated to determine which habitat factors were responsible for major differences in performance between tributaries. The hypothesis for this phase was that the performance characteristics of the fish would be similar across the upstream reference sites. Specifically, we proposed to determine whether there was similar performance of fish populations as measured by age distribution, growth rate, and reproductive/metabolic fitness in undeveloped reaches of all rivers. The fish collections have characterized growth rates, reproductive energy investments, and energy storage patterns. Detailed studies are also examining stable isotope ratios within the food web to characterize the feeding relationships at each site. Stable isotope ratios are a technique which uses changes in the ratios of the stable isotopes of carbon and nitrogen to identify energy origins and transfer between trophic levels.[23,24]

The initial reference sites for the latitudinal studies were selected to represent, as closely as possible, similar flows, depths, and habitats. The sites were

1. The Mattagami River immediately upstream from Smooth Rock Falls
2. The Kapuskasing River a short distance upstream from Kapuskasing
3. The Kapuskasing River approximately 40 km upstream of the mill
4. The Groundhog River at Fauqier
5. The Missinaibi River at Mattice

It was expected that if upstream mining development was affecting fish performance in the middle reaches, it would be detected during Phase I sampling as an altered performance of fish at that reference site. Although a hydroelectric facility was installed in late 1991 upstream of Fauqier at Carmichael Falls on the Groundhog River, this site was retained in the reference comparisons because of continuity with the 1991 samples. An upstream Groundhog River reference site above the Carmichael Falls facility was sampled in the fall of 1996.

Over the course of the project, reference samples have been collected during >20 sampling trips, including 13 reference sites. This data will eventually be incorporated to define expected performance levels. For phase I sampling, white sucker were collected in the fall of 1993, 1994, 1995, and 1996 (September) to determine year-to-year variability. White sucker were also collected during the prespawning (May) periods of 1993 and 1994 to examine seasonal variability. Trout-perch samples were collected in the fall of 1995 and 1996. In addition to the fish species collected, data were also collected to describe the following:

1. Basic habitat mapping at the selected sites, including some basic water chemistry measurements such as pH, conductivity, temperature, water depth, and velocity, as well as indications of dominant substrate classes and vegetation.
2. Basic water quality sampling at each of the selected sites, including triplicate water samples for (a) dissolved inorganic carbon, (b) phenols and pulp mill contaminants, (c) AOX analysis, (d) major ions, and (e) total nutrient analysis.
3. Qualitative collections of benthic invertebrates, consisting of 1 to 3 g of each major order of invertebrates for use in stable isotope analysis.
4. Samples of all easily available species of forage fish, frozen at and stored at temperatures less than $-15°C$, for stable isotope analysis.
5. An integrated sample of 0 to 2 cm of sediment from depositional zones.
6. Temperature probes (Optic Stowaway; Onset Computer Corp., Pocasset, MD 02559) were installed in fall of 1995, and data were recorded from each site at 15-min intervals. Data recording will continue into 1999.

## Phase I Preliminary Results

The habitat surveys showed that elevation, water depth, substrate, temperature profiles, etc. were fairly similar between sites (C. Portt, unpublished data).

White sucker collections showed that there were significant year-to-year variations within and between sites and that there was significant variability between seasons. In spite of the variability within a year and season, there was remarkable

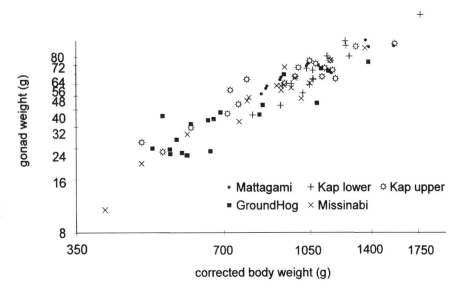

**FIGURE 5** Relationship between gonad size and body size for female white sucker collected during the fall of 1994. There were no statistical differences between the reference sites in Phase I sampling.

similarity between upstream reference sites. The 1993 and 1994 baseline white sucker collections showed much less latitudinal variability than was expected. Although limited numbers of male fish were collected during 1993, females showed few statistically significant differences between sites. Similar consistency was seen during 1994 periods, but the dramatic differences between seasons and years requires further analysis to determine the most stable predictors of performance (Figure 5). More detailed analysis is ongoing to evaluate methods for calibrating "reference" performance levels.

The similarity between tributaries suggested that upstream reaches of the river yield similar performance characteristics in white sucker despite some minor variability in habitats. The second phase of the studies involved determining the expected, normal downstream changes in performance in the absence of industrial development. Fish collections from the Missinaibi River during the fall of 1996 (Phase II) will be used for this purpose.

Stable isotope ratios were used to characterize the feeding relationships at each site to identify energy origins and transfer between food groups. The analysis of stable isotopes has been utilized in studies to better understand carbon flow in aquatic ecosystems.[17,18] The latitudinal comparison was undertaken to identify differences in food web dynamics via the use of carbon and nitrogen stable isotopes. Similar species were collected at each site, and relationships between stable carbon isotopes were used to identify carbon flow (Figure 6). Factors associated with changes in the flow of carbon in the system will be identified. Analysis of stable carbon isotopes for dissolved inorganic carbon, dietary items, and white sucker muscle tissue show differences among the reference sites (A. Farwell, unpublished data). Preliminary

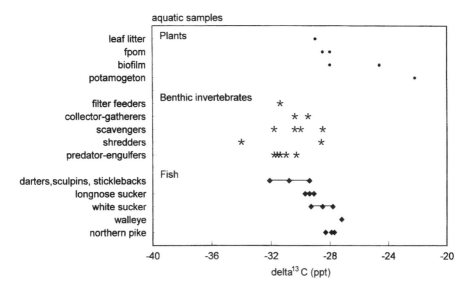

**FIGURE 6** Ratios of stable carbon isotopes for samples collected from the Missinaibi River during the fall 1993 sampling period. The data shows the normal shifts to enrichment in the upper portions of the food chain.

results suggest that the differences found in the carbon isotope signatures for white sucker tissue at different sites may be a function of the source of dissolved inorganic carbon and subsequent carbon recycling at the reference sites. This work is ongoing at this time. Data from Phase II sampling and follow-up experiments from the fall of 1996 will be used to complete the picture.

## PHASE II: LONGITUDINAL COMPARISON OF UNDEVELOPED SITES

Due to the length of the rivers in the Moose River basin, we would expect changes in productivity and fish performance as the width, depth, and flow of the rivers changes downstream.[25,26] All of the industrialized rivers require upstream reference sites and downstream exposed sites. Since the Missinaibi River is undeveloped, it can be used as an indicator of the expected downstream changes in the other tributaries for the Moose River basin had development not occurred. This phase of the study took place in the fall of 1996 and focused on longitudinal comparisons of the habitat variations and the performance characteristics of fish populations downstream on the Missinaibi River. The underlying hypothesis is that performance parameters will change in a predictable fashion as downstream habitat characteristics change. The stable isotope ratios should allow clear definition of the changes in food web structure and energy flow as the rivers progress downstream.

Analysis of dietary items and fish muscle tissue for samples collected during Phase I sampling at a middle reach of the Missinaibi River show stable carbon isotope signatures that are similar for sources of carbon (leaf material, fine particle organic matter, and biofilm) for benthic invertebrates and benthic fish species (white sucker, longnose sucker, darters, sculpins, and sticklebacks) (Figure 6). This finding is to be expected since carbon isotope signatures have been shown to alter very little from one level of the food chain to higher levels. This reference site is important to the understanding of energy flow in the tributaries of the Moose river where hydroelectric dams and pulp mill effluent have potentially altered the carbon flow and energetics. Future work includes intensive sampling at sites further downstream on the Missinaibi river to complete the understanding of carbon flow as predicted by the river continuum theory where autochthonous material increases in importance.

The collection sites for Phase II were selected to be

1. The Missinaibi River upstream of Mattice and downstream of the junction with the Mattawitchewan River
2. The Missinaibi River at Mattice (Sampling was conducted here in both 1993 and 1994, and the 1996 samples will allow further temporal comparison.)
3. The Missinaibi River upstream of the confluence with the Mattawishkwia River from Hearst
4. The Missinaibi River downstream of Thunderhouse Falls

The third phase of data collection will allow two levels of comparison at developed sites: to the upstream reference sites and to the lateral reference sites on the undeveloped tributary. During the course of these studies, there will have been >20 reference site collections during the fall sampling period at a total of 13 different reference sites. The data will be integrated and compared for various methods of calculating background performance levels.

## PHASE III: LONGITUDINAL COMPARISON AT DEVELOPED SITES

Phase III represents the studies designed to attempt to separate the impacts associated with hydroelectric development from those responses associated with the pulp mill effluent discharges.

The main Phase III study sites were located at

1. Kapuskasing River (a) immediately downstream of the pulp mill discharge and (b) downstream of the discharge, but upstream of Freddy Flat bridge (approximately 16 km downstream)
2. Mattagami River (a) approximately 500 m to 2 km downstream of the outfall and (b) downstream of Cyprus Falls (approximately 40 km downstream)
3. Groundhog River at Fauqier (This location originally served as a reference site for the 1991 collections prior to the completion of the Carmichael Falls hydroelectric complex and as a reference site during the 1993 collections. An additional site upstream of the Carmichael Falls facility was added during the fall 1996 sampling period.)

The original study design intended for the first two phases to be complete prior to initiating Phase III studies, but several circumstances required earlier initiation of Phase III sampling. Both the pulp mills in the study area (Kapuskasing and Smooth Rock Falls) were in the middle of process changes and the installation of secondary treatment facilities which would be finalized in late 1994. The bleached kraft mill at Smooth Rock Falls installed an aerated stabilization basin for secondary treatment, which became operational in late October 1994. A number of other process changes took place since 1991, including increasing the level of chlorine dioxide substitution from 15 to 100%. The thermomechanical pulp mill at Kapuskasing installed an activated sludge system for waste treatment, which became operational in late April, 1995. Furthermore, the Kapuskasing mill modernized their pulping process and converted their pulping procedures from a mixed process (thermomechanical, groundwood, and magnetite pulping) to thermomechanical pulping.

Collections conducted in 1993 and 1994 would provide a series of baseline studies to evaluate the efficacy of process changes and the impacts of installation of secondary waste treatment, for improving the performance of fish. These studies would also contribute to additional data on year-to-year variations within sites. It was necessary to understand the responses to installation of secondary treatment in order to help in the separation of hydroelectric and mill effluent responses.

## Preliminary Phase III Results

There have been some responses within the system to process and waste treatment changes at the pulp mill. While the reduced gonadal size originally reported at Smooth Rock Falls has persisted through process changes, fall gonad size in female white sucker at Kapuskasing has shown considerable recovery (Figure 7). The improvements took place prior to completion of the waste treatment system. Similar improvements have been seen with liver size (Figure 8), although there are still some site differences evident within the system.

The growth data for the system has shown considerable year-to-year variability (Table 2). In order to eliminate some of the variability, we are currently back-calculating size at age and growth rates (L. Ruemper, unpublished data).

In 1995, trout-perch were collected from five sites including two reference sites upstream on the Kapuskasing River, one site downstream on the Kapuskasing River, one site downstream on the Mattagami, and one site at the Groundhog River. Additional sites on the Missinaibi River were added in the fall of 1996, and the Kapuskasing sites were resampled to look at temporal variability. Approximately 70 to 100 trout-perch were collected from each site, including mature males, mature females, and immature fish. The samples will be used to examine the suitability of small fish species for monitoring environmental effects in river systems. We have previously used forage fish in both the Fraser[27] and the Athabasca River systems.[28]

Whole organism data for the trout-perch for two Kapuskasing reference sites were similar, except for male condition factor. Individuals captured downstream of the Kapuskasing pulp mill were younger, but larger and showed faster growth (W.N. Gibbons, K.R. Munkittrick et al., unpublished data). Male fish showed mixed-function oxygenanse enzymes (MFO) induction and higher steroid hormone

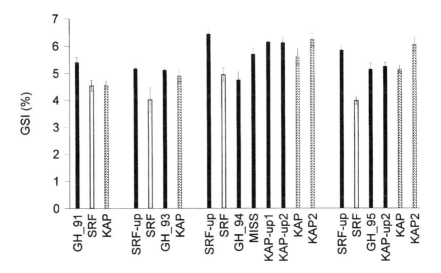

**FIGURE 7** Relative gonad size (GSI = 100 * gonad weight/body weight) for female white sucker collected during the fall sampling periods from 1993 to 1995. The abbreviations represent the Kapuskasing River (two downstream sites at 2 and 16 km and two upstream sites), Mattagami River (Smooth Rock Falls site), Groundhog River, and Missinaibi River. An additional six sites are expected in the fall of 1996.

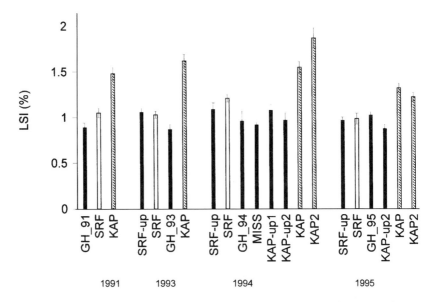

**FIGURE 8** Relative liver size (LSI = 100 * liver weight/body weight) for male white sucker collected during the fall sampling periods from 1993 to 1995. The abbreviations represent the Kapuskasing River (two downstream sites at 2 and 16 km and two upstream sites), Mattagami River (Smooth Rock Falls site), Groundhog River, and Missinaibi River.

## TABLE 2
## Characteristics of Female White Sucker Collected During Fall Sampling Periods Downstream of the Pulp Mills and Dams on the Kapuskasing and Mattagami Rivers

| River | Site | Year | n | Slope | INT | $R^2$ | Length | Weight |
|---|---|---|---|---|---|---|---|---|
| Kapuskasing | TB | 1995 | 25 | 0.532 | 37.364 | 0.433 | 42.3 ± 0.41 | 127 ± 34 |
| | FF | 1994 | 24 | 0.573 | 37.765 | 0.52 | 41.7 ± 0.5 | 1084 ± 35 |
| | FF | 1991 | 19 | 0.672 | 36.454 | 0.628 | 40.9 ± 0.5 | 1051 ± 41 |
| | TB | 1991 | 16 | 0.933 | 35.67 | 0.665 | 43.8 ± 0.6 | 1194 ± 40 |
| | TB | 1993 | 15 | N/A | N/A | N/A | 43.1 ± 0.9 | 1225 ± 64 |
| | FF | 1995 | 27 | 0.942 | 34.856 | 0.692 | 43.5 ± 0.7 | 1251 ± 65 |
| | TB | 1994 | 12 | 1.293 | 34.022 | 0.884 | 42.6 ± 0.9 | 1170 ± 91 |
| Reference | KAP1 | 1994 | 14 | 0.599 | 38.486 | 0.492 | 43.1 ± 0.7 | 1185 ± 68 |
| | KAP2 | 1994 | 19 | 1.006 | 32.782 | 0.569 | 40.6 ± 0.9 | 1018 ± 69 |
| | KAP2 | 1995 | 25 | 1.181 | 29.07 | 0.63 | 39.2 ± 0.6 | 874 ± 44 |
| Groundhog | | 1991 | 17 | 0.378 | 37.041 | 0.369 | 40.7 ± 0.5 | 995 ± 41 |
| | | 1993 | 10 | N/A | N/A | N/A | 42.0 ± 1.0 | 1135 ± 89 |
| | | 1995 | 31 | 0.835 | 32.297 | 0.259 | 39.4 ± 0.5 | 915 ± 39 |
| | | 1994 | 21 | 2.424 | 24.826 | 0.629 | 38.4 ± 0.8 | 788 ± 54 |
| Mattagami | | 1994 | 13 | 0.679 | 37.268 | 0.353 | 42.3 ± 0.8 | 1156 ± 71 |
| | | 1991 | 15 | 1.124 | 32.559 | 0.648 | 42.2 ± 1.0 | 1177 ± 100 |
| | | 1993 | 16 | N/A | N/A | N/A | 41.7 ± 0.8 | 1181 ± 59 |
| | | 1995 | 19 | 1.539 | 27.018 | 0.616 | 40.5 ± 1.0 | 958 ± 67 |
| Reference | | 1993 | 21 | N/A | N/A | N/A | 40.8 ± 0.6 | 1013 ± 50 |
| | | 1994 | 12 | 0.69 | 36.37 | 0.394 | 42.4 ± 0.9 | 1154 ± 74 |
| | | 1995 | 25 | 0.793 | 34.908 | 0.569 | 42.7 ± 0.5 | 1155 ± 39 |

production, but females showed no differences in either parameter. Downstream of the Smooth Rock Falls mill, trout-perch were older, smaller, showed increased condition factor, and decreased liver size relative to their reference site on the Groundhog River. MFO activity was induced in both sexes, and females showed lower levels of steroid hormone production (W.N. Gibbons, K.R. Munkittrick et al., unpublished data).

Physiological testing with white sucker has also shown reduced steroid hormone level and elevated MFO activity in male fish.[29] Preliminary analyses from the 1995 collections also suggest that steroid differences between sites are not as strong as in 1993 and 1994. Laboratory screening of effluents for their potential to disrupt steroid hormone production has been ongoing at both pulp mills on a regular basis, before and after the installation of waste treatment (L.S. McCarthy and K.R. Munkittrick, unpublished data).

The usefulness of stable carbon isotope analysis is dependent on the ability to isotopically differentiate between natural vs. anthropogenic sources of carbon. Initial results from the stable isotope analysis show dramatic changes between upstream and downstream sites. Some collections were conducted downstream of the pulp mills and hydroelectric dams at Kapuskasing and Smooth Rock Falls during 1993 and 1994 (unpublished data). These collections were necessary as baseline studies because of the installation of new waste treatment facilities at the pulp mills. The data show clear differentiation of the ratios of carbon isotopes between upstream and downstream sites, and the factors associated with these differences will be identified.

The data will be integrated into a model for the Moose River system. The model will adapt and further develop the response framework outlined by Gibbons and Munkittrick.[11]

## PHASE IV: MODEL DEVELOPMENT

It is obvious from the collections that different species within the system show different responses to the changes within the ecosystem. The trout-perch and white sucker show very different population-level changes. Phase IV will develop a population-level response framework based on previous work[5,9-12] to assist in the interpretation of responses. Preliminary analysis suggests that even though there are differences in the whole organism characteristics, the stressors the populations are responding to are similar. For example, downstream of the pulp mill/hydroelectric facility at Kapuskasing, white sucker are larger downstream, while trout-perch are smaller.[30,31] However, both species seem to be responding to similar stressors within the system.[31] These preliminary analyses have only been conducted for the Kapuskasing River. Phase IV will concentrate on expanding the analysis to include the other river systems, and on integrating the philosophical approach and the reference site data into a performance-based model using the characteristics of fish collected at the reference sites along the heritage waterway and at headwater sites. The model will account for variability in fish performance characteristics attributable to habitat and anthropogenic developments. Habitat characteristics will include assessments of food availability and usage, physical characteristics (substrate, water depth, and velocity), and availability of spawning habitat. Indices of physiological performance will include measures of energy stores and expenditure along with measures of reproductive fitness (age and size at maturity, fecundity, egg size).

The second tier of comparisons will be to test the applicability of the model. At this stage it would be premature to provide a detailed analysis, since the final field work will not be completed for another year. However, we envisage testing the predictability of the model in explaining the performance characteristics of fish populations from both developed and undeveloped rivers. This could include rivers with a variety of development scenarios including hydroelectric, pulp and paper, mining, or various combinations.

## PHASE V: EVALUATION OF CAUSE AND EFFECT RELATIONSHIPS

This phase involves describing cause and effect relationships and will involve characterizing the habitat and anthropogenic effects on fish performance characteristics.

This phase focuses on the relationships between effects, the mechanisms underlying these effects, and the chemical or habitat responsible for the changes. It involves subdivision of reaches to evaluate response gradients associated with habitat and development variables. The process will attempt to quantify the relative contribution of single/multiple stressors to the overall response patterns. The ultimate goal of the project is to develop an information base which would allow for some level of predictability within the Moose River basin and to outline a framework and sampling strategy which could be utilized to allow predictability in other watersheds.

## CONCLUSIONS

Most of the studies described in this chapter are still under analysis. The long-term objectives are to provide a characterization of performance parameters in fish inhabiting undeveloped and developed reaches of the Moose River basin, key indicators which reflect the performance of fish populations within the Moose River watershed, the identity of habitat changes and environmental stressors associated with performance differences between sites, a common methodology and an interpretation framework for assessing cumulative effects as it applies to the performance characteristics of fish within a watershed affected by a variety of developmental scenarios, and a model to make predictions of the effect of proposed industrial developments on fish populations.

## ACKNOWLEDGMENTS

John Nickle passed away tragically in March 1995 in the middle of these studies. Additional data on laboratory evaluations of the pulp mill effluents studied in this project are being provided by the Department of Fisheries & Oceans Green Plan Toxics Fund (Dr. L.H. McCarthy, P.L. Luxon, J.J. Jardine, and B.E. Blunt). The authors would like to also acknowledge the assistance of Dr. S. McKinley and Dr. W. Taylor (Department of Biology, University of Waterloo); the cooperation of the Large River Ecosystem Unit and the regional offices of the Ontario Ministry of Natural Resources; and Marty Bergmann, Department of Fisheries & Oceans, Winnipeg, Manitoba. Co-funding for this project has been received from the Canadian Electricity Association, Technology Development, Montreal, Quebec (T. Kingsley, T. Glavicic-Théberge); the Department of Fisheries and Oceans (PERD funding, GLLFAS A-base funding); Environment Canada (AECB, NWRI); Ontario Hydro Technologies, Environmental Sciences, Toronto, Ontario (C. Dawson, R. Sheehan, S. Griffiths); the Department of Zoology, University of Guelph, (Dr. G. Van Der Kraak); Department of Environmental Biology, University of Guelph (Dr. K. Solomon); and Department of Biology, University of Waterloo (Dr. R.S. McKinley).

For questions or further information regarding the work described in this chapter, please contact:

**Kelly R. Munkittrick**
Dept. of Biology
Room 141, Loring Bailey Hall
University of New Brunswick
Fredericton, NB E3B 5A3
e-mail: krm@unb.ca

## REFERENCES

1. Environment Canada. 1997. *Sustainable Development Strategy*. Minister of Public Works and Government Services, Ottawa, En-21-162/1997E (ISBN 0-662-25711-1).
2. Munkittrick, K.R., G.J. Van Der Kraak, M.E. McMaster, C.B. Portt, M.R. van den Heuvel and M.R. Servos. 1994. Survey of receiving water environmental impacts associated with discharges from pulp mills. 2. Gonad size, liver size, hepatic EROD activity and plasma sex steroid levels in white sucker. *Environmental Toxicology and Chemistry* 13: 1089–1101.
3. Greig, L.A., J.K. Pawley, C.H.R. Wedeles, P. Bunnell and M.J. Rose. 1992. Hypotheses of effects of development in the Moose River Basin. Report for Dept. Fisheries & Oceans, ESSA, Richmond Hill, ON L4C 9M5.
4. ESSA Technologies Ltd. 1996. Planned and potential future development activities in the Moose River Basin. Prepared for the Environmental Information partnership, Northeast Region, Ontario Ministry of Natural Resources, Cochrane, Ontario, 58 pp. + appendices.
5. Munkittrick, K.R. 1992. A review and evaluation of study design considerations for site-specifically assessing the health of fish populations. *Journal of Aquatic Ecosystem Health* 1: 283–293.
6. Munkittrick, K.R. and L.S. McCarty. 1995. An integrated approach to ecosystem health management: top-down, bottom-up or middle-out? *Journal of Aquatic Ecosystem Health* 4: 77–90.
7. Adams, S.M., ed. 1990. *Biological Indicators of Stress in Fish*. American Fisheries Society Symposium 8, American Fisheries Society, Bethesda, MD.
8. Huggett, R.J., R.A. Kimerle, P.M. Mehrle, Jr. and H.L. Bergman, eds., 1992. *Biomarkers: Biochemical, Physiological and Histological Markers of Anthropogenic Stress*. Lewis Publishers, Boca Raton, FL, 347 p.
9. Munkittrick, K.R. and D.G. Dixon. 1989a. An holistic approach to ecosystem health assessment using fish population characteristics. In M. Munawar, G. Dixon, C.I. Mayfield, T. Reynoldson and M.H. Sadar (eds.), *Environmental Bioassay Techniques & Their Application*. Kluwer Academic Publishers, The Netherlands; *Hydrobiologia* 188/189: 122–135.
10. Munkittrick, K.R. and D.G. Dixon. 1989b. Use of white sucker (*Catostomus commersoni*) populations to assess the health of aquatic ecosystems exposed to low-level contaminant stress. *Canadian Journal of Fisheries and Aquatic Sciences* 46: 1455–1462.

11. Gibbons, W.N. and K.R. Munkittrick. 1994. A sentinel monitoring framework for identifying fish population responses to industrial discharges. *Journal of Aquatic Ecosystem Health* 3: 227–237.
12. Munkittrick, K.R., C.B. Portt, G.J. Van Der Kraak, I.R. Smith and D.A. Rokosh. 1991. Impact of bleached kraft mill effluent on population characteristics, liver MFO activity and serum steroid levels of a Lake Superior white sucker (*Catostomus commersoni*) population. *Canadian Journal of Fisheries and Aquatic Sciences* 48: 1371–1380.
13. Brousseau, C.S. and G.A. Goodchild. 1989. Fisheries and yields in the Moose River basin, Ontario. p 145–158 In D.P. Dodge (ed.), Proceedings of the International Large River Symposium. Canadian Special Publications of Fisheries and Aquatic Sciences, 106.
14. Card, K.D. and B.V. Sanford. 1989. Geology, Timmins, Ontario-Quebec; Geological Survey of Canada, Geological Atlas, map NM-17-G, scale 1:1,000,000, sheet 1 of 5, The National Earth Science Series.
15. Environment Canada. 1982. *Canadian Climate Normals*, Volume 2, Temperature, 1951–1980. Minister of Public Works and Government Service, Ottawa.
16. Environment Canada. 1982. *Canadian Climate Normals*, Volume 3, Precipitation, 1951–1980. Minister of Public Works and Government Service, Ottawa.
17. Rounick, J.S. and Winterbourne, M.J. 1986. Stable carbon isotopes and carbon flow in ecosystems measuring $^{13}$C to $^{12}$C ratios can help trace carbon pathways. *Bioscience* 36: 171–177.
18. Peterson, B.J. and B. Fry. 1987. Stable isotopes in ecosystem studies. *Annual Review of Ecological Systems* 18: 293–320.
19. McMaster, M.E., G.J. Van Der Kraak, C.B. Portt, K.R. Munkittrick, P.K. Sibley, I.R. Smith and D.G. Dixon. 1991. Changes in hepatic mixed function oxygenase (MFO) activity, plasma steroid levels and age at maturity of a white sucker (*Catostomus commersoni*) population exposed to bleached kraft pulp mill effluent. *Aquatic Toxicology* 21: 199–218.
20. Munkittrick, K.R., G.J. Van Der Kraak, M.E. McMaster and C.B. Portt. 1992. Response of hepatic mixed function oxygenase (MFO) activity and plasma sex steroids to secondary treatment and mill shutdown. *Environmental Toxicology and Chemistry* 11: 1427–1439.
21. Munkittrick, K.R., M.E. McMaster, C.B. Portt, G.J. Van Der Kraak, I.R. Smith and D.G. Dixon. 1992. Changes in maturity, plasma sex steroid levels, hepatic MFO activity and the presence of external lesions in lake whitefish exposed to bleached kraft mill effluent. *Canadian Journal of Fisheries and Aquatic Sciences* 49: 1560–1569.
22. Van Der Kraak, G.J., K.R. Munkittrick, M.E. McMaster, C.B. Portt and J.P. Chang. 1992. Exposure to bleached kraft pulp mill effluent disrupts the pituitary-gonadal axis of white sucker at multiple sites. *Toxicology and Applied Pharmacology* 115: 224–233.
23. Hecky, R.E. and R.H. Hesslein. 1995. Contributions of benthic algae to lake food webs as revealed by stable isotope analysis. *Journal of the North American Benthological Society* 14(4): 613–653.
24. Hesslein, R.H., K.A. Hallard, and P. Ramlal. 1993. Replacement of sulfur, carbon, and nitrogen in tissue of growing broad whitefish (*Coregonus nasus*) in response to a change in diet traced by $\delta^{34}$S, $\delta^{13}$C, and $\delta^{15}$N. *Canadian Journal Fisheries Aquatic Science* 50: 2071–2076.

25. Vannote, R.L., G.W. Minshall, K.W. Cummins, J.R. Sedell and C.E. Cushing. 1980. The river continuum concept. *Canadian Journal Fisheries Aquatic Science* 37: 130–137.
26. Gagnon, M.M., J.J. Dodson, P.V. Hodson, G. Van Der Kraak and J.H. Carey. 1994. Seasonal effects of bleached kraft mill effluent on reproductive parameters of white sucker (*Catostomus commersoni*) populations of the St. Maurice River, Quebec, Canada. *Canadian Journal Fisheries Aquatic Science* 51: 337–347.
27. Gibbons, W.N., W.D. Taylor and K.R. Munkittrick. 1995. Suitability of small fish species for monitoring the effects of pulp mill effluent on fish populations of the Fraser River. Environment Canada Report DOE FRAP 1995-11, North Vancouver, B.C., 37 p. + appendices.
28. Gibbons, W.N., K.R. Munkittrick and W.D. Taylor. 1996. Characteristics of spoonhead sculpins and lake chub in the Athabasca River near sites which discharge pulp mill effluents. Northern Rivers Study Board Report#100, Project 2353-D1 (Forage fish collections in the Athabasca River).
29. Nickle, J., M.E. McMaster, K.R. Munkittrick, C. Rumsey, C. Portt and G.J. Van Der Kraak. 1997. Reproductive effects of primary-treated bleached kraft and thermomechanical pulp mill effluents on white sucker in the Moose River basin. 3rd Internat. Conf. Environmental Fate and Effects of Pulp and Paper Mill Effluents, Rotorua, NZ, November 9–13, 1997.
30. Gibbons, W.N. 1997. Suitability of small species for monitoring the effects of pulp mill effluents on fish populations. Ph.D. thesis, University of Waterloo, Waterloo, Ontario, 209 p.
31. Gibbons, W.N., K.R. Munkittrick and W.D. Taylor. 1998. Monitoring aquatic environments receiving industrial effluents using small fish species. 2. Comparison between responses of trout perch and white sucker downstream of a pulp mill. *Envir. Toxicol. Chem.* 17: 2238–2245.

# 9 Air Pollutants and Forests: Effect at the Organismal Scale

*Teresa W.-M. Fan and Richard M. Higashi*

## CONTENTS

Introduction ............................................................................................................. 175
An Overview of Ozone Effects ............................................................................... 176
    Growth and Morphological Effects ................................................................. 176
    Physiological Effects ........................................................................................ 176
    Biochemical Effects .......................................................................................... 178
Current Development of In Situ Biochemical Assessment of Ozone Effects ...... 180
    Pine Needle Studies ......................................................................................... 180
    Historical-Scale Methods ................................................................................. 182
Concluding Remarks .............................................................................................. 187
Acknowledgments .................................................................................................. 187
References .............................................................................................................. 188

**Keywords:** *dendrobiochemistry, Rubisco, lignin, ozone, historical-scale, pyrolysis GC/MS, SDS-PAGE*

## INTRODUCTION

Among the known air pollutants, ozone has been considered to be of principal concern, and its effects on ponderosa and Jeffrey pine have been well-documented both in laboratory and field studies.[1] Despite this concern and widespread efforts to reduce its precursors, ozone concentrations have not been abated in sensitive forest areas of the U.S. (e.g., southern Sierra Nevada range).[2] In addition to ozone, other air pollutants including $SO_2$, $NO_x$, acids, and heavy metals may interact with ozone to contribute to tree decline, but this aspect is not well defined.[3] Thus, in this chapter we will focus on the effects of ozone on forests at the organismal level, while in Chapter 13, Paul Miller and colleagues will discuss effects at the landscape level.

The observed effects of ozone on forest trees range from reduced growth, accelerated senescence, and morphological aberrations (e.g., chlorotic mottles) to specific biochemical changes such as alterations in levels of antioxidants, amino acids, pigment, proteins, and enzyme activities. The growth and morphological effects can be utilized as markers to survey large areas of natural forests for ozone-elicited injuries, which is valuable in linking ozone pollution to the decline in tree population, and its linkage to physiological and biochemical changes can provide a quantitative detection of ozone and injury, for example.[4] Taken together, all of these approaches facilitate a mechanistic understanding of injury symptoms and adaptation strategy to better understand the cause-and-effect relationship between ozone exposure and forest decline. For example, these different levels of effects have been useful in linking ozone pollution to the decline of conifer population in Southern California forests.[1,5]

An overview of each level of effects will be discussed in this chapter, along with our recent findings on biochemical changes in California pines naturally exposed to ozone.

## AN OVERVIEW OF OZONE EFFECTS

### Growth and Morphological Effects

These measurements are used widely in studies of landscape-level effects covered in Chapter 13 by Miller et al. and will be mentioned only briefly here.

Early abscission of foliage is a commonly observed effect of ozone exposure in different species of pines abundant in North American forests. In controlled-environment studies of seedlings, tree height, stem diameter, and biomass of the foliage, roots, and stems have been most frequently used for growth assessment, for example.[6-8] Among these variables, stem diameter appeared to be the most consistent and stable indicator of growth response.[6] A related parameter, radial growth obtained from tree ring analysis, proved to be useful for evaluating detailed growth effects on mature stands in natural forests.[8] However, the annual changes in rings are affected by biological, chemical, and physical factors affecting tree growth and depend on complex assumptions and extensive statistical analyses (M. Arbaugh, personal communication).

Morphological effects have been useful in assessing ozone injuries over large areas of forests in southern and central California.[5,9,10] In particular, the development of chlorotic mottles and reduced needle retention of pines have long been associated with ozone injuries.[11,12] As with growth effects, morphological effects of ozone are modified by other environmental factors such as drought and acid deposition. The complex patterns of interactions among these factors have been demonstrated for ozone responses of ponderosa pine seedlings.[7]

### Physiological Effects

A number of physiological parameters have been investigated in association with ozone treatments, including, but not limited to, rate of photosynthesis, stomatal conductance, dark respiration, allocation and reallocation of N and carbohydrates, and mycorrhizal colonization. The following is a brief overview of the types of responses that have been observed.

Reduced rate of photosynthesis is a typical response observed for ozone-exposed forest trees both in controlled fumigation studies[13-15] and in the natural environment.[16] Other responses measured that are directly connected to photosynthesis include pigment content and stomatal conductance of gases. A dose-dependent decrease in chlorophylls/carotenoids[15] and stomatal conductance[16,17] was reported for ozone-exposed loblolly and ponderosa pines. The decrease in stomatal conductance did not account for the extent of inhibition in photosynthesis, which suggests that other components along the $CO_2$ diffusion pathway were also affected by ozone exposure.[16] It should also be noted that although reduced photosynthesis has been amply documented for ozone-treated plants, stimulation was reported on current-year needles of ozone-fumigated ponderosa pines, which was interpreted as a compensatory response to ozone exposure.[18] Consequently, the expression of ozone effects on photosynthesis and associated pathways can be greatly altered by interactions among different whorls of needles.

To interpret ozone effects on net photosynthesis, it is important to measure the effect on dark respiration, that is, the oxidation of carbohydrates fixed by photosynthesis to $CO_2$. However, reported effects of ozone exposure on dark respiration have been inconsistent, ranging from suppression or no effect to stimulation in different pine species.[14,19-21] Interspecies and tissue organ variations, as well as ozone dosage, are likely to be responsible for this inconsistency.

Changes in soluble sugars and storage carbohydrates represent a downstream response from photosynthesis and respiration in ozone-exposed plants. These changes can be linked to carbon allocation and reallocation and, ultimately, to biomass production. A decrease in foliar starch concentration has been reported in a number of studies.[22-24] This change in starch content was correlated with a reduced $^{14}C$ partitioning into starch and increased partitioning into organic acids, insoluble residues, lipids, and pigments.[25] The decrease in starch content was also correlated with an increase in nontranslocatable sugars (reducing sugars such as glucose) in shoot tissues.[22,26] These effects are consistent with a reduced C export from shoots to roots,[27] reduced root to shoot ratios in favor of new leaf and stem growth,[28] and increased C allocation to repair of injured foliage or defense.[22,25]

In close association with C allocation is N metabolism, which is likely to be perturbed by ozone exposure due to the effect of ozone on accelerating leaf senescence.[5] Increases in contents of current-year foliar N,[29] soluble proteins and amino acids, and activity of main N metabolizing enzymes[30] have been observed in pines in response to ozone treatment. These effects reflected an increased N turnover and were explained by the scenario of N reallocation from senescing needles to new growth.[29] The increased N allocation into current-year needles may, in turn, be linked to the increased C partitioning and net photosynthesis in young foliage.[30]

The decreased C supply to roots in ozone-exposed pines is likely to affect mycorrhizal colonization,[3] since the latter is dependent on the carbohydrate status of the host roots.[31] A decrease in mycorrhizal colonization could lead to less nutrient availability from soils and eventually reduced growth. The observed effects of ozone exposure on mycorrhizal infection in pine roots has been inconsistent, ranging from a reduction[23,32,33] to no significant effect.[34]

In all cases, connections of *in situ* physiologic parameters to the established morphologic ozone injury parameter—chlorotic mottle in pines—is crucial to linking the laboratory studies to the field. For example, principal components analysis of multiple morphologic and physiologic parameters in the field hold the promise of leading to a more quantitative measure of ozone injury in ponderosa pine, as compared to chlorotic mottle only.[4] In turn, these "calibrated" physiological parameters can be vital in connecting to laboratory and *in situ* biochemical studies.

## BIOCHEMICAL EFFECTS

As with all toxicants, ozone exerts its effects through chemical reactions. Therefore, effects at the physiological levels are expected to originate from specific biochemical alterations, just as effects on the growth can be ascribed to physiological changes. Some ozone fumigation studies have been inititated to investigate the biochemical changes which are connected to some of the physiological effects described earlier.

Inhibition of photosynthesis in forest foliage has been a major physiological effect observed from ozone exposure. At least several biochemical/metabolic changes can contribute to this effect. As stated earlier, changes in stomatal conductance alone cannot explain the degree of photosynthetic inhibition, which means that some other mechanism(s) must also be at work. One likely candidate involves the $CO_2$ fixation enzyme, ribulose-1,5-bisphosphate carboxylase (Rubisco). Indeed, there have been reports on the decline of Rubisco concentration and activity in crop plants and poplar trees subjected to relatively short-term ozone treatment.[35-37] However, no apparent changes were noted in Rubisco content in Eastern white pine over one growing season of ozone fumigation.[38] There has been very little information on the cumulative effect of ozone exposure on Rubisco in pines and other forest species over relatively long-term ozone exposure, especially in their natural setting.

In addition to $CO_2$ fixation, Rubisco represents a major N sink in plant leaves, and its decline in ozone-treated leaves of crop species may be due to an enhanced sensitivity to proteolysis.[38] This increased turnover of Rubisco is likely to have a pronounced effect on N metabolism. A decrease in protein and increase in total amino acid content were reported for ozone-fumigated bean leaves,[39,40] which is consistent with an increase in the degradation of proteins including Rubisco. However, an opposite response in protein content was observed for conifer species, along with an increase in glutamine synthetase and glutamate dehydrogenase activities.[30,41] Stimulation of these two enzyme activities was accompanied by an increase in the pools of glutamine, glutamate, aspartate, and alanine. These changes were interpreted as an increased export and N reassimilation of protein breakdown products from older senescing leaves under ozone exposure.[30] In addition, elevated pools of asparagine, glycine, and serine were occasionally observed in ozone-fumigated leaves,[39,42,43] which may be indicative of enhanced photorespiration[43,44] resulting from the decrease in stomatal conductance. As a whole, there appeared to be a substantial variability in the response of N metabolism to ozone treatment, depending on the plant type, duration and dosage of exposure, and other factors.

As a powerful oxidant, ozone is expected to elicit changes in the antioxidation machinery in plants. Antioxidant compounds such as ascorbate and glutathione have been shown to be depleted in some ozone-fumigated crop plants,[45,46] while an increase in ascorbate, glutathione, vitamin E, and catechin content has also been reported in some other plants under ozone fumigation.[47-49] These opposite responses of antioxidant levels may be attributable to a differential ozone tolerance among plants[48] or to different periods of adaptation. Closely linked to the antioxidant response was the observed stimulation of antioxidant enzyme activities including ascorbate peroxidase and superoxide dismutase.[50,51] However, such stimulation was not always observed in conifers, depending on other conditions including light and temperature.[52] This variation in peroxidase response may be related to ethylene production resulting from other interacting environmental stresses. Ethylene appeared to promote ascorbate peroxidase activity in mung beans and peas, which was associated with protection against ozone-induced injuries.[51] However, ethylene formation and increased ascorbate peroxidase activity were also linked to ozone sensitivity in pines.[50,53] This apparent contradiction will need to be resolved before ethylene and ascorbate peroxidase can be useful as indicators of defense against ozone damage.

In addition to the ascorbate/ascorbate peroxidase system, two other oxyradical scavenger systems involving phenolic compounds and polyamines have been investigated. Levels of free polyamines and polyamine conjugates in leaves of many plants were elevated upon ozone exposure[54-56] which was related to the induction of the biosynthetic enzyme (L-arginine decarboxylase) activity.[54,57] The changes in polyamine chemistry appeared to counteract the premature senescence effect of ethylene which was also associated with the development of ozone-induced foliar injuries.[54] Again, this finding contradicts the protective role of ethylene reported in another study[51] mentioned earlier.

Effects of ozone treatment on enzymes along the phenolic pathways were also investigated to some extent, including phenylalanine ammonia lyase (the gateway enzyme to phenolic biosynthesis), cinnamyl alcohol and coniferylalcohol dehydrogenases (key enzymes of the lignin pathway), chalcone synthase (key enzyme in the flavonoid pathway), and stilbene synthase (key enzyme of phytoalexin pathway). Stimulation of all of these enzyme activities has been reported in ozone-treated pine needles.[58-60] Changes in some of the enzyme activities were also accompanied by the product accumulation, including stilbene metabolites[58] and flavonoids.[60] Other than being good oxyradical scavengers, many of these secondary metabolites are cell wall and wood components which respond to wounding or fungal attack.[61] Thus, they may prove to be reliable secondary indicators of ozone impact on tree health.

Other plant defense mechanisms which responded to ozone exposure involve the induction of $\beta$-1,3-endoglucanase and endochitinase activities.[62] These two enzymes were shown to inhibit the growth of fungi that contain $\beta$-1,3-glucans or chitin as cell wall polymers, and a role of these hydrolases in plant defense against microbial infection was thus postulated.[63] In addition, a plant cell wall component ($\beta$-1,3-glucan callose) that has been known to respond to wounding and pathogen attack was found to accumulate in ozone-injured leaves.[62] These biochemical effects are consistent with the notion that ozone exposure predisposes plants to wounding and pathogen attack which divert resources away from growth or weaken the individual's ability to cope with other stresses. It is also possible that callose is formed from nonbiological injury.

This variety of biochemical changes in response to ozone treatment have facilitated a mechanistic and therefore cause-and-effect understanding of ozone impact on tree decline. However, since a vast majority of the studies were conducted using control or field open-top chambers and young seedlings, it is difficult to relate these changes to the effects on mature trees and naturally exposed stands. In natural environments, ozone-induced responses will vary as a function of multiple factors, including developmental stage or tree age, vigor, exposure conditions (dosage, application rate, duration, light, and temperature), soil/nutrient status, water availability, and presence of other pollutants. These factors must be taken into account for an *in situ* assessment of ozone impact on forest health and for implementing an effective measure for forest protection against ozone damages.

## CURRENT DEVELOPMENT OF *IN SITU* BIOCHEMICAL ASSESSMENT OF OZONE EFFECTS

One way to generate information for *in situ* assessment of air pollution impact is to explore the response of trees in natural settings. Such an approach is even more subject to individual difference and environmental variables than fumigation chamber studies where these variables can be carefully controlled. However, these drawbacks may be minimized if each individual tree can serve as its own control.

### PINE NEEDLE STUDIES

It is possible to examine the needle Rubisco content in natural stands of pines, much in the way that morphologic and physiologic characterization has been done. As an example, in our laboratory, we are investigating pine needles from several ozone-impacted sites in the Sierra Nevada range and in Southern California. Total proteins including Rubisco were extracted from lyophilized and pulverized (<3 μm size) needle powder using the SDS (sodium dodecyl sulfate) sample buffer (BioRad protocol) at boiling temperature for 5 min, followed by analysis on discontinuous SDS-PAGE (polyacrylamide gel electrophoresis).[64] Using a 12% separating gel, the large (L) and small (S) subunits of Rubisco were separated from other proteins as shown in Figure 1. The identity of L-Rubisco was confirmed by performing a Western blot on the same set of protein extracts using a polyclonal antibody raised against purified spinach Rubisco (gift of Dr. Steven Gutteridge) (Figure 2).

The research is in its infancy, but some preliminary results are presented here for illustrative purposes. As expected, the L- (53.7 kD) and S-Rubisco (13.4 kD) comprised the major fraction of the total proteins in pine needles. In addition, the L-Rubisco content increased approximately as a function of increasing needle age class in all ponderosa pine needles collected from two relatively mild ozone impact sites (1220 m elevation, White Cloud and sea level, Davis) (Figure 3). In contrast, the L-Rubisco content in ponderosa pine needles collected from a moderate ozone impact site (1830 m elevation, Mtn. Home) showed a slight decreasing trend with increasing needle age class, except for the 1994 growth year. A similar difference

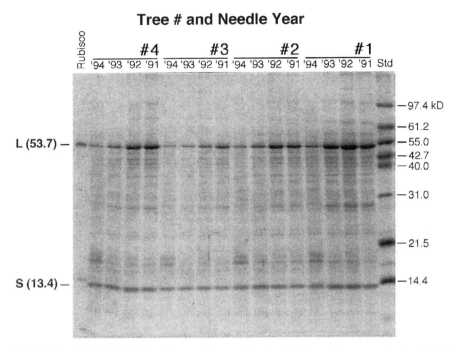

**FIGURE 1** SDS-PAGE of ponderosa pine needle extracts. Sample lanes are designated at the top; the needle whorl years from four individual trees; the leftmost lane that of purified Rubisco from spinach (Sigma), and the rightmost lane that of protein molecular weight markers (Promega). The vertical axis is the electrophoretic mobility. The pine needles were collected from low- to mid-ozone impacted sites (White Cloud) and pulverized frozen in liquid nitrogen to <3 μm particles in a Microdismembrator before lyophilization. About 15 to 20 mg of the lyophilized powder was extracted with BioRad sample buffer (containing 62.5 m$M$ Tris-HCl, 10% glycerol, 2% SDS, 5% 2-β-mercaptoethanol, and 12.5 ppm bromophenol blue, pH 6.8) at boiling temperature for 5 to 10 min and centrifuged to remove tissue debris. SDS-PAGE was performed on a 12% separating gel with a 4% stacking gel at 10°C on a slab gel apparatus (Owl Scientific). The gel was stained in a Coomassie blue G-250 staining solution (Sigma), air dried, and digitized for image analysis. The gel image shown here illustrates the protein patterns of different age class needles from four ponderosa pine stands located at White Cloud. Two prominent protein bands were present for all extract samples, which corresponded to the large (L, 53.7 kD) and small (S, 13.4 kD) subunits of Rubisco (compared with those of the spinach Rubisco, leftmost lane).

in the L-Rubisco distribution pattern was observed for Coulter pine needles collected from low (Davis) and high (830 m, Tanbark Flat) ozone impact sites (Figure 4).

Therefore, the needle age-dependent pattern of L-Rubisco appeared to be correlated with ozone gradients for naturally grown pine stands. Such correlation applied to both ozone-sensitive and tolerant pines. Assuming that foliar Rubisco content decreases with senescence, the observed L-Rubisco pattern for moderate and severe ozone impact sites is consistent with the effect of accelerated senescence. On the other hand, the ozone impact sites selected also differ in several other environmental factors, notably altitude, N deposition, soil conditions, and precipitation. In partic-

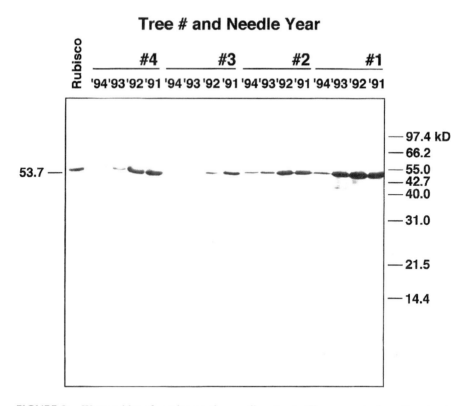

**FIGURE 2** Western blot of ponderosa pine needle extracts. The same set of needle extracts from Figure 1 along with spinach Rubisco (Sigma) was electrophoresed on a similar SDS gel, and the proteins in the gel were blotted onto a piece of nitrocellulose membrane using a semi-dry blotting apparatus (Owl Scientific) in a transfer buffer containing 25 m$M$ Tris-HCl, 192 m$M$ glycine, and 20% methanol, pH 8.25 for 4 h at a constant current of 400 mA. The blot was probed with a polyclonal antibody raised against purified spinach Rubisco according to the BioRad protocol. The presence of a 53.7-kD band in all needle extracts confirmed the identity of this band as the large subunit of Rubisco.

ular, differences in N deposition may be an important contributing factor to the large site-dependent variability in Rubisco content of pine needles (Figures 3 and 4). However, it is generally unclear how these factors influence the L-Rubisco pattern. Clearly, more systematic studies are needed to establish the L-Rubisco pattern as a cumulative indicator of ozone impact in natural pine populations.

## HISTORICAL-SCALE METHODS

For long-lived organisms such as pines, their tree rings harbor a record of their life histories. This record can be explored to trace the progressive changes elicited by the onset of air pollution. These changes will take into account all the incidental variables that are interacting with the air pollutant(s) of interest.

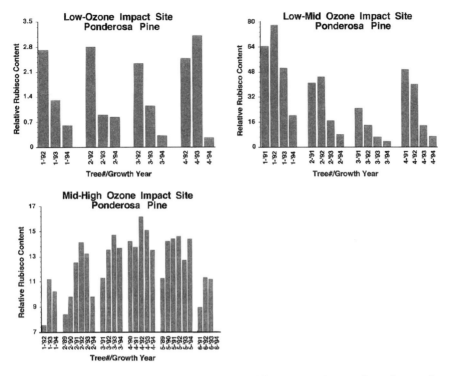

**FIGURE 3** Distribution patterns of Rubisco in different age classes of ponderosa pine needles collected from three ozone-impacted sites. The different age class pine needles were collected in the summer of 1994 from the low (Davis), low to mid (White Cloud), and mid to high (Mtn. Home) ozone-impacted sites. The large subunit of Rubisco (L-Rubisco) in the needle extract was separated by SDS-PAGE as in Figure 1 and digitized, followed by densitometric analysis by the NIH Image program (National Institutes of Health, Bethesda, MD). The relative Rubisco content was calculated from the densitometric quantitiy of L-Rubisco normalized against that of known quantity of a molecular weight standard. The age class-dependent distribution of L-Rubisco was established for four to six trees from each of the three sites.

Dendrochronological and dendrochemical analyses of naturally grown stands can be used to assess tree age, radial growth, soil conditions, water use efficiency, and metal pollution.[65] In particular, the radial growth parameter was used to assess the effect of elevated ozone on growth trends of ponderosa pine in natural forests of the Sierra Nevada.[8] The use of dendrochemistry in regional ecosystem health assessment has been utilized in several studies.[65]

Although "dendrochemistry"—defined as the analysis of nutrients and trace elements in wood[65]—is recognized as a useful retrospective indicator of forest conditions, the value of "dendrobiochemistry" has not been explored to any extent. It is reasonable to postulate that the tree biochemistry, represented by the biosynthesized organic constituents in rings, records at least some biochemical changes incurred in a tree's

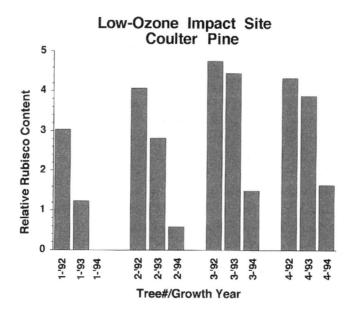
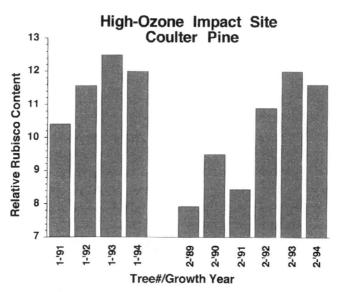

**FIGURE 4** Distribution patterns of Rubisco in different age classes of Coulter pine needles collected from low and high ozone-impacted sites. The different age class pine needles were collected in the summer of 1994 from the low (Davis) and high (Tanbark Flat) ozone-impacted sites. The relative Rubisco content was obtained similarly as in Figure 3.

lifetime. If understood, dendrobiochemistry can be a powerful tool for revealing the history of air pollution impact, thereby facilitating the prediction of future trends. It was for this purpose that we began the dendrobiochemical research.

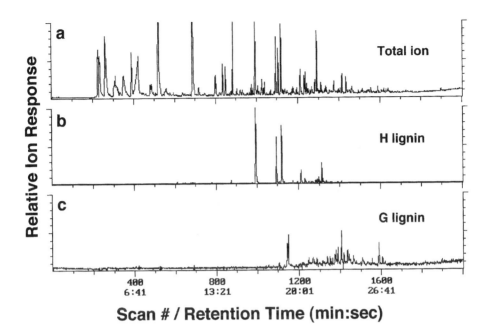

**FIGURE 5**  Pyrolysis/GC-MS chromatograms from a tree ring segment of ponderosa pine. Panel a shows the total data set (total ion chromatogram) obtained from a tree core sample, illustrating the complexity of the chemical structure data. One way of conceptualizing the data set is to view this as the sum of more than 300 "mass fragment channels" of chromatograms. Panels b and c show how six of each such "channels" can be dissected out to represent H- and G-type lignin substructures, respectively. The total area under all peaks in a given panel were summed for data in Figure 6.

Since the majority of the tree ring components are not amenable to extraction, we have adopted the pyrolysis/gas chromatography-mass spectrometry (Py/GC-MS) technique that was originally developed for the analysis of insoluble materials such as processed wood, polymers, and textile fibers. This technique generates thermolytic fragments from the sample under rapid heating and He atmosphere, which are subsequently separated by gas chromatography and structurally characterized by mass spectrometry. Py/GC-MS analyses require almost no sample preparation, and a wealth of biochemical information may be obtainable, in part because of the minimal sample manipulation.[66-68] This is illustrated by the usage of Py/GC-MS in the single-step analysis of cellulose, hemicellulose, lignocellulose, pectins, suberins, proteins, etc. directly from woody materials.[66,69-71] Most importantly, analytes do not have to be preselected, so the data can be revisited in the future by researchers armed with improved knowledge of wood biochemistry and air pollution.

In a collaborative project with M. Arbaugh (USDA Forest Fire Laboratory, Riverside, CA), we have been applying Py/GC-MS to analyze tree ring cores of ponderosa pines located in the San Bernardino National Forest, northeast and upwind of the Los Angeles basin in California. The studies are conducted at two well-established, high-ozone injury sites, Dogwood (DG) and Camp Paivika (CP), and

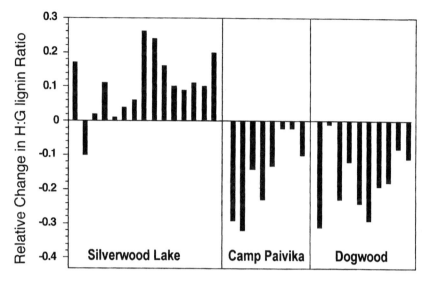

**FIGURE 6** Changes in the H-to-G lignin ratio of pre- and postindustrialization tree core segments from individual trees collected at three sites. The data illustrated in Figure 5 were used to calculate the ratio of H- and G-type lignin for tree ring segments representing pre- (1900 to 1940) and industrial boom (1950 to the present) in the Los Angeles basin. The relative changes in the H-to-G lignin ratio are plotted as bars. Here, all trees at the low ozone impacted site (SL), with the exception of one tree, showed declines in the H-to-G lignin ratio over time, while all trees at the two highly impacted sites (DG and CP) showed increases. This different trend may reflect ozone-induced metabolic changes which were effectively recorded in the tree ring segments. The variance between individual trees may result from differences in genetics and other environmental conditions.

one low-ozone injury site, Silverwood Lake (SL). Tree ring core segments representing periods of 1900 to 1940 (preindustrial boom) and 1950 to the present (industrial boom) for a given tree were analyzed using analytical methods very similar to that of Faix et al.[71] As such, dendrobiochemical changes as a consequence of industrialization or other factors can be established on individual trees without major interferences from intrinsic individual differences. Presently, the data complexity combined with a lack of sufficient understanding of the nature of the thermolytic fragments precludes a comprehensive analysis of the tree rings. Nevertheless, it is feasible to interpret changes in subcomponents of wood such as specific α- and β-glucan linkages of polysaccharides[70] and certain details of lignocellulosic structures and other phenolic compounds.[68,71]

Figure 5a illustrates a typical total ion chromatogram from Py/GC-MS of a tree ring segment of ponderosa pine. An example of wood substructure analysis is shown in Figures 5b and c, which was based on mass fragment assignments for H-type (hydroxyphenylpropenyl) and G-type (guaiacyl or monomethoxy-H-type) lignin substructures,[71] respectively. The sum of integrated peak areas of all H- and G-type lignin substructures was obtained to calculate the ratio of H to G lignin for each pre- and postindustrial ring segment. Figure 6 shows the results of H to G lignin ratio change for 15 trees at the low ozone-injury site (SL) and 18 trees at the two high ozone-injury sites (DG and CP), following heavy industrialization of the Los Angeles basin. At SL, 14 out of 15 trees showed a positive direction of change, while at DG and CP, all trees exhibited a negative direction of change. At the present time, we hypothesize that the *direction* of change in the lignin types is indicative of air pollution effects, particularly the ozone impact. It is possible that the H to G lignin ratio changes are related to the changes in lignin pathway enzymes, as discussed earlier. We feel that the variability in the *magnitude* of change is due to differences among the individual trees such as genetics, soil and nutritional factors, canopy position, local terrain, etc. This sensitivity to interactions among different factors is currently under investigation.[72]

## CONCLUDING REMARKS

Our current understanding of the different levels of air pollutant effects on forest trees is largely based on fumigation studies of seedlings, either in the laboratory or in field open-top chambers. In addition, most of the effects observed were dependent on the exposure conditions and developmental stage or growth year. Therefore, there is an urgent need to translate these findings to impacts on trees in natural settings. This aspect has been difficult due to environmental complexity and technical obstacles, but it will be required for decision making regarding air pollution, tree injury, and any associated forest decline. Since the majority of the effects on trees are dynamic and transient in nature, it has been difficult to utilize this knowledge to predict the long-term impact of air pollution on forest trees. In 1989, the National Research Council laid down a challenge that "to date no readily detectable, pollutant-specific, single marker for identifying the effects of air pollution on forests or trees has been identified."[73] Today, at the organismal level, *in situ* cumulative or integrative responses over long-term or even historical time scales are being explored to meet this challenge.

## ACKNOWLEDGMENTS

Some of the studies described were supported, in part, by the U.S. EPA funded (grant #R819658) Center for Ecological Health Research at the University of California, Davis. Although the information in this document has been funded partly by the U.S. Environmental Protection Agency, it may not necessarily reflect the views of the agency and no official endorsement should be inferred. We wish to acknowledge Dr. Andrew N. Lane for his expertise and the U.K. Medical Research Council Biomedical NMR Centre for use of NMR instrumentation. We would also like to acknowledge the two anonymous reviewers for their helpful comments.

For questions or further information regarding the work described in this chapter, please contact:

**Teresa Fan**
Dept. of Land, Air and Water Resources
One Shields Avenue
University of California
Davis, CA 95818
e-mail: twfan@ucdavis.edu

## REFERENCES

1. Miller, P.R., Arbaugh, M.J., Temple, P.J. 1996. Ozone and its known and potential effects on forest in the Western United States. In Sandermann, H. et al., eds., *Forest Decline and Ozone*. Springer-Verlag, Berlin.
2. Van Ooy, D.J., Carroll, J.J. 1995. The spatial variation of ozone climatology on the western slope of the Sierra Nevada. *Atmos. Environ.* 29:1319–1330.
3. Berrang, P., Meadows, J.S., Hodges, J.D. 1996. An overview of responses of Southern pines to airborne chemical stresses. In S. Fox and T.A. Mickler, eds., *Impact of Air Pollutants on Southern Pine Forests. Ecological Studies*, Vol. 118. Springer-Verlag, New York, pp. 196–243.
4. Grulke, N.E., Lee, E.H. 1997. Assessing visible ozone-induced foliar injury in ponderosa pine. *Can. J. For. Res.* 27:1658–1668.
5. Miller, P.R., McBride, J.R., Schilling, S.L., Gomez, A.P. 1989. Trend of ozone damage to conifer forests between 1974 and 1988 in the San Bernardino Mountains of southern California. In R.K. Olsen and A.S. Lefohn, eds., *Effects of Air Pollution on Western Forests*, APCA Transactions Series No. 16. AWMA, Pittsburgh, PA, pp. 309–323.
6. Flagler, R.B., Chappelka, A.H. 1996. Growth response of Southern pines to acidic precipitation and ozone. In S. Fox and R.A. Mickler, eds., *Impact of Air Pollutants on Southern Pine Forests. Ecological Studies*, Vol. 118. Springer-Verlag, New York, pp. 389–424.
7. Temple, P.J., Riechers, G.H., Miller, P.R., Lennox, R.W. 1992. Growth responses of ponderosa pine to long-term exposure to ozone, wet and dry acidic deposition, and drought. *Can. J. For. Res.* 32:59–66.
8. Peterson, D.L., Arbaugh, M.J., Robinson, L.J. 1989. Ozone injury and growth trends of ponderosa pine in the Sierra Nevada. In R.K Olson and A.S. Lefohn, eds., *Effects of Air Pollution on Western Forests*, APCA Transactions Series No. 16. AWMA, Pittsburgh, PA, pp. 293–307.
9. Miller, P.R., Millecan, A.A. 1971. Extent of oxidant injury to some pines and other conifers in California. *Plant Dis. Rep.* 55:555–559.
10. Duriscoe, D.M., Stolte, K.W. 1989. Photochemical oxidant injury to ponderosa pine (*Pinus ponderosa* Laws.) and Jeffrey pine (*Pinus jeffreyi* Grev., and Balf.) in the national parks of the Sierra Nevada of California. In R.K Olson and A.S. Lefohn, eds., *Effects of Air Pollution on Western Forests*, APCA Transactions Series No. 16. AWMA, Pittsburgh, PA, pp. 261–278.
11. Miller, P.R., Parmeter, Jr., J.R., Taylor, O.C., Cardiff, E.A. 1963. Ozone injury to the foliage of *Pinus ponderosa*. *Phytopathology* 53:1072–1076.

12. Taylor, O.C. ed., 1980. Photochemical oxidant air pollution effects on a mixed conifer forest ecosystem, Final Report, EPA-600/3-8-002, U.S. EPA, Corvallis, OR.
13. Miller, P.R., Parmeter, Jr., J.R., Flick, B.H., Martinez, C.W. 1969. Ozone dosage response of ponderosa pine seedlings. *J. Air Pollut. Control. Assoc.* 19:435–438.
14. Barnes, R.L. 1972. Effects of chronic exposure to ozone on photosynthesis and respiration of pines. *Environ. Pollut.* 3:133–138.
15. Richardson, C.J., Sasek, T.W., Fendick, E.A., Kress, L.W. 1992. Ozone exposure-response relationships for photosynthesis in genetic strains of loblolly pine seedlings. *For. Ecol. Manage.* 51:163–178.
16. Coyne, P., Bingham, C. 1981. Comparative ozone dose response of gas exchange in a ponderosa stand exposed to long-term fumigations. *J. Air Pollut. Control. Assoc.* 31:38–41.
17. Weber, J.A., Clark, C.S., Hogsett, W.E. 1993. Analysis of the relationships among $O_3$ uptake, conductance, and photosynthesis in needles of *Pinus ponderosa*. *Tree Physiol.* 13:157–172.
18. Beyers, J.L., Riechers, G.H., Temple, P.J. 1992. Effects of long-term ozone exposure and drought on the photosynthetic capacity of ponderosa pine (*Pinus ponderosa* Laws.). *New Phytol.* 122:81–90.
19. Edwards, N.T. 1991. Root and soil respiration responses to ozone in Pinus taeda L. seedlings. *New Phytol.* 118:315–321.
20. Hanson, P.J., McLaughlin, S.B., Edwards, N.T. 1988. Net $CO_2$ exchange of Pinus taeda shoots exposed to variable ozone levels and rain chemistries in field and laboratory settings. *Physiol. Plant.* 74:635–642.
21. Skarby, L., Troeng, E., Bostrom, C. 1987. Ozone uptake and effects on transpiration, net photosynthesis, and dark respiration in Scots pine. *For. Sci.* 33:801–808.
22. Paynter, V.A., Reardon, J.C., Shelburne, V.B. 1991. Carbohydrate changes in shortleaf pine (*Pinus echinata*) needles exposed to acid rain and ozone. *Can. J. For. Res.* 21:666–671.
23. Meier, S., Grand, L.F., Schoeneberger, M.M., Reinert, R.A., Bruck, R.I. 1990. Growth, ectomycorrhizae and nonstructural carbohydrates of loblolly pine seedlings exposed to ozone and soil water deficit. *Environ. Pollut.* 64:11–27.
24. Horton, S.J., Schoeneberger, M.M., Reinert, R.A., Shafer, S.R., Allen, H.L. 1989. Growth, carbohydrate reserves and nutrient content of loblolly pine seedlings exposed to ozone and simulated acidic rain. *Agronomy Abstr.*, ASA, Madison, WI, p. 305.
25. Friend, A.L., Tomlinson, P.T. 1992. Mild ozone exposure alters $^{14}C$ dynamics in foliage of Pinus taeda L. *Tree Physiol.* 11:215–227.
26. Tingey, D.T., Wilhour, R.G., Standley, C. 1976. The effect of chronic ozone exposures on the metabolite content of ponderosa pine seedlings. *For. Sci.* 22:234–241.
27. Edwards, G.S., Friend, A.L., O'Neill, E.G., Tomlinson, P.T. 1992. Seasonal patterns of biomass accumulation and carbon allocation in Pinus taeda seedlings exposed to ozone, acidic deposition, and reduced soil Mg. *Can. J. For. Res.* 22:640–646.
28. Cooley, D.R., Manning, W.J. 1987. The impact of ozone on assimilate partitioning in plants: a review. *Environ. Pollut.* 47:95–113.
29. Temple, P.J., Riechers, G.H. 1995. Nitrogen allocation in ponderosa pine seedlings exposed to interacting ozone and drought stresses. *New Phytol.* 130:97–104.
30. Manderscheid, R., Jäger, H.-J., Kress, L.W. 1992. Effects of ozone on foliar nitrogen metabolism of *Pinus taeda* L. and implications for carbohydrate metabolism. *New Phytol.* 121:623–633.
31. Hacskaylo, E. 1973. The Torrey symposium on current aspects of fungal development: IV. Dependence of mycorrhizal fungi on hosts. *Bull. Torrey Bot. Club* 100:217–223.

32. Adams, M.B., O'Neill, E.G. 1991. Effects of ozone and acidic deposition on carbon allocation and mycorrhizal colonization of *Pinus taeda* L. seedlings. *For. Sci.* 37:5–16.
33. McQuattie, C.J., Schier, G.A. 1992. Effect of ozone and aluminum on pitch pine (*Pinus rigida*) seedlings: anatomy of mycorrhizae. *Can. J. For. Res.* 22:1901–1916.
34. Simmons, G.L., Kelly, J.M. 1989. Influence of $O_3$, rainfall acidity, and soil Mg status on growth and ectomycorrhizal colonization of loblolly pine roots. *Water Air Soil Pollut.* 44:159–171.
35. Dann, M.S., Pell, E.J. 1989. Decline of activity and quantity of ribulose bisphosphate carboxylase/oxygenase and net photosynthesis in ozone-treated potato foliage. *Plant Physiol.* 91:427–432.
36. Pell, E.J., Pearson, N.S. 1983. Ozone-induced reduction in quantity of ribulose-1,5-bisphosphate carboxylase in alfalfa foliage. *Plant Physiol.* 73:185–187.
37. Nakamura, H., Saka, H. 1978. Photochemical oxidants injury in rice plants. *Jpn. J. Crop Sci.* 47:704–714.
38. Pell, E.J., Landry, L.G., Eckardt, N.A., Glick, R.E. 1994. Air pollution and RubisCO: effects and implications. In R.G. Alscher and A.R. Wellburn, eds., *Plant Responses to the Gaseous Environment*. Chapman & Hall, London, pp. 239–253.
39. Manderscheid, R., Bender, J., Weigel, H.-J., Jäger, J. 1991a. Low doses of ozone affect nitrogen metabolism in bean (*Phaseolus vulgaris* L.) leaves. *Biochem. Physiol. Pflanz.* 187:283–291.
40. Ito, W., Okano, K., Totsuka, T. 1986. Effects of $NO_2$ and $O_3$ alone or in combination on kidney bean plants (*Phaseolus vulgaris* L.): amino acid content and composition. *Soil Sci. Plant. Nutr.* 32:351–363.
41. Bender, J., Manderscheid, R., Jäger, H.-J. 1990. The Hohenheim long-term experiment. Analyses of enzyme activites and other metabolic criteria after five years of fumigation. *Environ. Pollut.* 68:331–343.
42. Dohmen, G.P., Koppers, A., Langebartels, C. 1990. Biochemical response of Norway spruce (*Picea abies* (L.) Karst.) towards 14-month exposure to ozone and acid mist: effects on amino acid, glutathionine and polyamine titers. *Environ. Pollut.* 64:375–383.
43. Ito, O., Mitsumori, F., Totsuka, T. 1985. Effects of $NO_2$ and $O_3$ alone or in combination on kidney bean plants (*Phaseolus vulgaris* L.). Products of [$^{13}CO_2$] assimilation detected by [$^{13}C$] nuclear magnetic resonance. *J. Exp. Bot.* 36:281–289.
44. Manderscheid, R., Jäger, H.-J., Schoeneberger, M.M. 1991b. Dose-response relationships of ozone effects on foliar levels of antioxidants, soluble polyamines and peroxidase activity of (*Pinus taeda* L.): assessment of the usefulness as early ozone indicators. *Angew Botanik* 65:373–386.
45. Price, A., Lucas, P.W., Lea, P.J. 1990. Age dependent damage and glutathione metabolism in ozone fumigated barley: a leaf section approach. *J. Exp. Bot.* 41:1309–1317.
46. Luwe, M.W., Takahama, U., Heber, U. 1993. Role of ascorbate in detoxifying ozone in the apoplast of spinach (*Spinacia oleracea* L.) leaves. *Plant Physiol.* 101:969–976.
47. Booker, F.L., Anttonen, S., Heagle, A.S. 1996. Catechin, proanthocyanidin and lignin contents of loblolly pine (*Pinus taeda* L.) needles after chronic exposure to ozone. *New Phytol.* 132:483–492.
48. Mehlhorn, H., Seufert, G., Schmidt, A., Kunert, K.J. 1986. Effect of $SO_2$ and $O_3$ on production of antioxidants in conifers. *Plant Physiol.* 82:336–338.
49. Lee, E.H., Jersey, J.A., Gifford, C., Bennett, J. 1984. Differential ozone tolerance in soybean and snapbeans: analysis of ascorbic acid in $O_3$-susceptible and $O_3$-resistant cultivars by high-performance liquid chromatography. *Environ. Exp. Bot.* 24:331–341.

50. Benes, S.E., Murphy, T.M., Anderson, P.D., Joupis, J.L.J. 1995. Relationship of antioxidant enzymes to ozone tolerance in branches of mature ponderosa pine (*Pinus ponderosa*) trees exposed to long-term, low concentration, ozone fumigation and acid precipitation. *Physiol Plant.* 94:124–134.
51. Mehlhorn, H. 1990. Ethylene-promoted ascorbate peroxidase activity protects plants against hydrogen peroxide, ozone and paraquat. *Plant Cell Environ.* 13:971–976.
52. Polle, A., Rennenberg, H. 1991. Superoxide dismutase activity in needles of Scots pine and Norway spruce under field and chamber conditions: lack of ozone effects. *New Phytol.* 117:335–343.
53. Telewski, F.W. 1992. Ethylene production by different age class ponderosa and Jeffery pine needles as related to ozone exposure and visible injury. *Trees* 6:195–198.
54. Langebartels, C., Kerner, K., Leonardi, S., Schraudner, M., Trost, M., Heller, W., Sandermann, Jr., H. 1991. Biochemical plant responses to ozone. I. Differential induction by polyamine and ethylene biosynthesis in tobacco. *Plant Physiol.* 95:882–889.
55. Sandermann, H., Schmitt, R., Heller, W., Rosemann, D., Langebartels, C. 1989. Ozone-induced early biochemical reactions in conifers. In J.W.S. Longhurst, ed., *Acid Deposition. Sources, Effects and Controls.* British Library, London, pp. 243–254.
56. Rowland, A.J., Borland, A.M., Lea, P.J. 1988. Changes in amino acids, amines, and proteins in response to air pollutants. In S. Schulte-Hostede, N.M. Darrall, L.W. Blank, and A.R. Wellburn, eds., *Air Pollution and Plant Metabolism*, Elsevier, London, pp. 189–221.
57. Rowland-Bramford, A.J., Borland, A.M., Lea, P.J., Mansfield, T.A. 1989. The role of arginine decarboxylase in modulating the sensitivity of barley to ozone. *Environ. Pollut.* 61:95–106.
58. Rosemann, D., Heller, W., Sandermann, Jr., H. 1991. Biochemical plant responses to ozone. II. Induction of stilbene biosynthesis in Scots pine (*Pinus sylvestris* L.) seedlings. *Plant Physiol.* 97:1280–1286.
59. Heller, W., Rosemann, D., Osswald, W.F., Benz, B., Schönwitz, R., Lohwasser, K., Kloos, M., Sandermann, Jr., H. 1990. Biochemical response of Norway spruce (*Picea abies* (L) Karst.) towards 14-month exposure to ozone and acid mist: effects on polyphenol and monoterpene metabolism. *Environ. Pollut.* 64:353–366.
60. Langebartels, C., Heller, W., Kerner, K., Leonardi, S., Rosemann, D., Schraudner, M., Trost, M., and Sandermann, Jr., H. 1990. Ozone-induced defense reactions in plants. *Environmental Research with Plants in Closed Chambers. Air Pollut. Res. Rep. Euro. Commun.* 26:358–368.
61. Kindl, H. 1985. Biosynthesis of stilbenes. In T. Higuchi, ed., *Biosynthesis and Biodegradation of Wood Components.* Academic Press, London, pp. 349–377.
62. Schraudner, M., Ernst, D., Langebartels, C., Sandermann, Jr., H. 1992. Biochemical plant responses to ozone. III. Activation of the defense-related proteins β-1,3-glucanase and chitinase in tobacco leaves. *Plant Physiol.* 99:1321–1328.
63. Bowles, D.J. 1990. Defense-related proteins in higher plants. *Annu. Rev. Biochem.* 59:873–907.
64. Laemmli, U.K. 1970. Cleavage of structural proteins during assembly of the head of bacteriophage T4. *Nature* 227:680–685.
65. Lewis, T.E., ed. 1995. *Tree Rings as Indicators of Ecosystem Health.* CRC Press, Boca Raton, FL.

66. Saiz-Jimenez, C., Boon, J.J., Hedges, J.I., Hessels, J.K.C., de Leeuw, J. 1987. Chemical characterization of recent and buried woods by analytical pyrolysis: comparison of pyrolysis data with 13C NMR and wet chemical data. *J. Anal. Appl. Pyrolysis* 11:437–450
67. Schulten, H.-R., Simmleit, N., Müller, R. 1989. Characterization of plant materials by pyrolysis-field ionization mass spectrometry: high-resolution mass spectrometry, time-resolved high-resolution mass spectrometry, and Curie-point pyrolysis-gas chromatography/mass spectrometry of spruce needles. *Anal. Chem.* 61:221–227.
68. Scheijen, M.A., Boon, J.J. 1991. Micro-analytical investigations on lignin in enzyme-digested tobacco lamina and midrib using pyrolysis-mass spectrometry and Curie-point pyrolysis-gas chromatography/mass spectrometry. *J. Anal. Appl. Pyrolysis* 19:153–173.
69. van der Kaaden, A., Haverkamp, J., Boon, J.J., de Leeuw, J.W. 1983. Analytical pyrolysis of carbohydrates I: chemical interpretation of matrix influences on pyrolysis-mass spectra of amylose using pyrolysis-gas chromatography-mass spectrometry. *J. Anal. Appl. Pyrolysis* 5:199–220.
70. van der Kaaden, A., Boon, J.J., Haverkamp, J. 1984. Analytical pyrolysis of carbohydrates. 2. Differentiation of homopolyhexoses according to their linkage type, by pyrolysis-mass spectrometry and pyrolysis-gas chromatography/mass spectrometry. *Biomed. Mass Spectrom.* 11:486–492.
71. Faix, O., Deitrich, M., Grobe, I. 1987. Studies on isolated lignins and lignins in woody materials by pyrolysis-gas chromatography-mass spectrometry and off-line pyrolysis-gas chromatography with flame ionization detection. *J. Anal. Appl. Pyrolysis* 11:403–416.
72. Higashi, R.M., Fan, T.W.-M. 1997. Cumulative biochemical markers of air pollution in pine tree rings, *Abstract Book of the 18th Annual Meeting of the Society for Environmental Toxicology and Chemistry*, San Francisco, CA, November 1997. SETAC Press, Pensacola, FL, p. 221.
73. National Research Council. 1989. *Biologic Markers of Air-Pollution Stress and Damage in Forests*. National Academy Press, Washington, D.C.

# 10 DNA Fingerprinting as a Means to Identify Sources of Soil-Derived Dust: Problems and Potential

*Mary Ann Bruns and Kate M. Scow*

## CONTENTS

Abstract ........................................................................................................................ 193
Introduction .................................................................................................................. 194
Materials and Methods ................................................................................................. 194
    Collection of Bulk Dust and Source Soils ............................................................. 194
    DNA Extraction and Purification .......................................................................... 195
    Polymerase Chain Reaction ................................................................................... 195
    Thermal Gradient Gel Electrophoresis .................................................................. 196
Results and Discussion ................................................................................................ 196
    DNA Yields ............................................................................................................ 196
    DNA Fingerprints .................................................................................................. 198
Summary and Conclusions .......................................................................................... 202
Acknowledgments ....................................................................................................... 203
References ................................................................................................................... 203

## ABSTRACT

We are developing methods in our laboratory to produce molecular fingerprints from soil microorganisms as a means to differentiate soils and possibly identify sources of soil-derived fugitive dust. The methods described here are based on the extraction of microbial DNA from soil and dust samples which contain billions of microorganisms per gram. DNA fingerprinting involves the following: (1) soil sampling, (2) treating the soil to break open microbial cells, (3) extracting and purifying the DNA that is released from cells, (4) using polymerase chain reaction (PCR) to copy specific gene fragments from the DNA, and (5) generating a fingerprint from the gene fragments by electrophoretic separation in an analytical gel. The number and diversity of gene fragments in a fingerprint will depend on the specificity of DNA "primers" used in PCR to copy DNA. In this chapter we present our preliminary results on the extraction and analysis of microbial DNA from agricultural soils and bulk dusts from California's Central Valley. We describe experimental methods, technical difficulties observed, and basic assumptions that need to be considered in evaluating whether DNA fingerprinting can be applied to the problem of identifying sources of soil-derived fugitive dust.

**Keywords:** *DNA, fingerprinting, PM10, microbial communities*

## INTRODUCTION

Air quality, an important component of ecosystem health, is significantly affected by respirable particulate matter. PM10, or particulate matter with an aerodynamic diameter of 10 µm or less, is respirable and may consist of dust, soot, crystalline chemicals, microbial aerosols, and other fine materials.[1] High PM10 levels indicate that human health can be adversely affected by respiratory irritation, inflammation, or allergic reactions.[2] Throughout the state of California, PM10 levels often exceed the federal standard of 50 µm/m$^3$ of air over a 24-h period.[3] In California's Central Valley, airborne soil dust constitutes the greatest fraction of PM10 in late summer and early fall. The principal source(s) of this fugitive dust, which may be generated by agricultural field operations, construction activities, wind erosion, and traffic on unpaved roads,[4] has not been determined.

As a means to differentiate soils and possibly identify sources of soil-derived fugitive dust, we are developing methods in our laboratory to produce molecular fingerprints from microorganisms in soils. Molecular fingerprinting is based on the fact that soils are a habitat for bacteria, fungi, and other microorganisms,[5] all of which contain biochemical material which can be extracted and analyzed.[6] Our eventual goal is to use fingerprinting methods as tools to characterize and monitor changes in soil microbial communities, which are responsible for many biogeochemical transformations affecting environmental quality. Potential applications in soil microbial ecology include pollutant remediation, waste degradation, agricultural nutrient retention, plant protection, and reduction of trace gases contributing to global climate change. Molecular fingerprinting is potentially applicable to the characterization of soil microorganisms present in fugitive dust,[7] thus serving as a means to link airborne dust samples to the soil(s) of origin. In this chapter we describe how soil and dust samples can be analyzed with DNA fingerprinting methods. We also discuss some of the problems associated with this application for DNA fingerprinting based on insights from preliminary results. Our research currently focuses on characterization of microbes in soil-derived dust, rather than dust from other sources such as plant surfaces.[8]

## MATERIALS AND METHODS

### COLLECTION OF BULK DUST AND SOURCE SOILS

Bulk dust fall (total particulates) and source soils were collected at two locations. The first location was a fallow field at the University of California, Davis (UCD)

Campbell Tract which was being land planed. Dust was collected over a period of 1 h in pans taped to a horizontal bar located immediately above the land plane (1 m height). Surface soils were sampled volumetrically to a depth of 15 cm along a 20-m transect in an area represented by one soil map unit (Reiff loam). At a site in Kings County, CA, bulk dust was collected from the rear horizontal surfaces of a disker which had gone over a half section of land. A wheat crop had been harvested from this section two days before. Since this section contained two different soil map units, Westhaven loam and Kimberlina fine sandy loam, surface soils were sampled across separate transects in each map unit. Soil samples were also taken from an adjacent cotton field that lay within the Westhaven loam map unit. On the same day at this site, PM10 samples were collected on 25-mm Teflon filters with IMPROVE (Interagency Monitoring for Protection of Visual Environments) samplers, having a flow rate of 16.7 L/min and fitted with EPA-certified inlets (Sierra-Anderson). The PM10 filters had been archived at ambient temperatures for six months prior to analysis. Source soils were air dried, mixed by passage through a 2-mm sieve, and stored in cardboard cartons at ambient temperature.

## DNA Extraction and Purification

For soil and bulk dust samples, DNA was extracted from 1- to 5-g triplicate subsamples by a direct lysis procedure adapted from Zhou et al. (1995), which involved high salt/heat treatment. Subsequent steps to purify DNA from cellular debris and soil humic acids were chloroform extraction, precipitation of DNA with isopropanol, and agarose gel electrophoresis of crude DNA extracts. Portions of gel containing DNA were excised, and residual agarose was removed from the DNA by digestion with Gelase enzyme (Epicentre Technologies, Madison, WI), followed by washing and concentration of the DNA in a centrifugal filtration unit with a 100,000-Da molecular weight exclusion limit (Micron Separations, Inc., Westborough, MA). For PM10 samples, we modified the DNA extraction procedure to incorporate more severe physical treatments to ensure as much cell breakage as possible. These treatments consisted of (1) freezing in liquid nitrogen, alternating with heat shocks at 70°C (three times); (2) bead beating with sterile glass beads in a Vortex mixer (three times for 1 min each time); and (3) addition of chloroform prior to bead beating as an aid in cell breakage. After bead beating, the extraction was repeated to promote more exhaustive DNA recovery. For the modified method, we used smaller amounts of reagents in 2-cm$^3$ microcentrifuge tubes, into which we inserted the PM10 filters. DNA was quantified by measuring its absorbance at 260 nm in a Lambda 10 UV/Vis spectrophotometer (Perkin-Elmer Applied Biosystems, Foster City, CA). DNA purity was checked by measuring absorbances at 230 and 280 nm, which indicate contamination with humic acids and proteins, respectively. The resultant DNA was of sufficient purity for PCR amplification.

## Polymerase Chain Reaction

We used a GeneAmp 2400 thermal cycler (Perkin-Elmer Applied Biosystems, Foster City, CA) to amplify purified DNA with primers complementary to 16S ribosomal

RNA (rRNA) genes in bacteria.[6] PCR amplification was carried out in 25-μL reaction volumes containing the following: 5 ng of purified community DNA in 1X PCR buffer (500 m$M$ KCl, 100 m$M$ Tris-HCl, pH 9, 1% Triton X-100); 2.5 m$M$ MgCl$_2$; 1.25 m$M$ deoxyribonucleotides, 10 pmol each of the forward and reverse primers; and 1.5 units of Taq DNA polymerase (Promega Corp., Madison, WI). The sequence of the forward primer was 5′-CGC CCG CCG CGC GGC GGG CGG GGC GGG GGC ACG GGG GG <u>CCT ACG GGA GGC AGC AG</u>, the underlined portion of which corresponds to *Escherichia coli* positions 341–357.[9] The 5′ portion of the forward primer consisted of a "GC clamp," 38 nucleotides long,[10] the purpose of which is to improve the resolution of PCR fragments in denaturing gradient gel electrophoresis. The sequence of the reverse primer was 5′-CCC CGT CAA TTC CTT TGA GTT T, which corresponds to *E. coli* positions 907–928.[9] The PCR program consisted of 30 cycles of denaturing at 94°C for 1 min; primer annealing at 55°C for 1 min; and DNA extension at 72°C for 1 min. The approximate lengths of the resulting PCR fragments were 625 base pairs.

### THERMAL GRADIENT GEL ELECTROPHORESIS

A thermal gradient gel electrophoresis (TGGE) system[11] was constructed in our laboratory from a vertical electrophoresis rig fitted with an aluminum block, which was coated with a heat-conducting, compressible pad. Glass plates (42 cm high × 30 cm wide) containing a denaturing polyacrylamide gel were clamped tightly against the block assembly. The upper portion of the aluminum block contained an internal channel (8 mm diameter) through which water was pumped using a Neslab RTE-111 circulating water bath (Neslab Corp., Portsmouth, NH). The bottom portion of the block contained an electrical heating strip connected to a temperature controller (Model 1500, Dwyer Instruments Incorporated, Michigan City, IN). The temperatures of the water bath (for the cooled upper block) and the controller unit (for the hotter lower block) could be adjusted to provide a linear temperature gradient of 38 to 60°C down a portion of the gel. The gels contained 6% polyacrylamide (37.5:1 ratio of acrylamide to bisacrylamide), 7 $M$ urea, 20% formamide, 0.1% ammonium persulfate, and 50 μL TEMED per 100 mL of gel. The gel and running buffer consisted of 0.5X TAE. Electrophoresis was done at 10 V/cm for 16 h. Gels were stained either with ethidium bromide or with more sensitive silver reagents[12] to visualize the bands. Gel images were recorded using a charge-coupled-device camera with the BP-M1/722 TWAIN digital imaging kit (Bioimage, Ann Arbor, MI) and evaluated with Photofinish and GPTools image analysis software.

## RESULTS AND DISCUSSION

### DNA YIELDS

Before comparing DNA fingerprints from bulk dusts and their source soils, we evaluated their relative DNA yields. Determination of total DNA yield per gram dry weight sample is a quality control measure which can provide information on the relative amounts of microbial biomass on which fingerprints will be based. Total

## TABLE 1
## DNA Yield (in Micrograms) from Soil and Dust Samples

| Sampling Site | Soil Map Unit | DNA Yield (per gram dry soil) | | DNA Yield (per gram dust) | | DNA Yield (per gram PM10) |
|---|---|---|---|---|---|---|
| | | Range | Mean | Range | Mean | |
| Harvested wheat field Kings County, CA | Westhaven loam | 1.3–8.9 | 5.4 (±3.8) | 4.1–8.3 | 5.5 (±2.4) | 10 (±2) |
| | Kimberlina fine sandy loam | 1.5–5.9 | 4.2 (±1.5) | | | |
| Fallow UCD Tract Yolo County, CA | Reiff silt loam | 2.3–6.0 | 4.3 (±1.4) | 11.3–17.9 | 15.4 (±3.5) | ND |

yield determinations also serve as a check on DNA extraction efficiency. Considerable variation in DNA yield was observed among field replicates taken from the same transects in each map unit (Table 1). The nine samples from the Westhaven soil map unit, for example, had nearly a sevenfold range of magnitude in DNA yield. Similar variation in DNA yields among analytical replicates have been observed in other studies,[13] and we hypothesize that this variation reflects spatial variability in the distribution of microorganisms in soils and the patchiness of plant material contributions to soil organic matter. There was no significant difference in mean DNA yields from the Westhaven, Kimberlina, or Reiff soils.

DNA yields per gram dry weight of bulk dust samples tended to be higher than DNA yields from soils. Due to the high variability in soil DNA yields, only Reiff bulk dust showed a significantly higher DNA yield than its corresponding soil ($p < 0.01$). We expected DNA yields from bulk dust to be higher than from soil, because dust contains a greater proportion of lightweight organic matter and fine clays, which tend to be more tightly associated with soil microorganisms.[14] As expected, DNA yield per gram of PM10 material was even higher (Table 1), due in part to the fact that PM10 samples were subjected to more severe cell lysis procedures in the protocol used to extract DNA. Using these procedures, we recovered approximately 6 ng of DNA from Teflon filters containing approximately 600 μg of PM10, indicating that a considerable amount of DNA remained undegraded during the 6-month storage period. DNA absorbance ratios at 260/280 and 260/230 nm were similar for all samples (1.6 and 1.2, respectively).

These results indicate that spatial variability in the microorganisms inhabiting soils will impose difficulties in obtaining representative samples of "source soils" for fingerprint comparisons. Table 2 lists other factors which affect reliability and reproducibility of fingerprints generated from soil microbial DNA. Differences in soil properties, for example, will influence recovery efficiency and representativeness of the DNA "starting material" for fingerprinting, even when the same DNA extraction and purification methods are used for all samples. DNA yields tend to be lower from soils with higher clay contents (>35%) because clay particles can protect tightly

## TABLE 2
## Factors Affecting Reproducibility and Reliability of Soil Microbial Community DNA Fingerprints

| | |
|---|---|
| DNA extraction procedure | Incomplete cell lysis and low recovery may yield DNA that is not representative of the total community. |
| DNA purification procedure | Presence of contaminants (e.g., soil humic acids) may preclude reliable DNA quantitation and interfere with polymerase activity during PCR. |
| Concentration of template DNA in PCR reaction | Excess template DNA can result in production of nonspecific PCR products that introduce artifacts into the fingerprint. |
| PCR primer selection | General primers (complementary to DNA of all organisms in broad groups) may produce excessively complex mixtures that are difficult to separate. |
| Stringency (specificity) of PCR conditions | Low stringency conditions (lower annealing temperatures, higher magnesium concentrations) can result in more nonspecific PCR products. |
| Separation and resolution of PCR products | Incorrect gradients during gel electrophoresis may result in poor or no separation of PCR products. |

bound microbes from being lysed.[14,15] Although more severe physical lysis methods, such as bead beating, freeze thawing, or sonication, can be applied to high-clay soils to obtain DNA yields comparable to coarser soils, severe physical treatments can also cause DNA shearing and degradation.[13] Furthermore, cells of different microorganisms vary greatly in their susceptibilities to being broken open. Fungi and Gram-positive bacteria are more difficult to lyse than Gram-negative bacteria because their cell walls are resistant to chemical digestion.[15] A potential solution for comparing fingerprints from diverse soils, therefore, may be to extract DNA only from the fine fractions (<2 µm) using a consistent procedure for all samples.

### DNA FINGERPRINTS

DNA analysis has been used extensively in clinical medicine, forensics, and plant breeding to differentiate individual members of a species and evaluate genetic variation among species.[16] We considered three kinds of DNA analysis as possible means to fingerprint microbial DNA from soils and dusts. All three approaches involve PCR,[17] in which a minute amount of DNA (one billionth of a gram or less) provides the template for a polymerase enzyme to copy DNA fragments from "target" sites, generating large amounts of PCR products for analysis. The three approaches differ according to which target sites are copied during PCR and how PCR products are processed. Table 3 lists fingerprinting methods based on these three approaches: (1) copying DNA fragments from multiple sites on microbial chromosomes, followed by direct separation of DNA fragments by size; (2) copying DNA fragments from specific sites on chromosomes (i.e., single genes or operons), followed by cutting the fragments into smaller pieces and separating them by size; and (3) copying DNA fragments from specific sites on chromosomes, followed by direct separation of fragments by size or sequence.

## TABLE 3
## Microbial DNA Fingerprinting Approaches Based on PCR

| Fingerprinting Based on DNA from: | Examples and References | Relative Lengths of DNA Fragments (in base pairs) | PCR Products Separated by Gel Electrophoresis in: |
|---|---|---|---|
| Multiple sites on chromosome | RAPDs, or Randomly Amplified Polymorphic DNA (Hadrys et al., 1992) | 1000–10,000 | Agarose |
| | REP-PCR, or Repetitive Extragenic Palindrome-PCR (deBruijn, 1992) | | |
| Specific gene or operon, cut by restriction enzymes | ARDRA, or Amplified Ribosomal DNA Restriction Analysis (Massol-Deya et al., 1995) | 50–3000 | Agarose or high-resolution agarose |
| Ribosomal RNA operon, uncut | RISA, or Ribosomal Intergenic Spacer Analysis (Borneman and Triplett, 1997) | 500–2500 | High-resolution agarose or nondenaturing polyacrylamide |
| | DGGE/TGGE, or Denaturing/Thermal Gradient Gel Electrophoresis (Muyzer et al., 1993; Heuer and Smalla, 1997) | 200–600 | Denaturing polyacrylamide |

We consider the third approach listed in Table 3 to be most appropriate for fingerprinting complex mixtures of DNA from different microbial species. With this approach, PCR can be used to generate DNA fragments from the ribosomal RNA (rRNA) genes of many different microorganisms in a single reaction.[6] This is possible because rRNA genes, possessed by all organisms, contain short stretches of "universal" DNA, which exhibit very little sequence variation, interspersed with stretches that are more variable and sequence specific.[18] PCR primers complementary to universal regions can initiate copying of the variable parts of the gene.[19] DNA fragments of the same size but different sequence can be separated in denaturing polyacrylamide gels having either a chemical gradient, as in DGGE,[20] or a thermal gradient, as in TGGE,[21] By varying the degree of specificity of PCR primers for rRNA genes, fingerprints can be generated from either broad or narrow groups within the community. Microbial ecologists have successfully applied this approach in studying the composition of specific bacterial populations in mixed cultures,[22] rhizosphere samples,[23,24] and soils.[10] DGGE has also been used to analyze relatively simple microbial assemblages in biofilms,[20] marine and fresh waters,[25,26] and microbial mats from hot springs.[27] More recently, Heuer and Smalla[21] have described the use of DGGE/TGGE for study of whole soil microbial communities.

We used TGGE to generate all-bacterial rRNA gene fingerprints from Central Valley soil and dust samples. Figure 1a is a photograph of the TGGE fingerprints we obtained with DNA from soil and dust samples using primers that generate PCR products 625 base pairs in length. Figure 1b is a diagram showing positions of the bands in different lanes and arrows indicating several band locations. Although PCR bands were observed in the gel between 42 and 54°C, only the portion from 42 to 48°C is shown, due to band smearing at temperatures above 48°C. The poor resolution of bands in Figure 1a indicates that we had not yet optimized the temperature gradient and electrophoresis conditions for these mixtures of DNA. Thus, the extensive time and experience required to optimize TGGE conditions, among other technical problems listed in Table 2, will hamper routine application of this type of fingerprinting to the analysis of large numbers of diverse samples.

TGGE fingerprints from field replicates samples (i.e., samples taken from different locations within one soil map unit in a given field) are shown in Lanes 1 and 2 for Westhaven loam in cotton, Lanes 3 and 4 for Kimberlina fine sandy loam after wheat harvest, and Lanes 5 through 8 for Westhaven loam after wheat harvest (Figure 1). Under the TGGE conditions we used, fingerprints from replicate soil samples were identical and reproducible, despite the fact that DNA yields from replicate samples were quite variable (Table 1). Second, fingerprints of soil samples from the same field but different map units were similar, but not identical. For example, fingerprints from wheat field soils in Lanes 3 and 4 (Kimberlina fine sandy loam) had four out of eight bands in common with Westhaven loam soils in Lanes 5 through 8 (bands a, b, c, and d). Although one band was common to all samples (band b), fingerprints for soils and dust were quite dissimilar. We consider the latter observation to be due to the dust samples being enriched with microorganisms which are more tightly associated with lighter-weight clays and organic material.

DNA bands can be cut out of these gels to determine the DNA sequence(s) of the fragments localized in the bands. The DNA sequences can then be compared to

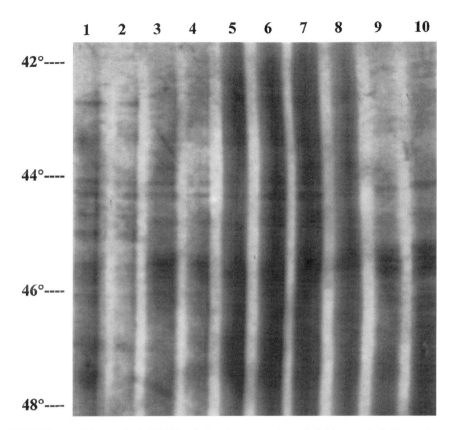

**FIGURE 1a**  Silver-stained TGGE gel showing separation of 625-base pair PCR products from microbial DNA extracted from source soils and bulk dust. Samples were as follows: Westhaven loam soil under cotton crop (Lanes 1 and 2); Kimberlina fine sandy loam soil after wheat harvest (Lanes 3 and 4); Westhaven loam soil after wheat harvest (Lanes 5 through 8); and bulk dust collected from disker following disking of harvested wheat field (Lanes 9 and 10).

other sequences in the Ribosomal Database Project, or RDP, a depository of hundreds of rRNA reference sequences accessible on the Internet for use in microbial identification.[28] The independent sequence data in the RDP help to check against experimental artifacts when analyzing uncharacterized sites. The RDP cannot support the fingerprinting methods listed under the first approach in Table 3 because PCR products in these methods are generated from many different and uncharacterized sites on the chromosome. Since slight differences in the species composition of mixtures could affect reproducibility of RAPD[29] and REP-PCR,[30] these methods are more appropriate for typing DNA from individual species than for fingerprinting DNA mixtures. The RDP would support methods listed under the second analysis approach in Table 3 when it targets the rRNA gene, as in ARDRA.[31] However, the PCR products in ARDRA must be cut into unique sets of smaller fragments by enzymes at sequence-specific sites before they produce any distinguishable pattern in agarose gels. ARDRA is more appropriate for characterizing isolates and simple

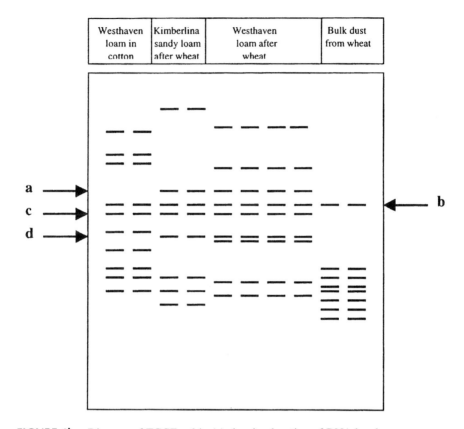

**FIGURE 1b** Diagram of TGGE gel in (a) showing location of DNA bands.

microbial assemblages, while RISA[32] and DGGE or TGGE[21] can be used for fingerprinting complex mixtures of microbial DNA.

## SUMMARY AND CONCLUSIONS

We have made the following observations about the feasibility of DNA fingerprinting to identify sources of soil-derived dust. First, DNA yields per gram of dry soil from the same field and soil map unit varied over a threefold range. Despite yield variability, DNA extracted from replicate soil samples gave identical and reproducible TGGE fingerprints. Fingerprints obtained from the same field but different soil map units were similar, but not identical, while fingerprints of source soils and bulk dust were quite dissimilar. A major obstacle to the application of TGGE fingerprinting to soil/dust analysis will be the time and experience needed to optimize electrophoresis conditions for large numbers of diverse samples. In addition, more information is needed on how spatial and temporal variability of soil microbial communities will affect the consistency and reproducibility of TGGE fingerprints.

Even if DNA fingerprinting is not found to be a feasible method for tracking PM10 sources, it should still provide a valuable research tool in studying soil microbial community changes which indicate environmental stress. Other colleagues in our laboratory are using fatty acid fingerprinting[33] to analyze the same dust and soil samples from this research project. If we can determine that the two approaches give similar results in differentiating soils, we will have good evidence that DNA fingerprinting is a valid and reliable method for characterizing soil and dust samples.

## ACKNOWLEDGMENTS

The authors would like to acknowledge the help of the following people in obtaining samples: Teresa James and Bob Matsumura of the Air Quality Group in the UCD Crocker Nuclear Laboratory, and Mike Mata of the UCD Agronomy Field Station. We would also like to acknowledge the laboratory assistance of Sandra Uesugi and the expertise of David Paige in designing and building the TGGE temperature gradient block.

For questions or further information regarding the work described in this chapter, please contact:

> Kate M. Scow
> Dept. of Land, Air and Water Resources
> University of California, Davis
> Davis, CA 95616-8627
> e-mail: kmscow@ucdavis.edu

## REFERENCES

1. Malm, W.C., J.F. Sisler, D. Hoffman, R.A. Eldred, and T.A. Cahill. 1994. Spatial and seasonal trends in particle concentration and optical extinction in the United States. *J. Geophys. Res.* 99:1347–1370.
2. Abbey, D.E., B.L. Hwang, R.J. Burchette, T. VanCuren, and P.K. Mills. 1995. Estimated long-term ambient concentrations of PM-10 and development of respiratory symptoms in a nonsmoking population. *Arch. Environ. Health* 50:139–152.
3. California Air Resources Board. Technical Support Division. 1986 to ongoing. *California Air Quality Data*, quarterly and annual summary reports, published by California Air Resources Board, Sacramento, CA.
4. Cunha, M., Jr. 1992. The role of agriculture in PM10 attainment in California. In J.C. Chow and D.M. Ono (eds.), *PM10 Standards and Nontraditional Particulate Source Controls*, Vol. 1, Air and Waste Management Association, Pittsburgh, PA, pp. 399–404.
5. Stotsky, G. 1997. Soil as an environment for microbial life. In J.D. van Elsas et al. (eds.), *Modern Soil Microbiology*, Marcel Dekker, Inc., New York, pp. 1–20.
6. Pace, N.R., D.A. Stahl, D.J. Lane, and G. Olsen. 1986. The analysis of natural microbial populations by ribosomal RNA sequences. *Adv. Microb. Ecol.* 9:1–55.
7. Kennedy, A.C. and A.J. Busacca. 1995. Microbial analysis to identify sources of PM-10 material. In Proceedings of the Air and Waste Management Association specialty conference on "Particulate Matter: Health and Regulatory Issues," Pittsburgh, PA, April 4–6, 1995, pp. 670–675.

8. Lighthart, B. 1997. The ecology of bacteria in the alfresco atmosphere. *FEMS Microbiol. Ecol.* 23:263–274.
9. Teske, A., C. Wawer, G. Muyzer, and N.B. Ramsing. 1996. Distribution of sulfate-reducing bacteria in a stratified fjord (Mariager Fjord, Denmark) as evaluated by most-probable-number counts and denaturing gradient gel electrophoresis of PCR-amplified ribosomal DNA fragments. *Appl. Environ. Microbiol.* 62:1405–1415.
10. Kowalchuk, G.A., J.R. Stephen, W. deBoer, J.I. Prosser, T.M. Embley, and J.W. Woldendorp. 1997. Analysis of ammonia-oxidizing bacteria of the β subdivision of the class *Proteobacteria* in coastal sand dunes by denaturing gradient gel electrophoresis and sequencing of PCR-amplified 16S ribosomal DNA fragments. *Appl. Environ. Microbiol.* 63:1489–1497.
11. Wartell, R.M., S.H. Hosseini, and J.D. Moran. 1990. Detecting base pair substitutions in DNA fragments by temperature-gradient gel electrophoresis. *Nucleic Acids Res.* 18:2699–2705.
12. Mitchell, L.G., A. Bodenteich, and C.R. Merril. 1994. Use of silver staining to detect nucleic acids. In A.J. Harwood (ed.), *Methods in Molecular Biology: Protocols for Gene Analysis*, Vol. 31, Humana Press, Inc., Totowa, NJ, pp. 197–203.
13. Zhou, J., M.A. Bruns, and J.M. Tiedje. 1995. DNA recovery from soils of diverse composition. *Appl. Environ. Microbiol.* 62:316–322.
14. Bakken, L.R. and R.A. Olsen. 1989. DNA content of soil bacteria of different cell size. *Soil Biol. Biochem.* 21:789–793.
15. Holben, W.E. 1994. Isolation and purification of bacterial DNA from soil. In R.W. Weaver et al. (eds.), *Methods of Soil Analysis Part 2 — Microbiological and Biochemical Properties*. Soil Science Society of America, Madison, WI, pp. 727–751.
16. Sensabaugh, G.F., D. Crim, and C. Von Beroldingen. 1991. The polymerase chain reaction: application to the analysis of biological evidence. In M.A. Farley and J.J. Harrington (eds.), *Forensic DNA Technology*, Lewis Publishers, Chelsea, MI, pp. 63–82.
17. Saiki, R.K., D.H. Gelfand, S. Stoffel, S.J. Scharf, R. Higuchi, G.T. Horn, K.M. Mullis, and H.A. Erlich. 1988. Primer-directed enzymatic amplification of DNA with thermostable DNA polymerase. *Science* 239:487–491.
18. Brosius, J., T.J. Dull, D.D. Sleeter, and H.F. Noller. 1981. Gene organization and primary structure of a ribosomal RNA operon from *Escherichia coli*. *J. Mol. Biol.* 148:107–127.
19. Lane, D.J. 1991. 16S/23S rRNA sequencing. In E. Stackebrandt and M. Goodfellow (eds.), *Nucleic Acid Techniques in Bacterial Systematics*, John Wiley & Sons, Ltd., New York, pp. 115–175.
20. Muyzer, G., E.C. De Waal, and A.G. Uitterlinden. 1993. Profiling of complex microbial populations by denaturing gradient gel electrophoresis analysis of polymerase chain reaction-amplified genes coding for 16S rRNA. *Appl. Environ. Microbiol.* 59:695–700.
21. Heuer, H. and K. Smalla. 1997. Application of denaturing gradient gel electrophoresis and temperature gradient gel electrophoresis for studying soil microbial communities. In J.D. van Elsas et al. (eds.), *Modern Soil Microbiology*, Marcel Dekker, Inc., New York, pp. 353–373.
22. Teske, A., P. Sigalevich, Y. Cohen, and G. Muyzer. 1996. Molecular identification of bacteria from a coculture by denaturing gradient gel electrophoresis of 16S ribosomal DNA fragments as a tool for isolation in pure cultures. *Appl. Environ. Microbiol.* 62:4210–4215.

23. Heuer, H., M. Krsek, P. Baker, K. Smalla, and E.M.H. Wellington. 1997. Analysis of actinomycete communities by specific amplification of genes encoding 16S rRNA and gel-electrophoretic separation in denaturing gradients. *Appl. Environ. Microbiol.* 63:3233–3241.
24. Kowalchuk, G.A., S. Gerards, and J.W. Woldendorp. 1997. Detection and characterization of fungal infections of Ammophila arenaria (marram grass) roots by denaturing gradient gel electrophoresis of specifically amplified 18S rDNA. *Appl. Environ. Microbiol.* 63:3858–3865.
25. Murray, A.E., J.T. Hollibaugh, and C. Orrego. 1996. Phylogenetic compositions of bacterioplankton from two California estuaries compared by denaturing gradient gel electrophoresis of 16S rDNA fragments. *Appl. Environ. Microbiol.* 62:2676–2680.
26. Ovreas, L., L. Forney, F.L. Daae, and V. Torsvik. 1997. Distribution of bacterioplankton in meromictic Lake Saelenvannet, as determined by denaturing gradient gel electrophoresis of PCR-amplified gene fragments coding for 16S rRNA. *Appl. Environ. Microbiol.* 63:3367–3373.
27. Ferris, M.J., G. Muyzer, and D.M. Ward. 1996. Denaturing gradient gel electrophoresis profiles of 16S rRNA-defined populations inhabiting a hot spring microbial mat community. *Appl. Environ. Microbiol.* 62:340–346.
28. Maidak, B.L., G. J. Olsen, N. Larsen, R. Overbeek, M.J. McCaughey, and C.R. Woese. 1997. The RDP (Ribosomal Database Project). *Nucleic Acids Res.* 25:109–111.
29. Hadrys, H., M. Balick, and B. Schierwater. 1992. Applications of random amplified polymorphic DNA (RAPD) in molecular ecology. *Mol. Ecol.* 1:55–63.
30. deBruijn, F.J. 1992. Use of repetitive (repetitive extragenic palindromic and enterobacterial repetitive intergeneric consensus) sequences and the polymerase chain reaction to fingerprint the genomes of *Rhizobium meliloti* isolates and other soil bacteria. *Appl. Environ. Microbiol.* 58:2180–2187.
31. Massol-Deya, A.A., D.A. Odelson, R.F. Hickey, and J.M. Tiedje. 1995. Bacterial community fingerprinting of amplified 16S and 16-23S ribosomal DNA gene sequences and restriction endonuclease analysis (ARDRA). In A.D.L. Akkermans et al. (eds.), *Molecular Microbial Ecology Manual*, Kluwer Academic Publishing, Dordrecht, The Netherlands, pp. 3.3.2/1–8.
32. Borneman, J. and E.W. Triplett. 1997. Molecular microbial diversity in soils from eastern Amazonia: evidence for unusual microorganisms and microbial population shifts associated with deforestation. *Appl. Environ. Microbiol.* 63:2647–2653.
33. Bossio, D.A., K.M. Scow, N. Gunapala, and K.J. Graham. In press. Management regime and seasonal effects on phospholipid fatty acid profiles of microbial communities from organic and conventional farming systems. *Microb. Ecol.*

# 11 Microbial Proteins as Biomarkers of Ecosystem Health

*Oladele A. Ogunseitan*

## CONTENTS

Abstract ................................................................................................................207
Introduction ..........................................................................................................208
Materials and Methods .........................................................................................209
　Environmental Samples ....................................................................................209
　Electrophoresed Protein Profiles ......................................................................212
　Extraction of Enzymatic Proteins from Water Samples ..................................214
Results and Discussion .........................................................................................215
Acknowledgment ..................................................................................................219
References ............................................................................................................220

## ABSTRACT

The microbial component of ecosystems represents the first tier of response to sustained systemic disturbance. Indices of microbial diversity and function are also relevant to the assessment of the potential for homeostatic ecosystem recovery. Research approaches based on molecular analysis of microbial diversity and physiology are critical for assessing ecosystem health indices at the fundamental level of structures and functions that are dependent on microorganisms such as bacteria, fungi, and viruses. This study provides two methods developed to analyze protein molecules extracted from natural aquatic microbial communities. The first method yields denatured protein molecules that can be resolved electrophoretically, and the second method yields protein molecules with preserved enzymatic activities. Electrophoretic resolution of proteins leads to the construction of profiles that provide information on population dynamics and ecological succession in response to specific disturbances. Direct analysis of microbial community enzymes provides qualitative and quantitative information on biogeochemical processes which sustain ecosystem health. In order to demonstrate the utility of these methods, aquatic samples collected from a pollution-prone estuarine ecosystem were processed to reveal variations in protein profiles and spectra of enzyme activities characteristic of microbial physiological responses to environmental factors.

**Keywords:** *microorganisms, proteins, enzymes, ecosystem, stress response*

## INTRODUCTION

Few investigators have attempted to produce an operational definition of ecosystem health because of the difficulties inherent in deriving internally consistent units of measurement for dynamic, hierarchical, multiscale systems. I present the following parametric model that is based on Costanza's homeostatic concept of ecosystem health as a set of indices that measure resilience, organization, and vigor according to the following equation.[1,2]

$$EH_i = G * V * O * R$$

where $EH_i$ refers to a system health index which is equivalent to a measure of sustainability. G, a new addition to the scheme, refers to the genetic potential of the system as determined by measures of intrinsic capacity for genetic innovation such as an integration of rates of spontaneous mutation within different hierarchical clusters. V is system vigor which is related to function and primary productivity with attributes that can be measured directly as kinetic rates of various metabolic activities. O refers to organization index on a scale of 0 to 1 scored through network analysis of species diversity and connectivity, including information exchange and consortia representation. R is resilience index, also on a scale of 0 to 1 scored through simulation modeling of values such as scope for growth and temporal scale of post-stress recovery.

Clearly, the parameters needed to produce a robust and predictive model of $EH_i$ are not all at the same level of scientific understanding, but advances in one segment are likely to lead to more intense activity in elucidating the contributions of other, less-developed factors. Quantitative analysis of microbial proteins that are useful as biomarkers of ecosystem function will ultimately contribute to better modeling of G, V, O, and R, but the current state of knowledge requires more finetuning of cause-and-effect relationships to support the integration of microbial diversity and functions in ecosystem health beyond the obvious contribution to *vigor* through recognized extensive metabolic capabilities. This chapter addresses some of the difficulties and challenges facing one approach toward improving the relevance of molecular precision and comprehensiveness in the assessment of ecosystem vigor at the fundamental level of microbial communities.

Systemic ecological problems result from overarching perturbations of populations and activities of consortia of species. Due to their small size, large numbers, and ubiquitous distribution in the environment, microbial species are particularly vulnerable to ecosystem disturbances. For the same reasons, microorganisms are also valuable indicators of the occurrence of disturbances attributable to exogenous physicochemical and biological stressors. At the organismic level, the mere presence of certain indicator bacteria and viruses can indicate sources of pollution into an aquatic environment,[3] but the molecular-level responses of autochthonous microorganisms to changes in ambient conditions are more critical for ecosystem health assessment. There is a wide array of molecules, including lipids, nucleic acids, and proteins, that is useful for diagnosing microbial responses to pollution and for monitoring environmental management strategies.[4-14] Protein molecules are particularly attractive because the initiation or cessation of their synthesis is often controlled by genetic-level responses to variable environmental conditions. For example, the presence of toxic chemicals in microbial ecosystems induces the synthesis of detoxifying or degradative enzymes and certain stress proteins.[15-19]

A conceptual framework of dose-response interactions between the species in a microbial community and toxic chemical pollutants is presented in Figure 1. Effects due to chemical toxicity on microbial population density are mediated by depletion of sensitive species and proliferation of resistant species that are capable of utilizing the excess chemicals as nutrients. Protein molecules mediate these effects by virtue of the ability of each species to synthesize degradative enzymes or otherwise engage in repair mechanisms through the activities of stress proteins and modified structural components. Monitoring these proteins provides information on toxic chemical *fates* (biodegradative enzymes), *effects* (toxicity-induced changes in protein profiles), and *risks* (mutations revealed by multilocus enzyme electrophoresis).

Several laboratory studies have identified specific proteins which are relevant to investigations of microbial responses to ecosystem perturbation (Table 1, and References 18 and 20 to 39). The research challenge is to translate such laboratory-based information to ecologically meaningful diagnostic tools for managing detrimental environmental conditions. Direct extraction and analysis of microbial community proteins are designed to contribute toward meeting this challenge (Figure 2). Radioisotope precursors of protein synthesis can be used in microcosm experiments to refine protein profiles and to pinpoint specific polypeptides useful for diagnosing microbial responses to specific environmental stimuli. Such experiments can be used to assess the readiness of microbial communities for remedial activities leading to the restoration of disturbed ecosystems.

## MATERIALS AND METHODS

### ENVIRONMENTAL SAMPLES

Water samples were obtained from an aquatic ecosystem composed of the San Diego Creek (SDC) which flows through the Upper Newport Bay Ecological Reserve (UNBER) and discharges into the Pacific Ocean (PO) at the Southern California Bight (Figure 3). The ecological reserve encompasses 752 acres of saltwater and freshwater marshland. The ecosystem harbors over 100 species of coastal and mudflat fish and more than 160 species of birds. The ecosystem is subject to perturbation from various sources, including surrounding urban development, migration of toxic chemical pollutants from groundwater beneath contaminated military bases, planned release of treated wastewater from the Irvine Ranch Water District facility, and agricultural runoff. The current research was initiated to determine whether systematic monitoring of microbial protein profiles and enzyme functions can be used to

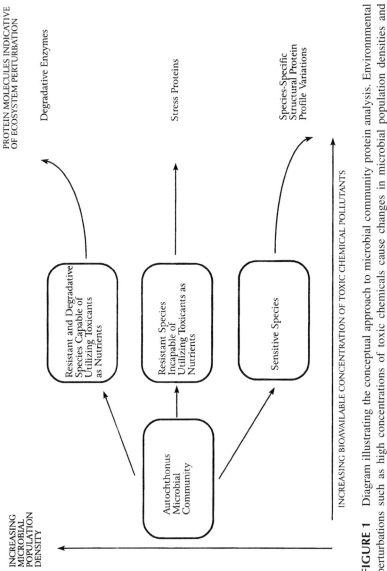

**FIGURE 1** Diagram illustrating the conceptual approach to microbial community protein analysis. Environmental perturbations such as high concentrations of toxic chemicals cause changes in microbial population densities and diversity. These population changes are reflected in the quality and quantity of protein molecules recoverable from the system. Resolution of such protein molecules produces distinct profiles and enzymatic activities that can be traced back to specific environmental conditions.

## TABLE 1
## Examples of Microbial Proteins Indicative of Ecosystem Health

| Ecosystem Condition | Organism | Protein Synthesized | Reference |
|---|---|---|---|
| | **Inducible Enzymes** | | |
| Mercury pollution | *Pseudomonas* sp. | Mercuric reductase and organomercurial lyase | 18, 20 |
| Polyaromatic hydrocarbon pollution | *Pseudomonas* sp. | Naphthalene dioxygenase | 21 |
| Chlorinated aliphatic hydrocarbon pollution | *Methanococcus* sp. | Methane monooxygenase | 22 |
| Methane production | *Methanobacterium* and *Methanosarcina* | Coenzyme F420 proteins | 23, 24 |
| Nitrogen cycling | *Pseudomonas* sp. and cyanobacteria | Nitrate reductase and nitrogenase | 25–27 |
| Phosphorus cycling | *Alcaligenes eutrophus* and cyanobacteria | Polyphosphate kinase and phosphatase | 28, 29 |
| Carbon cycling | *Phanerochaete chrysosporium* and phytoplankton | Lignin peroxidase, cellulase, and ribulose 1,5-bisphosphate carboxylase | 30 |
| | **Stress Proteins** | | |
| Hypersalinity | Bacillus subtilis | GroEL, DnaK, ClpP | 31 |
| Nutrient depletion | *Vibrio* sp. DW1 | 30 kDa protein | 32–36 |
| | *Bjerkandera* sp. BOS55 | Peroxidase | |
| | *Pycnoporus cinnabarinus* | Phenol oxidase | |
| Hyperthermal shift | *Halobacterium* sp. and *Escherichia coli* | Hsp 21-28, 75-105, GroE, DnaK | 37 |
| Oxygen tension | Bacteroides fragilis | Hsp 60, 90, 106 | 37, 38 |
| Ultraviolet light irradiation | *Phormidium laminosum* | Hsp33, 86, 89 | 37 |
| | *B. fragilis* | Hsp 95, 100 | |
| Virus infection | *E. coli* | DnaK, GrpE, UspA | 38 |
| Toxic chemical pollution | *E. coli* and *Photobacterium phosphoreum* | Proteins associated with loss of bioluminescence | 39 |

predict the responses of aquatic microbial communities in freshwater (SDC), estuarine (UNBER), and marine (PO) habitats to environmental pollution.

Duplicate 1-l water samples were collected from just beneath the surface at five locations within SDC, UNBER, and PO (Figure 3). The locations were selected to represent spatial variation in the quality of water samples from the least vulnerable (Site 1), toxic chemical pollution (Site 2), wastewater discharge (Site 3), agricultural and urban runoff (Site 4), and estuarine conditions (Site 5). At least three random point samples were collected from each site. To assess temporal variation in protein profiles from a single site, daily water samples were collected from Site 4 over a period of 2 weeks. Water samples were processed for protein extraction within 1 h of collection, and the remaining samples were kept at 4°C for not more than 24 h

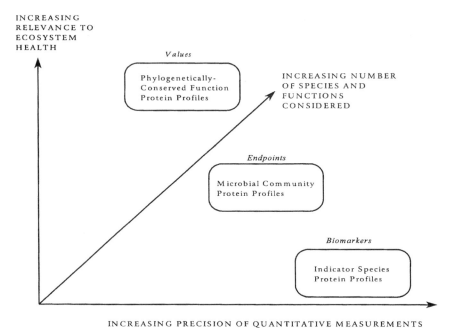

**FIGURE 2** The roles of biomarkers and community protein profiles toward the achievement of ecological health assessment. Development of precise indicators at the individual species level leads to generalizable assessment of phylogenetically conserved functions which are more directly associated with environmental management.

prior to the determination of microbial enumeration. Collection of microbial cells was routinely accomplished by centrifugation at 12,000 × $g$ for 10 min at 4°C. Protein extraction was performed according to the scheme presented in Figure 4. Preliminary experiments were performed to assess complementary physiological diversity in the microbial community by means of microbial identification strips (Biolog, Hayward, CA). The variability in response times and incubation period obtained with the Biolog technique limited the interpretation of the data, and the experiments were discontinued.

## ELECTROPHORESED PROTEIN PROFILES

Protein molecules were extracted directly from freshly collected water samples by the boiling method as previously described.[12] In order to increase the resolution of protein molecules and to facilitate the construction of profiles resulting from exposure to specific environmental pollutants, radiolabeled amino acids were incubated with 100-ml water samples in the presence of 0.1% (v/v) of benzene, toluene, xylene, and trichloroethylene prior to protein extraction.[15] Protein profiles were detected by silver staining of thin polyacrylamide gels and exposure of dried gels to X-ray films.[15]

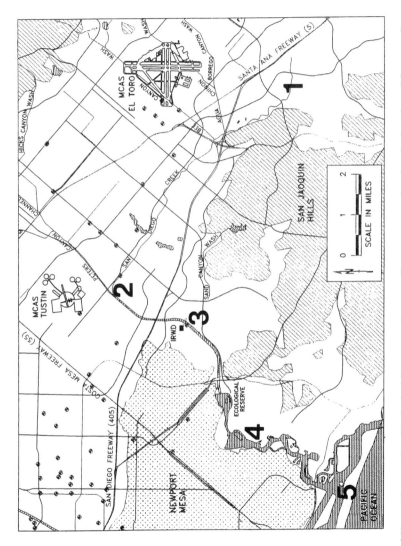

**FIGURE 3** Map of Upper Newport Bay Ecological Reserve and surrounding ecosystem showing the location of five sampling points. Water samples were collected from San Diego Creek (Sites 1, 2, and 3) and within the ecological reserve (Sites 4 and 5). Contaminants are introduced into the system from two polluted military bases (MCAS), urban runoff from shaded areas, and a wastewater treatment facility (IRWD). The partially filled, small circles represent the location of groundwater sampling wells.

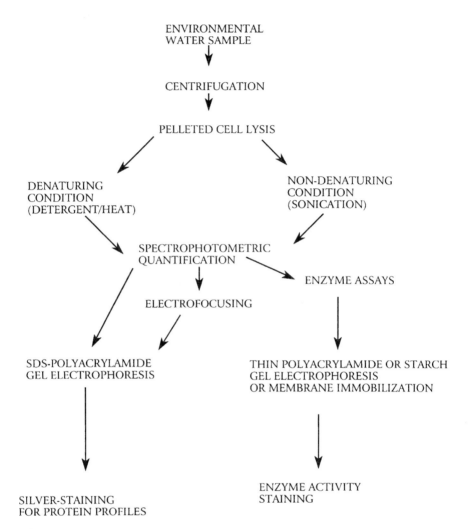

**FIGURE 4** Schematic diagram for processing water samples for recovery of proteins suitable for electrophoretic resolution and detection of enzyme activities.

### EXTRACTION OF ENZYMATIC PROTEINS FROM WATER SAMPLES

Protein molecules were released from aquatic microorganisms by sonication,[16-18] and crude extracts were immobilized on nitrocellulose membranes by means of a vacuum-blot apparatus (Schleicher & Schuell, Keen, NH). The membranes were stained to reveal specific enzyme activities by coupling electron transport requirements of catalysis to the production of colored dyes such as formazan.[41] The method has been used to detect numerous enzymes including peroxidases, xanthine dehydrogenase, catechol oxidase, and alkaline phosphatase.[16,17] A quantitative version of the technique has been used for direct analysis of mercuric reductase in aquatic environments.[41] Theoretically, the membrane method can be adapted to detect any enzyme activity that can be linked to redox reactions. Some enzymes (e.g., catechol oxidase) generate dark-stained products that are clearly visible on a white membrane background.

## RESULTS AND DISCUSSION

Quantitative assessment of ecosystem health requires unequivocal identification of relevant biomarkers, indicators, end points, and variable parameters useful for measuring perturbation and recovery (Figure 2). The present work is part of a comprehensive effort to develop molecular indicators of microbial response to toxic chemical pollutants and of changes in microbial diversity resulting from the introduction of exogenous species. Protein molecules were investigated as sensitive indicators of ecosystem status because of the dynamic functions these molecules perform in mitigating challenges to cellular physiology. The two methodological approaches presented here are (1) fingerprinting of environmentally responsive polypeptides in aquatic microbial communities exposed to toxic chemicals and (2) direct detection of enzymes involved in maintaining key biochemical functions at the microbial level. These two approaches cover the variety of microbial species found in aquatic samples and potentially circumvent confounding factors attributable to laboratory cultivation.

Variations in the profiles of protein molecules recovered from the five field sites are represented in Figure 5. The biological diversity of the samples corresponds to the variation observed in the molecular size and quantity of polypeptide bands. For example, the global pattern of protein profiles differed markedly among samples collected from spatially divergent sites (Figure 5A, Lanes a to j). The obvious interpretation of the data is that the diversity of microorganisms and their physiological status reflected in the protein profiles are characteristic of the prevalent environmental conditions at each site. This view is supported by data from a single site where water was collected on consecutive days (Figure 5A, Lanes k to t). In that experiment, the protein profiles generally remained within a recognizable pattern, with minor variation detected only between molecular size ranges 14 to 31 kDa. However, such simplistic interpretations are likely to be difficult to integrate into predictive models of ecological response because of the limited resolution of polypeptide bands and the extremely large number of variable environmental parameters associated with each site, and indeed each replicate sample. Subsequent radio-isotope-incorporation experiments were performed in order to enhance the discriminant function of protein profile analysis because the timeframe of analysis would presumably be more manageable.

The results of experiments using radioactive precursors of protein synthesis to trace the response of natural microbial communities to chemical stress are shown in Figures 5B and C. Figure 5B shows an autoradiograph of the dehydrated electrophoresis gel used to resolve radiolabeled polypeptides extracted from water samples collected from the five sites. The lane contents of Figure 5B correspond to those of Figure 5A. Clearly, several as yet uncharacterized factors led to the differences in the extent of protein radiolabeling observed for samples collected from different sites (Lanes a through j) and samples collected from the same site at different time

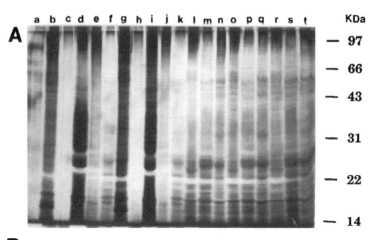
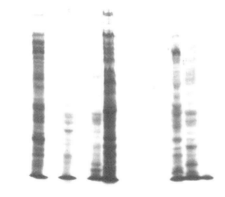
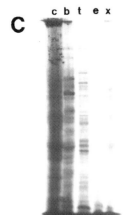

FIGURE 5

**FIGURE 5** (A) Silver-stained polyacrylamide gel used to resolve protein molecules extracted from nutrient-amended environmental water samples. Profiles illustrate both two samples each from Sites 1 through 5 (Figure 3), representing biologically diverse locations (Lanes a through j) and minor temporal variations from the biological uniform location of Site 4 (Lanes k through t). (B) Autoradiograph of electrophoretic gel showing assimilation of organic $^{35}$S (supplied as cysteine and methionine) into proteins by aquatic microorganisms in samples recovered from Sites 1 through 5. Lanes correspond to those shown in (A). Radioisotope incorporation varied according to both spatial and temporal origins of the water samples. (C) Autoradiograph showing variations in radiolabeled proteins synthesized after 60-min exposure of aquatic samples from Site 2 to toxic environmental pollutants. Lanes: c = control; b = benzene; t = toluene; e = trichloroethylene; x = xylene.

points (Lanes k through t). For example, in preliminary experiments, some samples incubated only with inorganic sources of radioactive sulfate did not show signals. This particular observation provides information about the rate and path of metabolic activities within the microbial ecosystem because prokaryotes and eukaryotes may utilize different precursors for *de novo* synthesis of proteins. In general, these results confirm the difficulty inherent in simple interpretations of microbial community protein profiles, but the challenge is to elucidate the contribution of various environmental factors to such differences in protein synthesis. An approach currently being pursued is to reconstruct the microbial community protein profiles by studying the profiles of individual isolates challenged with specific environmental stressors. However, this approach is limited by the fact that only a small proportion of microorganisms in a natural microbial community can be cultivated in the laboratory.

In order to further fine tune cause-and-effect responses for interpretation or microbial community protein profiles, water samples collected from Site 2 (the site most vulnerable to chemical pollution) were amended with petrochemicals (0.1% (v/v) benzene (b), toluene (t), trichloroethylene (e), or xylene (x)). The corresponding data from this experiment are shown in Figure 5C. When these chemical-amended microcosms are compared to the control experiment (c), it becomes apparent that (1) the chemicals caused a general decrease in protein synthesis and (2) the chemicals individually caused the synthesis of specific polypeptide bands that are not synthesized in the control samples. Although the specific identity of the polypeptides has not been ascertained, these key molecules may be induced enzymes or stress proteins that are synthesized in the presence of the toxic chemicals. Further studies must be conducted to demonstrate that such protein profiles can indeed be employed in diagnosing the bioavailability of chemical contaminants in polluted environments.[15]

Several investigators have attempted using qualitative detection of enzyme activities for assessing ecosystem function.[42-44] The spectra of enzyme activities detected in protein molecules extracted from water samples collected from Sites 1, 2, and 3 are represented in Figure 6. Immobilization of enzymes on membranes for colorimetric detection of catalytic activity facilitates quantitative and comparative assessment of inducible enzymes, potentially yielding information on chemical bioavailability and microbial diversity[16,17] (Figure 6). The six enzymes included in this study are all relevant to ecosystem functioning, and the three sites from which water samples are collected represent regions where microbial enzyme function would

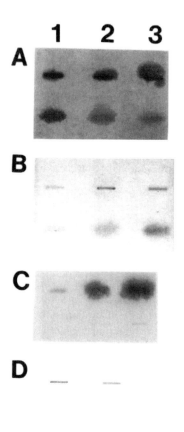

FIGURE 6

**FIGURE 6** Nitrocellulose membranes stained to reveal variations in enzyme activities of proteins extracted from aquatic samples collected on two occasions from Sites 1, 2, and 3 (identified in Figure 3), represented by columns 1, 2, and 3, respectively. A = alkaline phosphate; B = catechol oxidase; C = peroxidase; D = xanthine dehydrogenase; E = nitrate reductase; F = cellulase. The stain intensity corresponds to the activity of enzymes, reflecting the level of induction determined by the presence of substrates and population density of enzyme-producing organisms.

provide information on the intrinsic vigor of the freshwater system for natural remediation. For example, peroxidases, some of which are associated with cytochrome P450 family of proteins, have been implicated in the transformation of various environmental pollutants,[45] and these enzymes would be important in dealing with chemical pollutants from the contaminated military bases (Site 2). The variability observed in peroxidase activities at different sampling points could indicate the effects of different bioavailability factors influencing the distribution of specific toxicants in the ecosystem. Similarly, nitrate reductase activity can be used to assess the capacity of an aquatic ecosystem to avoid eutrophication which could otherwise result from nutrient loading from wastewater effluents (Site 3 in Figure 3 and Figure 6E). Further microcosm experiments are planned to test the effects of various concentrations of environmental chemicals on enzyme activities. The results of those experiments should provide quantitative data needed to support current multi-media and multipathway models used to assess chemical fates and the potential for *in situ* bioremediation in different compartments of a contaminated ecosystem.

In addition to facilitating closer understanding of the association between chemical pollution and microbiological activity, the recovery and immunological detection of specific microbial proteins can aid in surveillance studies of the transport and fate of viruses, such as the Norwalk agent, in extensive watersheds that receive treated wastewater. This approach is currently being employed in our laboratory to characterize the stability of recombinant Norwalk virus-like particles lacking nucleic acids in microcosms simulating groundwater recharge in the Southern California water basin.

## ACKNOWLEDGMENT

This work is funded in part by a grant (DEB 95-24481) from the National Science Foundation and Environmental Protection Agency partnership for environmental research.

For questions or further information regarding the work described in this chapter, please contact:

> **Oladele A. Ogunseitan**
> Dept. of Environmental Analysis & Design
> University of California
> Irvine, CA 92697-7070
> e-mail: oaogunse@uci.edu

## REFERENCES

1. Costanza, R. 1992. Toward an operational definition of ecosystem health. In R. Costanza, B.G. Norton, and B.D. Haskell, eds. *Ecosystem Health*. Island Press, Washington, D.C., pp. 239–256.
2. Haskell, B.H., B.G. Norton, and R. Costanza. 1992. What is ecosystem health and why should we worry about it? In R. Costanza, B.G. Norton, and B.D. Haskell, eds. *Ecosystem Health*. Island Press, Washington, D.C., pp. 3–20.
3. Metcalf, T.G., J.L. Melnick, and M.K. Estes. 1995. Environmental virology: from detection of virus in sewage and water by isolation to identification by molecular biology — a trip of over 50 years. *Annu. Rev. Microbiol.* 49: 461–487.
4. Sayler, G.S., M.S. Shields, E.T. Tedford, A. Breen, S.W. Hooper, K.M. Sirotkin, and J.W. Davis. 1985. Application of DNA-DNA colony hybridization to the detection of catabolic genotypes in environmental samples. *Appl. Environ. Microbiol.* 49: 1295–1303.
5. Stahl, D.A., D.L. Lane, G.J. Olsen, and N.R. Pace. 1985. Characterization of a Yellowstone hot spring microbial community by 5S ribosomal RNA sequences. *Appl. Environ. Microbiol.* 49: 1379–1384.
6. Selander, R.K., D.A. Caugant, H. Ochman, J.M. Musser, M.N. Gilmour, and T.S. Whittam. 1986. Methods of multilocus enzyme electrophoresis for bacterial population genetics and systematics. *Appl. Environ. Microbiol.* 51: 873–884.
7. Ogram, A., G.S. Sayler, and T. Barkay. 1987. The extraction and purification of microbial DNA from sediments. *J. Microbiol. Methods* 7: 57–66.
8. Steffan, R.J. and R.M. Atlas. 1988. DNA amplification to enhance detection of genetically engineered bacteria in environmental samples. *Appl. Environ. Microbiol.* 54: 2185–2191.
9. Ogunseitan, O.A., G.S. Sayler, and R.V. Miller. 1992. Application of DNA probes to analysis of bacteriophage distribution patterns in the environment. *Appl. Environ. Microbiol.* 58: 2046–2052.
10. Tsai, Y.-L., M. Park, and B.H. Olson. 1992. Rapid method for direct extraction of mRNA from seeded soils. *Appl. Environ. Microbiol.* 57: 765–768.
11. Wright, S.F. 1992. Immunological techniques for detection, identification, and enumeration of microorganisms in the environment. In M.A. Levin, R.J. Seidler, and M. Rogul, eds. *Microbial Ecology: Principles, Methods, and Applications*. McGraw Hill, New York, pp. 45–63.
12. Ogunseitan, O.A. 1993. Direct extraction of proteins from environmental samples. *J. Microbiol. Methods* 17: 273–281.
13. Ogunseitan, O.A. 1994. Biochemical, genetic, and ecological approaches to solving problems during in situ and off-site bioremediation. In D.L. Wise and D.J. Trantolo, eds. *Process Engineering for Pollution Control and Waste Minimization*. Marcel-Dekker, New York, pp. 171–192.
14. Ogunseitan, O.A. 1996a. Analytical prerequisites for environmental bioremediation. *Environ. Test. Anal.* 5: 36–40.
15. Ogunseitan, O.A. 1996b. Protein profile variation in cultivated and native freshwater microorganisms exposed to chemical environmental pollutants. *Microb. Ecol.* 31: 291–304.
16. Ogunseitan, O.A. 1997a. Extraction of proteins from aquatic environments. In A.D.L. Akkermans, J.D. van Elsas, and F.J. De Bruijn, eds. *Molecular Microbial Ecology Manual* 4.1.6. Kluwer Academic Publishers, The Netherlands, pp. 1–12.

17. Ogunseitan, O.A. 1997b. Direct extraction of catalytic proteins from natural microbial communities. *J. Microbiol. Methods* 28: 55–63.
18. Ogunseitan, O.A. 1998a. Protein method for investigating mercuric reductase gene expression in aquatic environments. *Appl. Environ. Microbiol.* 64: 695–702.
19. Ogunseitan, O.A. 1998b. Protein profile analysis for investigating genetic functions in microbial communities. In K. Cooksey, ed. *Molecular Approaches to the Study of the Ocean.* Chapman & Hall, London, Chapter 6.
20. Ogunseitan, O.A. and B.H. Olson. 1991. Potential for genetic enhancement of bacterial detoxification of mercury waste. In R. Smith and T. Mishra, eds. *Mineral Bioprocessing.* Engineering Foundation, New York, pp. 325–337.
21. Ogunseitan, O.A. and B.H. Olson. 1993. Effect of 2-hydroxybenzoate on the rate of naphthalene mineralization in soil. *Appl. Microbiol. Biotechnol.* 38: 799–807.
22. Bowman, J.P. and G.S. Sayler. 1994. Optimization and maintenance of soluble methane monooxygenase activity in methylosinus trichosporium OB3b. *Biodegradation* 5: 1–11.
23. Heine-Dobbernack, E., S.M. Schoberth, and H. Sahm. 1988. Relationship of intracellular coenzyme F420 content to growth and metabolic activity of *Methanobacterium bryanti* and *Methanosarcina barkeri. Appl. Environ. Microbiol.* 54: 454–459.
24. Moura, I., J.J.G. Moura, H. Santos, A.V. Xavier, G. Burch, H.D. Peck, and J. LeGall. 1983. Proteins containing factor F420 from *Methanosarcina barkeri* and *Methanobacterium thermoautotrophicum*: isolation and properties. *Biochim. Biophys. Acta* 742: 84–90.
25. Paerl, H.W., J.C. Priscu, and D.L. Brawner. 1989. Immunochemical localization of nitrogenase in marine *Trichodesmium* aggregates: relationship to nitrogen fixation potential. *Appl. Environ. Microbiol.* 55: 2965–2975.
26. Ward, B.B. and A.F. Carlucci. 1985. Marine ammonia and nitrite oxidizing bacteria: serological diversity determined by immunofluorescence in culture and in the environment. *Appl. Environ. Microbiol.* 50: 194–201.
27. Zehr, J.P., R.J. Limberger, K. Ohki, and Y. Fujita. 1990. Antiserum to nitrogenase generated from an amplified DNA fragment from natural populations of *Trichodesmium* spp. *Appl. Environ. Microbiol.* 56: 3527–3531.
28. Whitton, B.A. 1991. Use of phosphatase assays with algae to assess phosphorus status of aquatic environments. In D.W. Jefferey and B. Madden, eds. *Bioindicators and Environmental Management.* Academic Press, New York, pp. 295–310.
29. Chrost, R.J. and J. Overbeck. 1987. Kinetics of alkaline phosphatase activity and phosphorus availability for phytoplankton and bacterioplankton in Lake Plussee (North German eutrophic lake). *Microb. Ecol.* 13: 229–248.
30. Orellana, M.V., M.J. Perry, and B.A. Watson. 1988. Probes for assessing single cell primary production: antibodies against ribulose 1,5 bisphosphate carboxylase (RuBPCASE) and peridinin/chlorophyll *a* (PCP). In C.M. Yentsch, C. Mague, and P.K. Horan, eds. *Immunochemical Approaches to Coastal, Estuarine, and Oceanographic Questions. Lecture Notes on Coastal and Estuarine Studies.* Vol. 25. Springer-Verlag, New York, pp. 243–262.
31. Volker, U., H. Mach, R. Schmid, and M. Hecker. 1992. Stress proteins and crossprotection by heat shock and salt stress in *Bacillus subtilis. J. Gen. Microbiol.* 138: 2125–2135.
32. Jaan, A.J., B. Dahlof, and S. Kjelleberg. 1986. Changes in protein composition of three bacterial isolates from marine waters during short periods of energy and nutrient deprivation. *Appl. Environ. Microbiol.* 52: 1419–1421.
33. Watson, K. 1990. Microbial stress proteins. *Adv. Microb. Physiol.* 31: 183–216.

34. Kaal, E.E.J., E. De Jong, and J.A. Field. 1993. Stimulation of lignolytic peroxidase activity by nitrogen nutrients in the white rot fungus Bjerkandera sp. Strain BOS55. *Appl. Environ. Microbiol.* 59: 4031–4036.
35. Akamatsu, Y. and M. Shimada. 1996. Suppressive effect of L-phenylalanine on manganese peroxidase in the white-rot fungus Phanerochaete chrysosporium. *FEMS Microbiol. Lett.* 145: 83–86.
36. Jones, C.L. and G.T. Lonergan. 1997. Prediction of phenol oxidase expression in a fungus using the fractal dimension. *Biotechnol. Lett.* 19: 65–69.
37. Bukau, B., C.E. Donelly, and G.C. Walker. 1989. *DnaK* and *GroE* proteins play roles in *E. coli* metabolism at low and intermediate temperatures as well as high temperatures. In M.L. Pardue, J.R. Feramisco, and S. Lindquist, eds. *Stress-Induced Proteins*. Alan R. Liss, New York, pp. 27–36.
38. Ang, D., G.N. Chandrasekar, M. Zylicz, and C. Georgopoulos. 1986. *Escherichia coli grpE* gene codes for heat shock protein B25.3 essential for both lambda DNA replication at all temperatures and host growth at high temperature. *J. Bacteriol.* 167: 25–29.
39. Blom, A., W. Harder, and A. Matin. 1992. Unique and overlapping pollutant stress proteins of *Escherichia coli. Appl. Environ. Microbiol.* 58: 331–334.
40. Bulich, A.A. 1982. A practical and reliable method for monitoring the toxicity of aquatic samples. *Process Biochem.* 17: 45–47.
41. Manchenko, G.P. 1994. *Handbook of Detection of Enzymes on Electrophoretic Gels*. CRC Press, Boca Raton, FL.
42. Burns, R.G. 1978. *Soil Enzymes*. Academic Press, London, 380 pp.
43. Chrost, R.J. 1990. Microbial ectoenzymes in aquatic environments. In J. Overbeck and R.J. Chrost, eds. *Aquatic Microbial Ecology*. Springer-Verlag, New York, pp. 47–78.
44. Dermer, O.C., V.S. Curtis, and F.R. Leach. 1980. *Biochemical Indicators of Subsurface Pollution*. Ann Arbor Science, Ann Arbor, MI, 203 pp.
45. Thomas, P.E., D. Ryan, and W. Levin. 1976. An improved staining procedure for the detection of the peroxidase activity of cytochrome P-450 on sodium dodecyl sulfate polyacrylamide gels. *Anal. Biochem.* 75: 168–176.

# 12 Application of a Random Amplified Polymorphic DNA (RAPD) Method for Characterization of Microbial Communities in Agricultural Soils

*Padma Sudarshana, Jessica R. Hanson, and Kate M. Scow*

## CONTENTS

Abstract ..................................................................................................................223
Introduction ............................................................................................................224
Materials and Methods ..........................................................................................224
    DNA Extraction and Purification ...................................................................225
    Polymerase Chain Reaction Amplification ....................................................225
Results and Discussion .........................................................................................226
    DNA Isolation ................................................................................................226
    Comparison of Random Amplified Polymorphic DNA Fingerprints .............227
Acknowledgments .................................................................................................230
References .............................................................................................................230

## ABSTRACT

Effective monitoring of microbial populations in environmental samples requires highly sensitive analytical methods. Studies of microbial communities using traditional techniques are incomplete because of their inherent inability to identify and quantify all types of contributing populations. In this chapter we have used molecular genetic approaches to compare the responses of microbial communities to organic and conventional management practices. Soil DNA was isolated by the direct lysis method and subjected to random amplified polymorphic DNA (RAPD) analysis. Soil DNA yields ranged from 2.44 to 4.56 µg/g in conventional and 4.08 to 6.76 µg/g in organic samples. RAPD patterns generated were characteristic for organic and conventional soil types. Although some intermediate banding patterns were observed between the two soil types, major bands were distinguishable. The number of RAPD bands varied in both organic and conventional soil samples collected at 15-d intervals from April to July, which may indicate shifts in microbial populations. Thus, the RAPD technique has potential to detect changes in microbial population. However, the method requires processing of large number of replicate soil samples to generate reproducible polymorphic patterns.

**Keywords:** *soil DNA, RAPD, microbial diversity, soil management*

## INTRODUCTION

A variety of bacterial and fungal species are responsible for degradation of organic wastes, humification of organic complex polymers, and destruction of animal and plant pathogens. Analysis of microbial communities is important to understand the potential roles played by microorganisms in environmental processes and also to optimize the conditions for the growth and activity of indigenous microorganisms. Traditional methods for characterization of microbial communities require cultivation of microorganisms in the laboratory, and only a small percentage of the indigenous microorganisms can be isolated *in vitro*.[12] Methods based on microbial physiology such as phospholipid fatty acid analysis and BIOLOG have been shown to be useful for microbial community analysis.[1] These techniques indicate phenotypic changes in the microbial community structure, but cannot be used to measure genotypic changes. Also, BIOLOG is a culture-based method, and microbial fatty acid composition changes with environmental conditions.[13] Molecular techniques do not require culturing of microorganisms and may provide a less biased analysis of microbial communities.[7,11] In this chapter, we have adopted the random amplified polymorphic DNA technique (RAPD) to compare the responses of soil microbial population to organic and conventional management practices. The RAPD is a polymerase chain reaction (PCR) based technique that involves amplification of segments of target DNA using short oligonucleotide (random) primers. Amplified products are separated electrophoretically on an agarose gel to generate a banding pattern characteristic to the target DNA.[14,15] Banding patterns for DNA isolated from different samples are compared to calculate the similarity percentage.[4] The conventional soil management system involves application of pesticides and mineral fertilizers, whereas organic management relies on organic inputs such as cover crop and manure.[10] The objectives of this study were (1) to determine whether management practices and seasonal variations caused changes in soil microbial community and (2) to evaluate the sensitivity of the RAPD technique to detect changes in microbial community.

## MATERIALS AND METHODS

This study is part of the Sustainable Agriculture Farming Systems (SAFS) Project which is described in detail in several publications.[3,10] The experiment has a randomized, complete block split-plot design, with crop rotations as split plots within each farming system main plot, with four replications. The farming systems sampled for this study were organic and conventional 4-year rotations. The organic system relies on organic sources of nutrients obtained from a vetch winter cover crop, manure, seaweed, and fish powder. No pesticides are used, and the plots are managed according to California Certified Farmers requirements (California Certified Organic Farmers Inc., Santa Cruz, CA). Soils samples were collected from the tomato plots within each farming system. Soil sampling methods are described in detail elsewhere.[3] Thirty, 2.5-cm diameter soil cores taken randomly from each plot were pooled to obtain a representative surface soil (0 to 15 cm) sample. Samples were taken while cover crops were growing in the organic plot (April 4), after cover crop incorporation and manure application (April 18), 1 week prior to (May 9) and 1 week after (May 23) chemical fertilizer side dressing in a conventional system and twice more in the growing season (July 3 and July 28). Average soil temperatures measured at a depth of 15 cm were 14.8, 19.7, 21.1, and 25.2°C in April, May, June, and July months, respectively. Soil samples were air dried and stored at –20°C for DNA analysis.

## DNA Extraction and Purification

Soil DNA was extracted using the direct lysis method as described by Malik et al.[6] with a few modifications. One-gram soil samples were washed three times with phosphate buffer. Soil was suspended in 8 ml of solution containing 0.15 $M$ NaCl, 0.1 $M$ EDTA, and Lysozyme (10 mg/ml) and incubated at 37°C for 2 h. To this mixture, 8 ml of lysis solution containing 0.1 $M$ NaCl, 0.5 $M$ Tris HCl, and 10% SDS was added and mixed. Samples were subjected to freezing and thawing, repeated three times, and centrifuged. Supernatants were mixed with 2.7 ml of 5 $M$ NaCl and 2.1 ml of 10% cetyl trimethyl ammonium bromide (CTAB) and incubated at 65°C for 10 min. Later, an equal quantity of $CHCl_3$/isoamyl alcohol (24:1) was added and centrifuged. The resulted upper phase was mixed with 13% polyethylene glycol in 1.6 $M$ NaCl and incubated on ice for 90 min. The suspension was centrifuged, and the supernatant was discarded. The pellet was washed with 70% ethanol and resuspended in Tris-EDTA (TE). The sample was mixed with 1/10 volume of 10 $M$ ammonium acetate and incubated on ice for 2 h and centrifuged. The supernatant was collected and DNA was precipitated by adding two volumes of ethanol. The pellet was washed with 70% ethanol, dried, and resuspended in 50 µl of TE. DNA was further purified by electrophoresis through 0.75% low melting agarose gels containing 0.2% polyvinyl-pyrrolidone. Agarose gel slices containing DNA were excised and purified using prep-a-gene DNA purification kit (Biorad, Hercules, CA) following the methods suggested by the manufacturer. Purified DNA yielded an absorbance ratio of 1.7 (260/280 nm), and DNA concentration was measured at $A_{260}$.

## Polymerase Chain Reaction Amplification

For RAPD fingerprinting, PCR were conducted in a total volume of 25 µl containing 1X PCR buffer, 1.25 m$M$ $MgCl_2$, 2.5 units of Taq polymerase (Promega Inc., Madison, WI), 100 µ$M$ each of deoxy nucleoside triphosphates, 40 ng bovine serum albumin (BSA), and 20 pmol of decanucleotide primer (Table 1). The template

### TABLE 1
### Primers Used in This Study

| Primer | Sequence | Primer | Sequence |
|---|---|---|---|
| AA1 | AGACGGCTCC | AA11 | ACCCGACCTG |
| AA2 | GAGACCAGAC | AA12 | GGACCTCTTG |
| AA3 | TTAGCGCCCC | AA13 | GAGCGTCGCT |
| AA4 | AGGACTGCTC | AA14 | AACGGGCCAA |
| AA5 | GGCTTTAGCC | AA15 | ACGGAAGCCC |
| AA6 | GTGGGTGCCA | AA16 | GGAACCCACA |
| AA7 | CTACGCTCAC | AA17 | GAGCCCGACT |
| AA8 | TCCGCAGTAG | AA18 | TGGTCCAGCC |
| AA9 | AGATGGGCAG | AA19 | TGAGGCGTGT |
| AA10 | TGGTCGGGTG | AA20 | TTGCCTTCGG |

### TABLE 2
### DNA Yields From Soil Samples (mg/g ± SEM)

| Sample Date | Organic | Conventional |
|---|---|---|
| 4/4/95 | 6.76 ± 0.75 | 2.44 ± 0.70 |
| 4/18/95 | 6.43 ± 1.40 | 4.56 ± 0.90 |
| 5/9/95 | 3.94 ± 0.32 | 3.06 ± 0.40 |
| 5/23/95 | 4.00 ± 0.47 | 3.74 ± 0.73 |
| 6/13/95 | 4.34 ± 0.48 | 3.83 ± 0.34 |
| 7/3/95 | 4.09 ± 0.61 | 2.79 ± 0.50 |
| 7/23/95 | 3.98 ± 0.32 | 4.18 ± 0.22 |

concentration was 5 to 10 ng per reaction. An initial denaturation step was carried out at 94°C for 2 min followed by 40 cycles of 92°C for 1 min, 38°C for 1 min, and extension at 72°C for 2 min in a Perkin-Elmer Cetus Model 2400 thermocycler (Perkin Elmer, Norwalk, CT). At the end of the cycles, a final extension time period of 7 min at 72°C was carried out. Experiments were repeated at least two times to get reproducible fingerprint patterns. The products of PCR amplification were resolved by agarose gel electrophoresis (2% in Tris borate buffer).

## RESULTS AND DISCUSSION

### DNA Isolation

Isolated soil DNA was greater than 23 kb in size, as indicated by its migration relative to molecular weight size standards. Average DNA yields from samples collected at all sampling dates were higher in organic than in conventional soil samples ($p < 0.05$). DNA yields ranged from 4.08 to 6.76 µg/g in organic and from 2.44 to 4.56 µg/g in conventional soils (Table 2). These results are in agreement with previous studies on soil microbial biomass from SAFS experiments. Microbial biomass measures from 1992, 1993, and 1995 have demonstrated that organic systems maintained consistently higher biomass than conventional systems.[3,10] Microbial biomass carbon, microbial biomass nitrogen, and substrate-induced respiration were consistently lower in conventional than in organic plots.[2] Using the direct lysis method, the minimum soil sample size that could be used for DNA isolation was 200 mg. DNA extracted from soil samples below 200 mg was not detectable by spectrophotometry.

## COMPARISON OF RANDOM AMPLIFIED POLYMORPHIC DNA FINGERPRINTS

RAPD analysis was applied to community DNA isolated from the soil samples managed by organic and conventional methods. A comparison of major bands obtained from each soil with 15 different primers revealed that each primer generated distinguishable fingerprints except with primers AA5 and AA8 (Figures 1A and 1B). The absence of an RAPD pattern with primers AA5 and AA8 may indicate that these primer sequences had insignificant homology to the community DNA. Well-characterized primers with optimized cell lysis methods are likely to be important for the detection of microorganisms present in lower concentrations in a community.

RAPD fingerprints generated from DNA isolated at different sample dates from organic and conventional soil samples produced fingerprint patterns characteristic of each sample point. Examples of RAPD fingerprints generated using primer AA4 from organic and conventional soil samples on two sample dates 7-3-95 and 4-4-95 are presented in Figures 2, 3A, and 3B. Although some common bands were shared between the two soil types, some distinguishable bands were observed. Differences in RAPD patterns generated from soil samples of each treatment from different sample dates may indicate differences in microbial community composition associated with farming management practices. Although RAPD patterns between adjacent sample dates were similar, larger differences in band patterns were observed between April and July (Figures 2, 3A, and 3B). These observations are supported by phospholipid fatty acid analysis (PLFA) patterns measured for organic and conventional soil samples which were significantly different among sample dates.[2] Changes in soil properties such as altered carbon levels, pH, and moisture caused by organic and conventional management practices over time (e.g., incorporation of manure in organic or fertilizer application in conventional systems) may have caused changes in microbial population. An increase in soil temperatures during the period from April to July may also have affected the microbial population.

When RAPD fingerprints were generated using the same primer and DNA from four blocks of the same treatment, no major differences were noted, indicating that the field variability represented by blocks did not have a significant effect on RAPD patterns (Figure 2). On any given sample date, a higher number of bands were observed in organic than in conventional samples (Figures 3A and 3B). However, the differences between organic and conventional soils were not significant when RAPD patterns from all the sampling dates were pooled and analyzed ($p > 0.05$). Band intensities among samples were not considered for quantitative

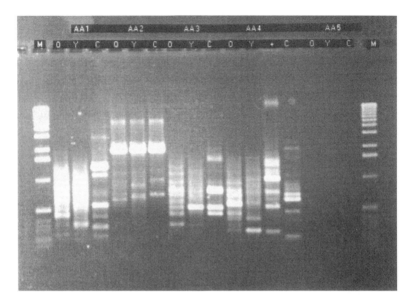

A

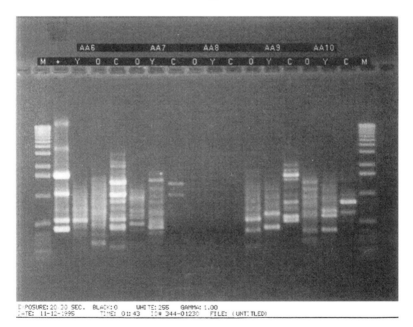

B

**FIGURE 1** (A, B) Agarose gel electrophoresis of RAPD fingerprints of DNA isolated from organic and conventional soil samples using primers AA1 through AA10. Each primer was used to develop RAPD patterns from organic (marked as "O") and conventional (marked as "C") soil samples.

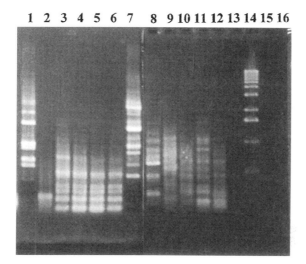

**FIGURE 2**  Agarose gel electrophoresis of RAPD fingerprints of DNA extracted from soil samples collected on 7-3-95. Primer AA4 was used. Lane 1—positive control (*Escherichia coli* DH5α DNA); 7 — positive control (soil DNA isolated and purified from samples collected from student farm, UC Davis); 13 — negative control; 2, 3, 4, 5, 6 — conventional soil samples; 8, 9, 10, 11, 12 — organic soil samples; 14 — DNA marker (1 kb ladder).

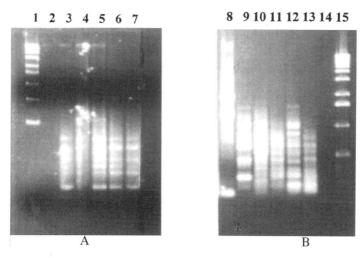

**FIGURE 3**  (A, B) RAPD fingerprints of DNA extracted from soil samples collected on 4-4-95 using primer AA4. Lanes 1, 15 — DNA marker (1 kb marker); 3, 4, 5, 6, 7 — conventional soil samples; 9, 10, 11, 12, 13 — organic soil samples; 8 — positive control; 2, 14 — negative control.

analysis of sequences because internal controls were not included. A major limitation in applying this technique to soil DNA analysis is the extreme sensitivity of RAPD reactions to the purity of target DNA. The presence of PCR inhibitory compounds in soil DNA can inhibit RAPD patterns. DNA isolation by the direct lysis method involves an alkaline lysis step which causes the co-extraction of humic acids and small phenolic compounds.[5] Humic acids are believed to strongly inhibit polymerase activity in PCR reactions.[5,8,9] Addition of BSA to the reaction mixture and dilution of target DNA (and contaminants) by 10- to 100-fold improved PCR amplification. However, dilution of target DNA reduced the sensitivity and also influenced the reproducibility of RAPD patterns. In contrast, other researchers have reported that PCR was not significantly inhibited by contaminants in the DNA extracted from compost.[6]

In comparison to other microbial community analysis techniques (e.g., cross-hybridization, DNA probes), RAPD is a simple, fast, sensitive technique for complex microbial community analysis. However, extensive sample purification and processing of large numbers of replicate soil samples are required to generate reproducible patterns. This study demonstrates that RAPD reaction conditions need to be optimized on a case-by-case basis to obtain less biased information on community analysis.

## ACKNOWLEDGMENTS

We thank A. Ogram for information on DNA extraction procedures and primers. Contributions of N. Gunapala and S. Uesugi are gratefully acknowledged. We are thankful to Mary Ann Bruns for suggestions.

For questions or further information regarding the work described in this chapter, please contact:

> **Padma Sudarshana**
> Dept. of Plant Pathology
> One Shields Avenue
> University of California
> Davis, CA 95616
> e-mail: plsudarshana@ucdavis.edu

## REFERENCES

1. Bossio, D. A., and K. M. Scow. 1995. Impact of carbon and flooding on the metabolic diversity of microbial communities in soils. *Appl. Environ. Microbiol.* 61:4043–4050.
2. Bossio, D. A., K. M. Scow, N. Gunapala and K. J. Graham. 1998. Determination of soil microbial communities: effects of agricultural management, season, and soil type on phospholipid fatty acid profiles. *Microb. Ecol.* 36:1–12.
3. Gunapala, N., and K. M. Scow. 1996. Dynamics of soil microbial biomass and activity in conventional and organic farming systems. *Soil Biol. Biochem.* 30:805–816.

4. Hadrys, H., M. Balick and B. Schierwater. 1992. Applications of random amplified polymorphic DNA (RAPD) in molecular ecology. *Mol. Ecol.* 1:55–63.
5. Holben, W. E., J. K. Jansson, B. K. Chelm and J. M. Tiedje. 1988. DNA probe method for the detection of specific microorganisms in the soil bacterial community. *Appl. Environ. Microbiol.* 54:703–711.
6. Malik, M., J. Kain, C. Pettigrew and A. Ogram. 1994. Purification and molecular analysis of microbial DNA from compost. *J. Microbiol. Methods* 20:183–196.
7. Ogram, A. V., and G. S. Sayler. 1988. The use of gene probes in the rapid analysis of natural microbial communities. *J. Ind. Microbiol.* 3:281–292.
8. Picard, C., C. Ponsonnet, X. Nesme and P. Simonet. 1992. Detection and enumeration of bacteria in soil by direct extraction and polymerase chain reaction. *Appl. Environ. Microbiol.* 58:2717–2722.
9. Ritz, K., and B. S. Griffiths. 1994. Potential application of a community hybridization technique for assessing changes in the population structure of soil microbial communities. *Soil. Biol. Biochem.* 26:963–971.
10. Scow, K. M., O. Somasco, N. Gunapala, S. Lau, R. Venette, H. Ferris, R. Miller and C. Shennan. 1994. Transition from conventional to low-input agriculture changes soil fertility and biology. *Calif. Agric.* 48: 20–26.
11. Steffan, R. J., J. Gokoyr, A. K. Bej and R. M. Atlas. 1988. Recovery of DNA from soils and sediments. *Appl. Environ. Microbiol.* 54:2908–2915.
12. Ward, D. M., R. Weller and M. M. Bateson. 1990. 16S rRNA sequences reveal numerous uncultured microorganisms in a natural community. *Nature* 345:63–65.
13. Wander, M. M., D. S. Hedrick, D. Kaufman, S. J. Traina, B. R. Stinner, S. R. Kehrmeyer and D. C. White. 1995. The functional significance of the microbial biomass in organic and conventionally managed soils. *Plant Soil* 170:87–97.
14. Williams, J. G. K., A. R. Kubilik, K. J. Livak, J. A. Ralfalski and S. V. Tingey. 1990. DNA polymorphisms amplified by arbitrary primers are useful as genetic markers. *Nucl. Acids Res.* 18:6531–6535.
15. Xia, X., J. Bollinger and A. Ogram. 1995. Molecular genetic analysis of the response of three soil microbial communities to the application of 2,4-D. *Mol. Ecol.* 4:17–28.

# 13 Air Pollution and Forests: Effects at the Landscape Level

*Paul R. Miller, John J. Carroll, Susan Schilling, and Raleigh Guthrey*

## CONTENTS

Abstract .................................................................................................................. 233
Introduction ........................................................................................................... 234
Background ........................................................................................................... 234
Methods ................................................................................................................. 237
Results ................................................................................................................... 239
    Ozone Injury Index (OII) Changes by General Location of
        Individual Plots Between Years 1991 and 1994 ..................................... 239
    Distribution of Ozone Injury Index in Relation to Crown Position Classes
        for All Locations in 1991 to 1994 ........................................................... 241
    Incidence of Other Abiotic Injury to Needles ............................................... 241
Discussion ............................................................................................................. 243
References ............................................................................................................. 246

## ABSTRACT

In the past, an approximate spatial characterization of the risks to California forests from ozone was developed by other investigators using existing forest resource and ozone databases and expert-estimated sensitivities for tree species. Potential impact was the primary product. Others have employed GIS-based risk assessment approaches; however, this approach requires much calibration and evaluation with on-the-ground measurements of mature tree response to ozone. In regard to the measured impacts of ozone on pines in the mixed conifer forest type in the Sierra Nevada and the mountains of southern California, there is a nearly 30-year record of the assessments at specific locations. Within the last 5 years, the assessment of ozone injury to ponderosa (*Pinus ponderosa* Dougl. ex Laws.) and Jeffrey (*P. jeffreyi* Grev. & Balf.) pines has been improved by a descriptive procedure called the ozone injury index (OII). This procedure has been used between 1991 and 1994 to assess the condition of ponderosa and/or Jeffrey pines using approximately 50 trees at each of 33 sites throughout the north to south extent of the Sierra Nevada, and including the San Bernardino mountains. Average ozone-caused foliar damage for all 4 years showed an increase from north to south in California. Crown position class had an influence on ozone injury amount; dominant (D) and open grown (OG) trees have generally lower OIIs (less injury) than codominant (CD) and intermediate (I) trees. The survey procedure confirmed ozone injury only if the distinctive chlorotic mottle symptom was present. However, the survey also recorded an upper leaf surface necrotic spotting called weather or winter fleck which is seen on needles that have been exposed to at least one winter period. The exact cause of winter fleck is unknown. Survey results show the intensity to be less in regions with relatively little photochemical pollutant exposure than in regions with moderate to high levels of pollutant exposure. The significance of winter fleck could be substantial, since it intensifies with each winter and diminishes the amount of healthy mesophyll tissue available for photosynthesis. Future assessments of the risk of ozone injury to the mixed conifer forest type should incorporate further developments in spatial analysis techniques, and ground truth should be provided by actual mature tree injury information from forest plots.

**Keywords:** *forest, landscape, ozone, pine, ponderosa, injury*

## INTRODUCTION

Precise data on the extent of forest effects from photochemical oxidant (or specifically ozone) air pollution transported from urban sources to the pine and mixed conifer forests of the western Sierra Nevada have been limited or nonexistant until recently. The purpose of this chapter is to point out the difficulties which have limited earlier estimates of air pollution effects, to evaluate a modern GIS-based approach for predicting injury, and to describe the results of a recent survey and its applications to understanding a number of aspects of air pollution effects primarily on ponderosa pine and Jeffrey pine, namely, the spatial extent of visible ozone injury, the presence of additional needle symptoms sometimes associated with ozone symptoms, and the influence of crown position class on the amount of tree injury.

## BACKGROUND

Data describing the effects of photochemical air pollutants on California forests are relatively scarce as are air quality measurements in rural forested areas.[1] Although the principle focus of this chapter is the Sierra Nevada, the forests of the entire state have been the subject of a risk analysis performed using a forest resource database including 91 variables in 3500 grid cells, each cell covering approximately 30,000 acres.[2] The database included tree species volume, elevation, soil type, land use, and interpolated ozone concentrations. The second major element of information was a ranking of the ozone sensitivity of California tree species based on literature values and expert judgment. A risk assessment was performed with these data and summarized by land ownership, landscape type, or administrative unit. Maps were produced to show the distribution of risk categories in California by grid cell for each of three ozone exposure scenarios. This process identified the most high risk cells in the central and southern Sierra Nevada; it tended to discount risk at even higher exposure levels in southern California because a lower species volume in this region became the main factor contributing to a low risk classification. The authors conclude that the components of data employed in the assessment have very wide margins of error leading to broad approximations in the assignment of risk to ozone damage. Certainly the 30,000-acre grid cell could encompass an enormous variety of species and landscape features, especially on steep elevational gradients. Thus, the assignment of a single risk category would be inappropriate for such a large cell.

Another ecological risk assessment approach has been tested for forests of the eastern U.S. and is under continuing development for eventual testing in the western states.[3] The method is designed to handle the spatial nature of tree and stand responses using a geographical information system (GIS). The three elements utilized by the assessment include estimated ozone exposure for the region in question, an estimate of tree and stand response to ozone, and spatially distributed intrinsic and extrinsic variables (genetics, environment) which influence tree and stand response to ozone. The characterization of species response is currently limited by its reliance on seedling response data and the lack of mature tree response information. Future efforts will feature process-based model simulations of 3- to 5-year growth responses of individual mature trees and a model simulated response of mixed species stands in response to appropriate environmental drivers in addition to ozone. This work is notable for its emphasis on the spatial representation of the ozone risk assessment. However, applying the method to western forests, which are characterized by rapid changes of forest type along elevational gradients requiring higher spatial resolution, will require additional refinements. Coordination between GIS-based spatial risk analysis methods and repeated measures of ozone injury in sample populations of mature trees in mixed stands will be an important next step in improving the risk analysis process.

Evergreen conifers have long been employed as a suitable biomonitor of changing ozone effects on foliage/crown condition.[4] The foliage of evergreen trees is potentially vulnerable to environmental stress because it is exposed simultaneously or sequentially to multiple stress agents during its lengthy functional life span. In the case of ponderosa pine and Jeffrey pine, needle whorls persist as long as 8 years, depending on the quality of the site and level of injury sustained from biotic and abiotic stresses. Ponderosa pine and Jeffrey pine have contiguous or slightly overlapping ranges of distribution in California.[5] Both species are sensitive to ozone air pollution, and chronic injury to both has been described in the San Bernardino Mountains and the Sierra Nevada.[6-8]

The characteristic leaf (needle) symptom used to distinguish ozone injury from other needle injuries is known as chlorotic mottle.[9] The appearance of this symptom on foliage of seedling trees is a predictable result of ozone fumigation.[10,11] Mottle appears on both upper and lower surfaces, but is easier to evaluate on the lower surface where other confounding symptoms like winter fleck are not present. Detection and measurement of ozone injury symptoms to ponderosa and Jeffrey pines in the Sierra Nevada[7] and subsequent surveys by the Forest Service, Forest Pest Management using 52 trend plots[12-15] provided the earliest data describing the extent of ozone injury and the early trends of the severity of injury. Pronos and Vogler[12] reported that between 1977 and 1980 the general trend was an increase in the amount of ozone symptoms present on pine foliage.

Peterson et al.[16] sampled crown condition and derived basal area growth trends based on cores collected from ponderosa pines at sites located from north to south in the Sierra Nevada, including Tahoe National Forest, Eldorado National Forest, Stanislaus National Forest, Yosemite National Park, Sierra National Forest, Sequoia-Kings Canyon National Park, and Sequoia National Forest. In July through August 1987, four symptomatic and four asymptomatic sites were visited in each location, and only sites with ponderosa pines greater than 50 years old were selected for sampling. The symptomatic sites generally indicated increasing levels of chronic ozone injury (reduced numbers of annual needle whorls retained and chlorotic mottle symptoms on younger age classes of needles) from north to south.

The results of Peterson et al.[16] document the regional nature of the ozone pollution problem originating primarily from the San Joaquin Valley Air Basin, as well as the San Francisco Bay Air Basin further to the west.[1] The study found no evidence of recent, large-scale growth changes in ponderosa pine in the Sierra Nevada mountains; however, the frequency of trees with recent declines of growth did increase in the southernmost units. Since these units have the highest levels of ozone (and more chlorotic mottle symptoms on needles of younger age classes), it was postulated from the weight of the evidence that ozone is one of the contributing factors to decline in basal area increase. Other factors limiting tree growth in this region include periodic drought, brush competition, and high levels of tree stocking.

Another tree ring analysis and crown injury study was focused on Jeffrey pines in Sequoia-Kings Canyon National Park.[17] This study suggested that decreases of radial growth of symptomatic, large, dominant Jeffrey pines growing on xeric sites (thin soils, low moisture holding capacity) and exposed to direct up-slope transport of ozone amounted to as much as 11% less over time (since the 1950s when the influence of air pollution was believed to begin) relative to adjacent trees without symptoms.

Both permanent plots and cruise surveys have been employed in Sequoia, Kings Canyon (SEKI),[18,19] and Yosemite (YOSE) National Parks to determine the spatial distribution and temporal changes of injury to ponderosa and Jeffrey pines within the parks.[8] Comparisons of the same trees at 28 plots between 1980 to 1982 and 1984 and 1985 in SEKI showed increases of ozone injury to many trees and increases of the total number of trees with ozone injury. Ozone injury was found to reach a maximum at around 1800 m elevation. The highest levels of tree injury in the Marble Fork drainage of the Kaweah River (at approximately 1800 m elevation) were associated with hourly averages of ozone frequently peaking at 80 to 100 ppb, but seldom exceeding 120 ppb.

A cruise survey in 1986 evaluated 3120 ponderosa or Jeffrey pines in SEKI and YOSE for ozone injury.[8] More than one third of these trees were found to have some level of chlorotic mottle. At SEKI symptomatic trees comprised 39% of the sample (574 out of 1470), and at YOSE they comprised 29% (479 out of 1650). Ponderosa pines were generally more severely injured than Jeffrey pines. The average Forest Pest Management (FPM) score (low score equals high injury) was 3.09 for ponderosa and 3.62 for Jeffrey.[20] These cruise surveys identified the spatial distribution of injury in SEKI and YOSE and indicated that trees in drainages nearest the San Joaquin Valley were most injured.

The Lake Tahoe Basin is an isolated air basin with an air quality situation distinct from other Sierra Nevada sites. In 1987 a survey of 24 randomly selected plots in the basin included a total of 360 trees of which 105 (29%) had some level of foliar injury.[21] Seventeen of these plots had Forest Pest Management (FPM) injury scores that fell in the slight injury category.[20] Of 190 trees in 16 cruise plots that extended observations to the east outside the basin, 22% had injury — less than in the basin.

In the San Bernardino mountains of southern California the levels of ozone injury to ponderosa and Jeffrey pines exceed those of the Sierra Nevada. However, chronic ozone injury is less than it was in the early 1970s as measured at 18 permanent forest plots oriented along a west to east gradient of decreasing ozone exposure.[4] There is evidence that stem growth of ponderosa and Jeffrey pines has been less than competing tree species that are less ozone sensitive.[22] Injury scoring methods used for the measurement of ozone injury to pines in permanent forest plots in California have not been entirely consistent. For example, the FPM index was used in the Sierra Nevada and an independently developed oxidant injury score was used in the San Bernardino mountains.

In recognition of the need for a uniform method for assessing ozone injury to ponderosa and Jeffrey pines in California and the western U.S., the Forest Ozone Response Study (FOREST) was started in 1991 by the Forest Service, National Park Service, and California Air Resources Board. Annual measurements were made at three, 50-tree plots located within 3 mi and 500 ft elevation from ozone monitoring stations at 12 locations until 1994. Occasionally, there were several trees omitted per plot because some trees were too tall to reach the lowest green branches with an extended pole pruner or sometime trees died due to bark beetle attack. Six of the ozone monitoring stations were established specifically for this purpose.[23] Field methods for tree measurements and calculation of an ozone injury index developed by the FOREST project are described in a protocol document.[9]

This chapter describes the results of the 1991 to 1994 annual evaluations at FOREST plots, including the development of a necrotic fleck symptom present on the same needles with ozone injury, as well as the relationship of crown position class and chronic ozone injury. This chapter also discusses the application of these results to the spatial analysis of tropospheric ozone risk to the mixed conifer forest type in California.

## METHODS

Ozone injury has been determined at FOREST plots by trained teams representing the interagency task group annually since 1991. The sample included 1600 to 1700 trees distributed in 36 individual plots located from Lassen Volcanic Park (north) to Mountain Home (south) in the Sierra Nevada, including 3 plots in the San Bernardino mountains.[24,25] The locations of plots included Lassen Volcanic, Yosemite, and Sequoia National Parks and six other Sierra Nevada sites (White Cloud, Sly Park,

**FIGURE 1** Location of Forest Ozone Response Study (FOREST) plots of ponderosa and Jeffrey pines.

Five-Mile Learning Center, Jerseydale, Shaver Lake, and Mountain Home) (Figure 1). Ozone monitoring stations were maintained near each site[23] or additional ozone records were obtained from nearby Forest Service or National Park Service monitoring sites. The calculated ozone injury indexes between years for the 1991 to 1994 period are reported as means for all trees at each monitoring site. A higher mean value indicates more injury. The index varies from 0 (no injury from ozone) to 100 (the most severe ozone injury). Trees that died during this period were not included in the comparison. Tree data from Jerseydale did not meet quality control standards, and most were excluded from analysis.

In order for a tree to qualify as having ozone injury it must be possible to find chlorotic mottle symptoms on needles of at least one annual needle whorl of at least one of five branch samples pruned from the lower crown. Branch samples were inspected by hand in bright sunlight.

Biotic and abiotic injuries to needles of handheld sample branches were also recorded; for example, the upper surface necrotic flecks (winter fleck) were counted as present if at least 1% of the surface area was involved. Other tree characteristics including stem diameter, percent live crown, and crown position class were obtained for the FOREST relational database.[9,24,25]

## RESULTS

### OZONE INJURY INDEX (OII) CHANGES BY GENERAL LOCATION OF INDIVIDUAL PLOTS BETWEEN YEARS 1991 AND 1994

The plot locations are illustrated in north to south sequence from left to right in Figure 2. The 4-year mean OII for each set of three, approximately 50-tree plots, tended to increase from 5.5 at Lassen to 45.7 at Barton Flats in the south. Grant Grove and Giant Forest in Sequoia National Park had the highest index value (highest injury) for the south central Sierra group. The 4-year average OII is shown in relation to ozone exposure in Table 1. In 1994, the OII was determined at an additional long-term plot originally established in 1974.[22] The plot at Camp Paivika (near Crestline) in the San Bernardino mountains had a mean OII of 65.7, the highest injury observed at any location.

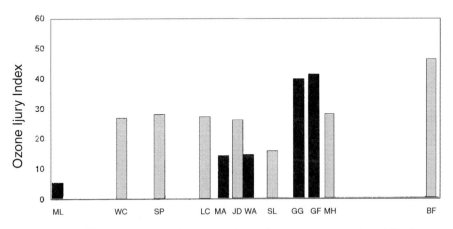

**FIGURE 2** The ozone injury index (OII) as a function of the north-south location of FOREST plots. The bars represent an average the OII for 4 years, namely, 1991 to 1994.

## TABLE 1
## OII Values for 1991 to 1994 at Sierra Nevada and San Bernardino Mountain Plots in Relation to Ozone Exposure

| Plot | Mean OII | 24-h Average Ozone (ppb)[a] |
|---|---|---|
| Lassen | 5.5 | 41.4 |
| White Cloud | 27.0 | 61.9 |
| Sly Park | 28.0 | 52.6 |
| Learning Center | 27.1 | 64.6 |
| Mather | 14.2 | 45.8 |
| Wawona | 14.5 | 40.2 |
| Jerseydale | Excluded[b] | 64.4 |
| Shaver Lake | 15.7 | 54.9 |
| Grant Grove | 38.0 | 63.7 |
| Giant Forest | 41.3 | 66.3 |
| Mountain Home | 28.0 | 71.2 |
| Barton Flats | 45.7 | 64.5 |

[a] June–September 1992–1994.
[b] Data excluded because bark beetle-caused mortality reduced tree numbers.

The proportion of the total number trees with confirmed evidence of any ozone injury is definitely lowest at the lower ozone pollution sites (Lassen) and highest at the highest ozone sites (Barton Flats) (Table 2).

Generally, year-to-year changes in OII and proportion of trees with symptoms were not large. This would be consistent with similar ozone exposures each year. The smallest yearly differences also appear mainly at those sites where the evaluation crew did not change composition each year.

The 1992 to 1994, June to September, 24-h ozone average in Table 1 shows the correspondence with the OII at each site. In the Sierra Nevada, the northern plots have a lower OII (or a lower proportion of affected trees) than do the plots in the southern Sierra Nevada where ozone averages are higher. At plots with the highest OII and highest ozone averages, namely, Grant Grove, Giant Forest, Mountain Home, and Barton Flats, the injury is not in exact proportion to average ozone. Of these four, Mountain Home has the highest ozone and the lowest OII (Table 1). Giant Forest has slightly higher ozone than Barton Flats and a lower OII than Barton Flats. We propose that these apparent discrepancies are explainable on the basis of actual ozone uptake by trees at a site. At Barton Flats in 1992 a significant increase in OII was due in part to a more favorable soil moisture supply, which in turn could possibly increase stomatal conductance. This mechanism may have resulted in the larger flux of ozone to foliage compared to other years.[26] The soil moisture holding capacity and yearly rainfall differences among the four plots with higher OII values should be evaluated further. The most useful method and data set needed for estimating ozone flux to foliage is the measurement of stomatal conductance at least once weekly during the entire summer season. The subsequent calculation of ozone flux will lead to a more

**TABLE 2**
**Changes in the Proportion of Ponderosa or Jeffrey Pines with Ozone Symptoms from 1991 to 1994 Indicating Significant Differences ($p \leq 0.05$) Compared to the Preceding Year**

| Location | 1991 | 1992 | 1993 | 1994 |
|---|---|---|---|---|
| Lassen Park | — | 0.200 | 0.187 | 0.393*a |
| White Cloud | 0.701 | 0.673 | 0.653 | 0.753 |
| Sly Park | 0.628 | 0.590 | 0.533 | 0.840* |
| Learning Center | 0.620 | 0.624 | 0.678 | 0.767* |
| Mather-Yosemite | 0.520 | 0.300* | 0.313 | 0.513* |
| Wawona-Yosemite | 0.490 | 0.317* | 0.159* | 0.593* |
| Jersey-dale | Excluded[a] | Excluded | Excluded | Excluded |
| Shaver Lake | 0.514 | 0.469 | 0.655* | 0.240* |
| Grant Grove | 0.946 | 0.953 | 0.913 | 0.940 |
| Giant Forest | 0.924 | 0.903 | 0.890 | 0.972* |
| Mountain Home | 0.813 | 0.535* | 0.721* | 0.680 |
| Barton Flats | 0.976 | 0.992 | 1.000 | 0.992 |

* Significantly different from the preceding year. (p = 0.05)
** Significantly different from the preceding year. (p = 0.01)
[a] Beetle-caused mortality.

meaningful ozone dose-response relationship.[26] Unfortunately, this task requires an intensive effort which is difficult to do routinely because of cost constraints.

## DISTRIBUTION OF OZONE INJURY INDEX IN RELATION TO CROWN POSITION CLASSES FOR ALL LOCATIONS IN 1991 TO 1994

Table 3 shows that dominant (D) and open grown (OG) trees have generally lower OIIs (less injury) than codominant (CD), intermediate (I), and suppressed (S) trees in 1991, 1992, 1993, and 1994. Intermediate trees are mostly younger trees in smaller stem diameter classes. In Table 3, the numbers of trees in the I category averages almost one third of the total population sampled. These data indicate that estimates of tree injury could easily be biased in either direction, depending on the sample distribution into different crown position classes. For example, it would be a mistake to sample only D or OG trees or only I trees to represent the entire tree population. I trees are younger and grow more rapidly than D or OG trees, thus ozone uptake may be higher for this class. Also, I trees may be more densely spaced and compete more for water, nutrients, and light — the resources needed to recover from chronic ozone exposure.

## INCIDENCE OF OTHER ABIOTIC INJURY TO NEEDLES

The other principal abiotic injury was upper-surface necrotic flecking or winter fleck. At least 1% of the needle surface must be occupied by fleck before it is counted as

## TABLE 3
### Ranking of OII in Relation to Crown Position Class

| Crown Position Class | Average OII | N | Mean Separation |
|---|---|---|---|
| Open grown, 1991 | 23.8 | 87 | a |
| Dominant, 1991 | 26.6 | 258 | a |
| Suppressed, 1991 | 28.6 | 107 | ab |
| Codominant, 1991 | 29.5 | 600 | ab |
| Intermediate, 1991 | 31.8 | 505 | b |
| Open grown, 1992 | 18.5 | 97 | a |
| Dominant, 1992 | 19.7 | 278 | a |
| Suppressed, 1992 | 21.2 | 91 | ab |
| Codominant, 1992 | 25.9 | 575 | b |
| Intermediate, 1992 | 28.1 | 454 | b |
| Dominant, 1993 | 18.4 | 295 | a |
| Open grown, 1993 | 18.4 | 103 | ab |
| Suppressed, 1993 | 25.0 | 92 | abc |
| Codominant, 1993 | 24.8 | 614 | bc |
| Intermediate, 1993 | 26.3 | 454 | c |
| Open grown, 1994 | 19.5 | 100 | a |
| Dominant, 1994 | 20.4 | 284 | a |
| Suppressed, 1994 | 26.3 | 85 | ab |
| Codominant, 1994 | 29.2 | 576 | b |
| Intermediate, 1994 | 30.6 | 437 | b |

*Note:* Single letters indicate a significant ($p = 0.05$) difference in mean compared to last year; multiple letters indicate insignificant mean from the last year.

present. The cause(s) of this symptom remains unclear, except that a comparison of the three highest ozone exposure sites with the three lowest ozone exposure sites shows a higher frequency of fleck at the high ozone sites (Figure 3). The corresponding frequency of chlorotic mottle is shown in Figure 4. One can see that there is a higher frequency of ozone injury (chlorotic mottle) accompanied by a higher frequency of winter fleck. No causal connection between the two foliar injury symptoms, however, has been established.

The phase during which defoliation was the most rapid coincided with the intensification of necrotic flecking. This defoliation has always been attributed to the increasing level of chlorotic mottle, which actually has a much lower incidence of occurrence at high ozone sites than necrotic flecking (Figures 3 and 4). The reason that chlorotic mottle incidence does not continue to increase to percent levels higher than 30% is that injured needles abscise earlier. There is no evidence that needle necrotic tissue (winter fleck) may lead to reduced vigor of trees, but the possibility cannot be entirely dismissed.

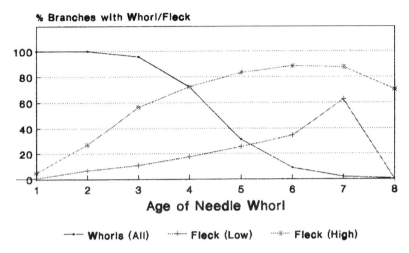

**FIGURE 3** Incidence (percent) of about 1600 ponderosa or Jeffrey pines that retain needle whorls aged 1 to 8 years compared with the average incidence of weather fleck on different whorls from three locations with the lowest ozone and three locations with the highest ozone exposure (Guthrey et al. 1993).

## DISCUSSION

The assessment of the actual amount of ozone injury or the risk of ozone injury to particular forest mixtures emphasized spatial distribution.[2,3] The scarcity of air quality data in remote forested regions and the lack of adequate species response information imposes severe limits on this approach. On the other hand, the intensive repeated measures at permanent forest plots as reported here for the Sierra Nevada and San Bernardino mountain plots do not alone lend themselves well to spatial analysis because plots are not intended to be randomly distributed on the landscape. The data from permanent plots are most helpful for understanding something about the spatial trend of injury and the influence of crown position class when standard methods for evaluating injury are employed.[9] Data from trees in permanent plots are also useful for analyzing the ozone exposure responses of pine foliage following several consecutive years of exposure[27,28] and for studying incremental increases in needle symptoms as a function of actual ozone uptake by foliage.[26] Ozone flux estimates to mature trees would lend support to the process model-supported-GIS approach being developed by Hogsett et al.[3]

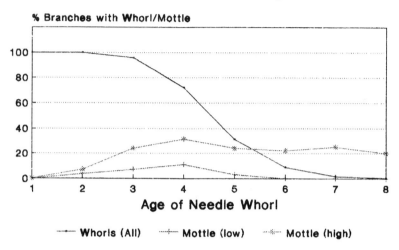

**FIGURE 4** Incidence (percent) of about 1600 ponderosa or Jeffrey pines that retain needle whorls aged 1 to 8 years compared with the average incidence of chlorotic mottle on different whorls from three locations with the lowest ozone and three locations with the highest ozone exposure (Guthrey et al. 1993).

Other kinds of needle injury may play an important role in determining the productivity and longevity of needles, particularly the condition referred to as "winter fleck" or "weather fleck" because it generally appears after the newest whorl of needles has been exposed to a winter season. Microscopic examinations showed that injury was evidenced by entire blocks of dead mesophyll cells extending from the epidermis beneath the necrotic fleck to the endodermis.[29]

Fink[30] described "the widespread phenomenon of necrotic spots on conifer needles in the field." This damage could not be attributed to acid rain because it was not characterized by damage to epidermal cells. Initial damage was to mesophyll cells, but not in the same pattern as with ozone injury. As described on Norway spruce (*Picea abies*), the cause still remains speculative, but it most closely resembles "winter fleck injury" as described on conifer needles by Miller and Evans[29] and by Brennan[31] in New Jersey. Winter injury to red spruce (*P. rubens* Sarg.) is characterized by a reddish-brown discoloration of the most recent foliage which later abscises in the spring.[32] Although this symptom is morphologically different from winter fleck on western pines, the spatial distribution of it is similar to winter fleck by being most prominent on the upper and southern-facing surfaces of needles and by not being present on needles that are deeply shaded.

Perkins and Adams[32] suspect that the rapid temperature fluctuations of red spruce foliage along with high solar loading during windless periods (causing needles to be 15°C hotter than air temperature) are contributing causes to winter injury. Similarly, Strimbeck[33] observed that during mild winters injury was concentrated on needles exposed to solar radiation. Large leaf-air temperature differentials and rapid rates of temperature change are suggested as contributing factors. Observations of increased winter injury to conifer foliage following summer exposures to ozone have been reported for Norway spruce (*P. abies* (L.) Karst.) seedlings.[34]

The dry deposition of nitric acid, one of the principal sources of nitrate, may have localized direct effects on leaf cuticles, which in turn may predispose underlying mesophyll cells to injury from low temperatures or desiccation. Exposures of ponderosa pine seedlings for 48 h to about 75 ppb of $HNO_3$ indicated drastic changes in chemistry of epicuticular waxes — proportion of estolides decreased with a simultaneous increase of the fatty acids fraction. The needles of the dark-fumigated pines showed increased nitrate reductase activity and uptake of $N^{15}$.[35] These findings suggest that $HNO_3$ can easily penetrate through the cuticle into the pine needle interior.

Direct injury by exposure to UV-B radiation varies widely between conifers and herbaceous plants. Biggs and Kosuth[36] categorized ponderosa pine as sensitive to UV-B, however, Day et al.[37] studied mature conifer needles (more than 1 year old) of nine species and found that all were particularly effective at screening UV-B. Additional work with *Pinus pungens* Lamb confirmed the cuticle of this conifer was a much more spatially uniform filter of UV-B than that of herbaceous plants.[37] In general, UV-B did reduce the net photosynthesis of loblolly pine (*Pinus taeda* L.) and did affect the buildup of methanol-extractable, UV-B absorbing compounds.[38]

In summary, an assortment of surveys and repeated measures in the Sierra Nevada mountains and San Bernardino mountains have focused on the mixed conifer forest type because detectable ozone injury was absent on most species of other forest and woodland types. For example, the highest levels of tree injury at Sequoia National Park were at 1800 m elevation in the Marble Fork drainage of the Kaweah River. Injury was associated with hourly averages of ozone frequently peaking at 80 to 100 ppb, but seldom exceeding 120 ppb.[8] Both risk analysis[2] and earlier studies[16] point to the southern Sierra Nevada as the region with the most foliar ozone injury to ponderosa and Jeffrey pines.

The FOREST results reported here confirm that injury increases from north to south in the Sierra Nevada and that injury is even higher in the San Bernardino mountains of southern California. The FOREST results also show that injury levels to trees within plots are significantly different depending on the crown position class, i.e., open grown, dominant, codominant, intermediate, and suppressed. Intermediate trees were significantly more injured than open grown or dominant trees. Intermediate trees comprise almost one third of the trees in the FOREST data-base. This within-species difference could approach the magnitude of difference between sensitive (ponderosa pine) and moderately sensitive species like white fir (*Abies concolor* Lindl.) Furthermore, there are suggestions from the FOREST data that landform or position of trees on the landscape could also influence amount of injury. However, additional sampling would be required to test this suggestion. Forest managers may find it useful to take into account crown position and possibly landform as important elements that could modify decisions about silvicultural treatments to be applied. Future surveys, efforts to establish permanent plots, and spatially oriented risk assessment procedures should be aware of the importance of crown position class and possibly landform as additional variables to consider.

For questions or further information regarding the work described in this chapter, please contact:

**Paul R. Miller**
5520 Via San Jacinto
Riverside, CA 92506-3651
e-mail: millerpr@aol.com

## REFERENCES

1. Bytnerowicz, A., J.J. Carroll, B.K.Takemoto, P.R. Miller and M.E. Fenn. 1996. *Integrated Assessment of Ecosystem Health,* Ann Arbor Press, Chelsea, MI.
2. Peterson, D.C. and C. Daly. 1989. Risks to California forests due to regional ozone pollution: a data base and ranking of relative sensitivity. In: Olson, R.K. and A.S Lefohn (eds) Effects of Air Pollution on Western Forests. Transactions Series, No.16, Air and Waste Management Association, Pittsburgh, pp 247–260.
3. Hogsett, W.E., A.A. Herstrom, J.A. Laurence, J.E. Weber, E.H. Lee and D.T. Tingey. 1996. An approach for characterizing tropospheric ozone risk to forests. *Environ. Manage.* 21:105–120.
4. Miller, P.R., J.R. McBride, S.L. Schilling and A.P. Gomez. 1989. Trend of ozone damage to conifer forests between 1974 and 1988 in the San Bernardino Mountains of southern California. In: Olson, R.K. and A.S. Lefohn (eds) Effects of Air Pollution on Western Forests. Transactions Series, No.16, Air and Waste Management Association, Pittsburgh, pp 309–324.
5. Critchfield, W.B. and E.L. Little, Jr. 1966. Geographic Distribution of Pines of the World. USDA, Misc. Pub. 991, Washington, D.C., 97 pp.
6. Miller, P.R., J.R. Parmeter, Jr., O.C. Taylor and E.A. Cardiff. 1963. Ozone injury to the foliage of *Pinus ponderosa. Phytopathology* 53:1072–1076.
7. Miller, P.R. and A.A. Millecan. 1971. Extent of oxidant air pollution damage to some pines and other conifers in California. *Plant Dis. Rep.* 55:555–559.
8. Duriscoe, D.M. and K.W. Stolte. 1989. Photochemical oxidant injury to ponderosa (*Pinus ponderosa Dougl.* ex Laws) and Jeffrey pine (*Pinus jeffreyi Grev. and Balf.*) in the national parks of the Sierra Nevada of California. In: Olson, R.K. and A.S. Lefohn (eds) Effects of Air Pollution on Western Forests. Transactions Series, No.16, Air and Waste Management Association, Pittsburgh, pp 261–278.
9. Miller, P.R., K.W. Stolte, D. Duriscoe and J. Pronos. 1996a. Evaluation of ozone air pollution injury to pines in the western United States. Pacific Southwest Research Station, Forest Service, USDA, Gen. Tech. Report, GTR-PSW 155. In Press.
10. Miller, P.R., G.J. Longbotham and C.R. Longbotham. 1983. Sensitivity of selected western conifers to ozone. *Plant Dis.* 67:1113–1115.

11. Temple, P.J., G.H. Riechers and P.R. Miller. 1992. Foliar injury responses of ponderosa pine seedlings to ozone, wet and dry acidic deposition, and drought. *Environ. Exp. Bot.* 32:101–113.
12. Pronos, J. and D.R. Vogler. 1981. Assessment of ozone injury to pines in the southern Sierra Nevada, 1979/1980. USDA Forest Service, Pacific Southwest Region, Forest Pest Management Report 81-20, 13pp.
13. Allison, J.R. 1982. Evaluation of ozone injury on the Stanislaus National Forest. USDA Forest Service, Pacific Southwest Region, Forest Pest Management Report 82-07, 7pp.
14. Allison, J.R. 1984a. An evaluation of ozone injury to pines on the Eldorado National Forest. USDA Forest Service, Pacific Southwest Region, Forest Pest Management Report 84-16, 10pp.
15. Allison, J.R. 1984b. An evaluation of ozone injury to pines on the Tahoe National Forest. USDA Forest Service, Pacific Southwest Region, Forest Pest Management Report 84-30, 10pp; Barnes, J.D. and A.W. Davison. 1988. The influence of ozone on the winter hardiness of Norway spruce (*Picea abies* L.). *New Phytol.* 108:159–165.
16. Peterson, D.L., M.J. Arbaugh and L.J. Robinson. 1991. Regional growth changes in ozone-stressed ponderosa pine (*Pinus ponderosa*) in the Sierra Nevada, California, USA. *Holocene* 1:50–61.
17. Peterson, D.L., M.J. Arbaugh and L.J. Robinson. 1989. Ozone injury and growth trends of ponderosa pine in the Sierra Nevada. In: Olson, R.K. and A.S. Lefohn (eds) Effects of Air Pollution on Western Forests. Transactions Series, No.16, Air and Waste Management Association, Pittsburgh, pp 293–308.
18. Wallner, D.W. and M. Fong. 1982. Survey Report, National Park Service, Three Rivers, CA.
19. Warner, T.E., D.W. Wallner and D.R. Vogler. 1982. Ozone injury to ponderosa and Jeffrey pines in Sequoia-Kings Canyon National Parks. In: van Riper, C., L.D. Whittig, and M.L. Murphy (eds) Proceedings of the Conference on Research in California's National Parks, University of California, Davis, pp 1–7.
20. Pronos, J., D.R. Vogler and R.S. Smith. 1978. An evaluation of ozone injury to pines in the southern Sierra Nevada. USDA Forest Service, Pacific Southwest Region, Forest Pest Management Report 78-1, 13pp.
21. Pedersen, B.S. 1989. Ozone injury Jeffrey and ponderosa pines surrounding Lake Tahoe, California and Nevada. In: Olson, R.K. and A.S. Lefohn (eds) Effects of Air Pollution on Western Forests. Transactions Series, No.16, Air and Waste Management Association, Pittsburgh, pp 279–292.
22. Miller, P.R., J.R. McBride and S.L. Schilling. 1991. Chronic ozone injury and associated stresses affect relative competitive capacity of species comprising the California mixed conifer forest type. In: Memorias Del Primer Simposio Nacional Agricultura Sostenible: Una opcion para el Desarrola sin Deterioro Ambiental. Comision de Estudios Ambientales C. P. y M. O. A. International, pp 161–172.
23. Van Ooy, D.J. and J.J. Carroll. 1995. The spatial variation of ozone climatology on the western slope of the Sierra Nevada. *Atmos. Environ.* 29:1319–1330.
24. Guthrey, D.R., S.L. Schilling and P.R. Miller. 1993. Initial Progress Report of an Interagency Forest Monitoring Project: Forest Ozone REsponse STudy (FOREST). Pacific Southwest Research Station, Riverside, CA. 42pp.
25. Guthrey, D.R., S.L. Schilling and P.R. Miller. 1994. Second Progress Report of an Interagency Forest Monitoring Project: Forest Ozone REsponse STudy (FOREST). Pacific Southwest Research Station, Riverside, CA. 22pp.

26. Temple, P.J. and P.R. Miller. 1996. Seasonal influences of ozone uptake and foliar injury to ponderosa and Jeffrey pines at a southern California site. Proceedings of the International Symposium on Air Pollution and Climate Change Effects on Forest Ecosystems. Albany, CA, Pacific Southwest Research Station, Gen. Tech. Report. PSW-GTR 166.
27. Miller, P.R., R. Guthrey, S. Schilling and J. Carroll. 1996. Ozone injury responses of ponderosa and Jeffrey pine in the Sierra Nevada and southern California mountains. Proceedings of the International Symposium on Air Pollution and Climate Change Efffects on Forest Ecosystems. Albany, CA, Pacific Southwest Research Station, Gen. Tech. Report. PSW-GTR 166.
28. Miller, P., A. Bytnerowicz, M. Fenn, M. Poth, P. Temple, S. Schilling, D. Jones, D. Johnson, J. Chow and J. Watson. 1998. Multidisciplinary study of ozone, acidic deposition, and climate effects on a mixed conifer forest in California, USA. *Chemosphere* 36:1001–1006.
29. Miller, P.R. and L.S. Evans. 1974. Histopathology of oxidant injury and winter fleck injury on needles of western pines. *Phytopathology* 64:801–806.
30. Fink, S. 1993. Microscopic criteria for the diagnosis of abiotic injuries to conifer needles. In: Huettl, R. and D. Mueller-Dombois (eds) *Forest Decline in the Atlantic and Pacific Region*, Springer-Verlag, Berlin, pp 175–188.
31. Brennan, E. 1988. Winter spot: a common symptom on coniferous evergreens in New Jersey. *Shade Tree* 61:44–45.
32. Perkins, T.D. and G.T. Adams. 1992. Winter injury and decline of red spruce in the northeast. Proceedings of the Conference: The effects of air pollution on terrestrial and aquatic ecosystems in New England and New York, October 19–21, 1992, USDA Forest Service, Northeastern Forest Experiment Station.
33. Strimbeck, R. 1992. Short-term fluctuations in midwinter needle temperature of red spruce. Proceedings of the Conference: The effects of air pollution on terrestrial and aquatic ecosystems in New England and New York, October 19–21, 1992, USDA Forest Service, Northeastern Forest Experiment Station.
34. Barnes, J.D. and A.W. Davison. 1988. The influence of ozone on winter hardiness of Norway spruce [*Picea abies* (L) Karst.]. *New Phytol.* 108:159–166.
35. Krywult, M., J. Hom, A. Bytnerowicz and K. Percy. 1994. Deposition of gaseous nitric acid and its effects on foliage of ponderosa pine (Pinus ponderosa L.) seedlings. Proceedings of the 16th International Meeting for Specialists in Air Pollution Effects on Forest Ecosystems "Air Pollution & Multiple Stress," IUFRO, Fredericton, Canada, September 7–9, 1994. Unpublished.
36. Biggs, R.H. and S.V. Kosuth. 1978. Impact of solar UV-B radiation on crop productivity. Effects of ultraviolet-B radiation enhancements on eighty-two different agricultural species. Final Report, Vol. II, SIRA File No. 142.23, EPA-IAG-D6-0168, USDA-EPA Stratospheric Impact Research and Assessment Program (SIRA), US Environmental Protection Agency, Washington, D.C., 79 pp.
37. Day, T.A., T.C. Vogelmann and E.H. DeLucia. 1992. Are some plant life forms more effective than others in screening out ultraviolet-B radiation? *Oecologia* 92:513–519.
38. Sullivan J.H. and A.H. Teramura. 1989. The effects of ultraviolet-B radiation on loblolly pine. I. Growth, photosynthesis and pigment production in greenhouse-grown seedlings. *Physiol. Plant.* 77:202–207.

# 14 Mercury in Lower Trophic Levels of the Clear Lake Aquatic Ecosystem, California

*Thomas H. Suchanek, Bradley A. Lamphere, Lauri H. Mullen, Cat E. Woodmansee, Peter J. Richerson, Darell G. Slotton, Lee Ann Woodward, and E. James Harner*

## CONTENTS

Abstract ........................................................................................................................249
Introduction ..................................................................................................................250
Methods ........................................................................................................................251
    Analytical Laboratory Procedures .........................................................................252
    Data Reduction and Statistical Analyses ...............................................................252
Results ..........................................................................................................................253
    Plankton .................................................................................................................253
    Benthic Invertebrates .............................................................................................257
Discussion ....................................................................................................................259
Summary ......................................................................................................................265
Acknowledgments ........................................................................................................265
References ....................................................................................................................266

## ABSTRACT

Total and methyl mercury were analyzed in plankton and benthic invertebrates from Clear Lake, CA, an aquatic ecosystem contaminated from mining at the Sulphur Bank Mercury Mine over an 84-year period. Sediment total (primarily inorganic) mercury concentrations exceed 180,000 ng/g near the mine. Total mercury in Clear Lake biota (up to 855 ng/g in plankton, 41,671 ng/g in oligochaetes, and 27,686 ng/g in chironomids) was found to reflect the concentration of mercury in the organisms' surroundings (water or sediment). Methyl mercury, however (up to 67 ng/g in plankton, 19.9 ng/g in oligochaetes, and 61.9 ng/g in chironomids), typically was not correlated with inorganic mercury concentrations in water and sediment. Anomalously high concentrations of methyl mercury in biota at great distances from the point source of inorganic mercury suggests either (1) methyl mercury is produced *in situ* at sites with low inorganic mercury or (2) methyl mercury is produced in regions with high inorganic mercury and transported to other regions of Clear Lake by wind-driven currents. An increasing ratio of methyl to total mercury with increasing trophic level supports a bioaccumulation model for methyl mercury dynamics in Clear Lake biota and suggests that bioavailable mercury increases as a function of distance from the mine. Compared with other contaminated sites, Clear Lake's plankton are relatively low in both total mercury and methyl mercury, despite the fact that sediments in this system are some of the most highly contaminated in the world. Benthic invertebrates at Clear Lake exhibit the highest total mercury values (by one order of magnitude higher than any other reported sites), yet have the lowest methyl mercury concentrations of any known contaminated sites.

**Keywords:** *Clear Lake, mercury, mining, invertebrates, aquatic*

## INTRODUCTION

Concern over mercury (Hg) bioaccumulation has stimulated substantial interest in Hg contamination in natural ecosystems and the relative efficiency of methyl Hg production and transport through trophic pathways.[1] While there has been considerable research on Hg in higher trophic level species such as predatory fishes and birds, little information exists on Hg in lower trophic levels. This information is critical to understanding the dynamics of Hg bioaccumulation, predicting the flow of Hg through various trophic levels, modeling systems contaminated with Hg, and providing potential solutions to problems associated with natural and anthropogenic sources of Hg. In addition, since lower trophic level organisms are more easily sampled than individuals at higher trophic levels, and since such collections have little impact on the populations of these organisms, it seems logical to use them as indicators of Hg contamination in the ecosystem as a whole. Here we present data on total and methyl Hg in plankton and benthic invertebrates from a highly contaminated aquatic ecosystem in Clear Lake, CA.

Clear Lake is a shallow, polymictic lake in the Northern California Coast Range. Mining of Hg ore from the Sulphur Bank Mercury Mine (periodically from about 1873 to 1957) resulted in the deposition of an estimated 100 metric tons of Hg into the Clear Lake aquatic ecosystem.[2] See Suchanek et al.[3-5,35] and Richerson et al.[6] for a more complete description of Clear Lake. Clear Lake is highly eutrophic with abundant phytoplankton (especially scum-forming cyanobacteria), zooplankton, and benthic invertebrates. In a preliminary study by the U.S. Environmental Protection Agency (USEPA), Hg levels in wet phytoplankton ranged from 0.6 to 160 ng/g, or about 160 to 700 ng/g on a dry weight basis.[7] To our knowledge, no previous study has assessed Hg levels in Clear Lake zooplankton or benthic invertebrates. Fishes from Clear Lake, specifically largemouth bass and channel catfish, also have been shown to contain Hg levels that exceed California state and federal health standards for safe consumption (i.e., $\geq 0.5$ mg/kg).[8,9]

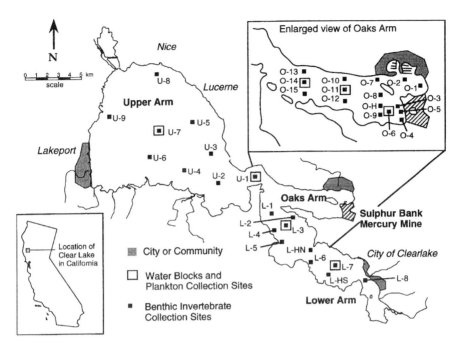

**FIGURE 1** Map of Clear Lake and sampling sites.

The primary objective of this chapter is to describe the distribution of total and methyl Hg within the lower trophic levels, primarily phytoplankton, zooplankton, and benthic invertebrates (oligochaetes and chironomids) of the Clear Lake aquatic ecosystem. These results represent only a portion of a baseline study conducted as a part of an Ecological Assessment of the USEPA Region IX Sulphur Bank Mercury Mine Superfund Site.[3,35] Concentrations of Hg within the abiotic components (i.e., sediments and water) of this ecosystem are given in Suchanek et al.,[5] while Slotton et al.[9] present Hg concentrations in Clear Lake fishes. Effects of Hg on invertebrate populations and community structure are presented in Suchanek et al.[4] Taken collectively, these data represent a snapshot in time of the distribution of Hg in both abiotic and biotic components.

## METHODS

All collections were made during July to November 1992. Plankton were collected from 7 water block stations throughout Clear Lake, while benthic invertebrates were collected from 35 benthic stations (Figure 1). Descriptive statistics (distance from the mine, depth, sediment grain size, total organic carbon, total Hg, methyl Hg, methyl to total ratio in sediments and water, plus standard limnological parameters) on these same sampling stations are provided in Suchanek et al.[3,5,35]

Plankton were collected (mesh size = 160 µm) over a 0.5-km² area centered on each of the seven stations until approximately 100 g was obtained. The resulting samples consisted of a mixture of approximately 90% phytoplankton (mostly blue-green "algae" = cyanobacteria) and 10% zooplankton. A subsample of this "total plankton" (the mixture of phytoplankton and zooplankton) was analyzed at each site. Zooplankton samples for Hg analysis were obtained by separating out zooplankton from the remainder of the total plankton samples by washing the total plankton samples through a 300-µm mesh screen using deionized water. During the plankton sampling periods, phytoplankton consisted primarily of *Anabeana, Gleotrichia,* and *Microcystis,* whereas the zooplankton consisted primarily of *Daphnia, Bosmina, Ceriodaphnia,* and *Diaphanosoma* (Norm Anderson, Lake County Mosquito and Vector Control, personal communication). Both total plankton and zooplankton samples were transferred to acid-washed bottles, dried in an evacuating drying oven at <60°C, and weighed before and after drying. Dried samples were homogenized with Teflon™ utensils, resulting in uniform powders.

Benthic invertebrates (oligochaetes and chironomids) were collected using 6-in. Ekman grab samples and sieved with a 0.5-mm mesh screen. Samples were sorted and placed in acid-washed glass containers with water and shipped live (on ice) for Hg analysis. Because of taxonomic complexities, oligochaetes were not identified to species, but previously collected specimens from Clear Lake are known to include numerous genera: *Aulodrilus, Bothrioneurum, Branchiura, Dero, Ilyodrilus, Limnodrilus, Potamothrix, Tubifex,* and *Varichaetadrilus* (Art Colwell, Lake County Mosquito and Vector Control, personal communication). Chironomid larvae consisted primarily of several species of the genus *Chironomus* (*C. plumosus, C. frommeri,* and *C. decorus*).

For comparisons of Hg concentrations in biota with water from various locations within the lake, raw (unfiltered) water included all particulates. Filtered water was passed through a 0.45-µm filter (see Suchanek et al.[5] for details).

## Analytical Laboratory Procedures

Plankton were analyzed for total Hg at all seven water block stations; methyl Hg was only analyzed for a subset of four stations. Total Hg in benthic invertebrates was analyzed at all 35 stations; methyl Hg was analyzed for a subset of 10 stations. Total Hg and methyl Hg for all tissues were analyzed by Brooks Rand Ltd. (Seattle, WA). Total Hg in plankton and benthic invertebrates was analyzed by dual amalgamation/cold vapor atomic fluorescence spectrometry.[10] Methyl Hg was analyzed in biotic samples utilizing aqueous phase ethylation, followed by cryogenic gas chromatography with cold vapor atomic fluorescence detection.[19]

Field collection and analytical methods for Hg in unfiltered and filtered water are described in detail in Suchanek et al.[5] All raw data for abiotic and biotic components are presented in Suchanek et al.[3]

## Data Reduction and Statistical Analyses

All data analysis was performed on an Apple Macintosh® platform. Preliminary plotting and calculation of simple regression statistics for curve-fitting routines were accomplished using the data management and statistical visualization program JMP® (SAS Institute, Inc.). Curve-fitting and $R^2$ values were calculated using the presentation graphics program Delta Graph® (DeltaPoint, Inc.). Step-wise multiple linear regression analyses were performed using the statistical package DataDesk® (Data Description, Inc.).

Data for independent variables on physical environmental parameters (such as concentrations of Hg in water and sediments) used in multiple regression analyses (see Table 1) were taken from Suchanek et al.[5]

## RESULTS

### PLANKTON

Plankton exhibited total Hg concentrations (69 to 855 ng/g for total plankton and 80 to 661 ng/g for zooplankton) about four to five orders of magnitude higher than the water from which they were collected (Figure 2A). Total Hg in both types of plankton declined with distance from the mine, though concentrations at the site most distant from the mine were slightly higher than expected. Using curve-fitting routines, this decline was best described by second order polynomials, which explained about 73% of the variability in the zooplankton data and about 93% of the variability in the total plankton data. Multiple regression analyses revealed that for total Hg in both total plankton and zooplankton, a positive relationship was observed with methyl Hg in unfiltered surface water (Table 1). This relationship was much stronger in the total plankton ($p < 0.001$) than in zooplankton ($p < 0.05$). Zooplankton total Hg concentrations were also found to be positively correlated to surface water conductivity values ($p < 0.05$).

Methyl Hg concentrations in plankton (10 to 29 ng/g for total plankton and 27 to 67 ng/g for zooplankton) were five to six orders of magnitude higher than the unfiltered water from which they were collected (Figure 2B). These data show only a slight decline with increasing distance from the mine, with the site furthest from the mine showing slightly elevated methyl Hg concentrations as compared to those sites at moderate distances, a trend similar to that of the total Hg data. As a result of this trend, plankton methyl Hg data was best described by second-order polynomials, explaining about 93% of the variability in zooplankton and 98% of the variability in total plankton. Multiple regression analysis on methyl Hg in total plankton yielded no significant relationship to any of the independent variables tested (Table 1). Total Hg in zooplankton, however, was found to have a moderately negative correlation with total suspended solids in surface waters ($p < 0.01$) and a weak positive correlation with surface conductivity.

Although total Hg concentrations in zooplankton and total plankton were comparable, methyl Hg levels were considerably higher in zooplankton, likely a reflection of their higher trophic status. The ratio of methyl to total Hg ranged from 0.04 to 0.18 (4 to 18%) in total plankton and from 0.09 to 0.46 (9 to 46%) in zooplankton (Figures 2C and 4). Both sets of ratios were several orders of magnitude higher than the ratios of methyl to total Hg in surface waters (0.004 to 0.017 ng/L in filtered water and 0.003 to 0.007 ng/L in unfiltered water), and both increased significantly with distance from the mine. Results of the multiple regression analysis showed a moderately negative correlation ($p < 0.01$) of the ratio of methyl to total Hg in total plankton with total Hg in filtered surface water (Table 1). Zooplankton Hg ratios were found to be most significantly (and negatively) correlated with the depth of a site ($p > 0.01$).

## TABLE 1
### Results of Multiple Regression Analyses

| Dependent Variables | Number of Samples Collected | Number of Samples in Model | Whole Model Significance | Whole Model R² | Site | | Sediment | | | | | Water | | | | | | | | | | | |
|---|---|---|---|---|---|---|---|---|---|---|---|---|---|---|---|---|---|---|---|---|---|---|---|
| | | | | | Distance from Mine | Depth | Total Hg | Methyl Hg # | Methyl/Total Hg Ratio | Grain Size | TOC | Surface Total Hg — Unfiltered | Surface Total Hg — Filtered | Surface Methyl Hg — Unfiltered | Surface Methyl Hg — Filtered | Surface TOC — Unfiltered | Surface TOC — Filtered | Surface TSS | Surface Temperature | Surface pH | Surface Conductivity | Surface D.O. | Surface Eh |
| Total plankton-total Hg | 7 | 7 | *** | 0.98 | ns | ns | ○ | ○ | ○ | ○ | ○ | ns | ns | ***/+ | ns | ns | ns | ns | ns | ns | ns | ns | ns |
| Total plankton-methyl Hg | 4 | 4 | ns | 0.73 | ns | ns | ○ | ○ | ○ | ○ | ○ | ns | ns | ns | ns | ns | ns | ns | ns | ns | ns | ns | ns |
| Total plankton-methyl/total Hg ratio ◊ | 4 | 4 | ** | 0.99 | ns | ns | ○ | ○ | ○ | ○ | ○ | ns | **/− | ns | ns | ns | ns | ns | ns | ns | ns | ns | ns |
| Zooplankton-total Hg | 7 | 7 | * | 0.75 | ns | ns | ○ | ○ | ○ | ○ | ○ | ns | ns | */+ | ns | ns | ns | ns | ns | ns | */+ | ns | ns |
| Zooplankton-methyl Hg | 4 | 4 | * | 1.00 | ns | ns | ○ | ○ | ○ | ○ | ○ | ns | ns | ns | ns | ns | ns | **/− | ns | ns | */+ | ns | ns |

# Mercury in Lower Trophic Levels of Clear Lake Aquatic Ecosystem

| Variable | | | | | | | | | | | | | | | | | | | |
|---|---|---|---|---|---|---|---|---|---|---|---|---|---|---|---|---|---|---|---|
| Zooplankton-methyl/total Hg ratio ◊ | 4 | 4 | ** | 0.99 | ns | **/− | ○ | ○ | ○ | ○ | ns | ns | ns | ns | ns | ns | ns | ns | ns |
| Chironomids-total Hg # | 35 | 31 | *** | 0.56 | ns | ns | ***/+ | ns | ns | ns | ns | ns | ns | ns | ns | ○ | ○ | ○ | ○ |
| Chironomids-methyl Hg | 10 | 10 | ns | 0.26 | ns | ns | ns | ns | ns | ns | ns | ns | ns | ns | ns | ○ | ○ | ○ | ○ |
| Chironomids-methyl/total Hg ratio ◊ | 10 | 7 | ** | 0.97 | ns | †/+ | ns | */+ | **/+ | ns | ns | ns | ns | ns | ns | ○ | ○ | ○ | ○ |
| Oligochaetes-total Hg | 35 | 33 | *** | 0.37 | ns | ns | ***/+ | ns | ns | ns | ns | ns | ns | ns | ns | ○ | ○ | ○ | ○ |
| Oligochaetes-methyl Hg # | 10 | 7 | * | 0.58 | ns | ns | ns | */+ | ns | ns | ns | ns | ns | ns | ns | ○ | ○ | ○ | ○ |
| Oligochaetes-methyl/total Hg Ratio ◊ | 10 | 7 | ** | 0.68 | ns | **/+ | **/+ | ns | ns | ns | ns | ns | ns | ns | ns | ○ | ○ | ○ | ○ |

*Note:* Significance levels from multiple regression analysis where † = $0.05 < p < 0.10$; * = $0.01 < p < 0.05$; ** = $0.001 < p < 0.01$; *** = $p < 0.001$; ns = not significant; ○ = not tested; + = positive correlation; − = negative correlation; ◊ = variable logit transformed; # = variable log transformed.

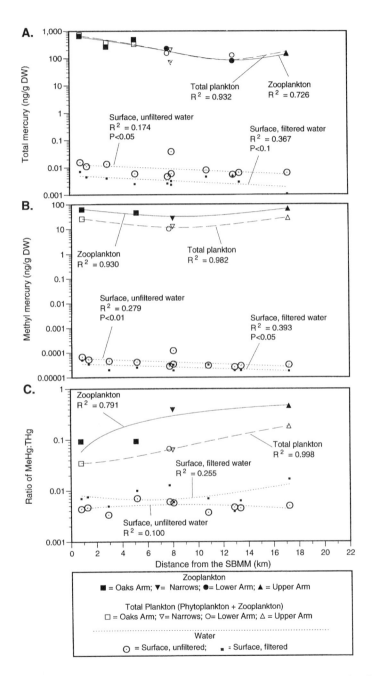

**FIGURE 2** Plankton results: (A) total mercury; (B) methyl mercury; and (C) ratio of methyl to total mercury in zooplankton, total plankton, and raw and filtered surface water plotted against distance from the Sulphur Bank Mercury Mine. Note log scales. Curve fits are second-order polynomials for all. Water Hg data from Suchanek et al.[5]

## Benthic Invertebrates

Total Hg concentrations in oligochaetes and chironomids were statistically indistinguishable from each other, each exhibiting a significant exponential decline as a function of distance from the mine (Figure 3A). These values spanned over two orders of magnitude, from 316 ng/g at distant sites to 42,000 ng/g for sites less than 1 km from the mine, and were about one order of magnitude lower than the sediments from which these organisms were obtained. Because individual specimens were not purged before analysis, some of the total Hg could have been the result of sediments remaining in their guts at the time of analysis. However, the entire body burden of Hg, including gut contents, is probably a more realistic value to use when evaluating bioaccumulation pathways. Using multiple regression analysis, concentrations of total Hg in both chironomids and oligochaetes were best explained by the concentrations of total Hg in the sediments from which they were derived ($p < 0.001$), and no other factors tested were significant.

Methyl Hg ranged from 5.3 to 19.9 ng/g in oligochaetes and from 5.2 to 61.9 ng/g in chironomids (Figure 3B). In oligochaetes, this trend was best explained by an exponential decline as a function of distance from the mine, but it was not significant and exhibited a very low $R^2$ (0.186). In contrast, methyl Hg in chironomids was best fit with a second order polynomial curve ($R^2 = 0.677$), though this fit depends greatly upon the single datum in the Upper Arm at U-7, the furthest sampling station from the mine (Figure 3B). Unfortunately, this was the only sampling site for methyl Hg in the Upper Arm. Invertebrate methyl Hg concentrations were, at most, only about one order of magnitude higher than the methyl Hg values in the sediments from which they were derived. Multiple regression analysis on methyl Hg in chironomids yielded no significant relationship to any of the independent variables tested. Oligochaete methyl Hg concentrations showed only a weak positive correlation with sediment methyl Hg concentrations.

The percentage of total Hg as methyl in both invertebrate taxa were considerably lower than those for plankton and not much different than those values for water and sediment (see Suchanek et al.[5]), with values ranging from 0.3 to 1.4% for oligochaetes and from 0.6 to 6.0% for chironomids (Figures 3C and 4). Interestingly, the highest methyl to total Hg ratio was observed at U-7, a site over 17 km from the mine. Since no attempts were made to void oligochaete guts before Hg analyses, it is unclear whether the ratio of methyl to total Hg in these species was significantly influenced by the total Hg levels of sediment remaining in their guts. Chironomid larvae, benthic filter feeders, also reflected sediment levels of total Hg in their environment. The percentage of methyl Hg in chironomids, however, was generally higher (Figures 3C and 4) than in oligochaetes. This is probably a reflection of (1) their higher trophic position and (2) the fact that Hg levels in oligochaetes may have been influenced by residual sediments in their guts. Results of multiple regression analyses on the ratio of methyl to total Hg in chironomids revealed a moderately positive correlation ($p > 0.01$) with the ratio of methyl to total Hg in sediment and a weaker positive correlation ($p < 0.05$) with sediment methyl Hg (Table 1). The ratio of methyl to total Hg in oligochaetes was found to have a moderately positive correlation ($p < 0.01$) with sediment total Hg concentrations.

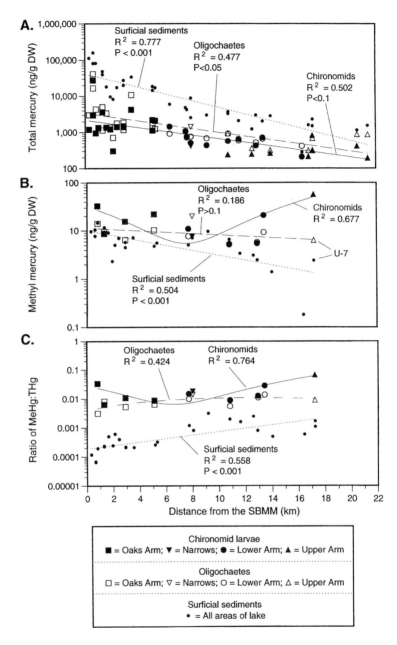

**FIGURE 3** Benthic invertebrate results: (A) total mercury; (B) methyl mercury; and (C) ratio of methyl to total mercury in chironomids, oligochaetes, and surficial sediments. Note log scales in A and B. Curve fits are as follows: (A) all are exponential; (B) sediment and oligochaete data are fit with exponential fits, and chironomid data are second-order polynomial; and (C) sediment data are fit with an exponential fit and oligochaete and chironomid data with second-order polynomial. Sediment data from Suchanek et al.[5]

## DISCUSSION

The declines of total Hg with distance from the mine in Clear Lake plankton and benthic invertebrates probably result from comparable declines in sediment and water concentrations of inorganic Hg. As shown in Suchanek et al.[5] abiotic components of the ecosystem (sediment and water) from Clear Lake exhibited classic point source distributions of total and methyl Hg with maximum concentrations occurring at sites closest to the mine. The correlation of biotic total Hg levels with total Hg levels in the abiotic environment is particularly evident in the results of the multiple regression analyses presented in Table 1. Methyl Hg burdens in Clear Lake biota, on the other hand, were not correlated with inorganic Hg concentrations in the abiotic environment. The relatively high methyl Hg concentrations in biota collected at sites distant from the mine indicate that methyl Hg burdens may be better explained by considering factors relating to bioavailability or methylation potential. Further evidence that methyl Hg is more readily available at sites further from the mine was seen in the studies of the abiotic components of Clear Lake.[5] That study found that the ratio of methyl to total Hg in sediment and the particulate fraction of water increased with distance from the mine. The conclusions of Suchanek et al.[5] that there was either (1) differential downgradient transport of methyl Hg or (2) preferential production of methyl Hg at sites distant from the mine are consistently reflected in the biotic data presented here.

When comparing Hg concentrations in lower trophic biota from Clear Lake to those from other systems throughout the world, we note some interesting trends (Table 2). Total Hg in Clear Lake zooplankton, which ranged from 80 to 661 ng/g, was relatively low compared to other lake systems where total Hg ranged over 3000 ng/g.[14,15] It should be noted that, like plankton, Hg concentrations (0.005 to 0.05 µg/L) in Clear Lake water, in comparison with other studies (0.0003 to 0.39 µg/L), are also at the low end of the range.[5] Though fewer studies have measured methyl Hg in plankton, Clear Lake plankton, with methyl Hg concentrations of 10 to 29 ng/g in total plankton and 27 to 67 ng/g in zooplankton, are typically much lower than those values reported in most other studies, documenting 32 to 220 ng/g in phytoplankton[16,17] and 260 ng/g in zooplankton.[17] Clear Lake plankton typically ranged from 3 to 18% methyl Hg for total plankton and 4 to 83% methyl Hg for zooplankton, much lower than those values typically reported from other contaminated and uncontaminated sites worldwide which often ranged from 70 to 100% methyl Hg.[20,24,32]

In contrast to the plankton, total Hg in Clear Lake oligochaetes (416 to 41,671 ng/g) and chironomids (194 to 27,686 ng/g) were much higher than the typical range (71 to 4900 ng/g) found in other studies, both freshwater and marine.[12,16-20] The elevated inorganic Hg in these organisms could be a reflection of the higher sediment Hg concentrations in Clear Lake (270 to 183,000 ng/g) relative to the other study sites (0 to 30,000 ng/g). Methyl Hg concentrations in Clear Lake oligochaetes (5.3 to 19.9 ng/g) and chironomids (5.2 to 61.9) were about one order of magnitude lower than the typical range (52 to 1000 ng/g) for benthic invertebrates from other sites.[16,17,29] Interestingly, Clear Lake sediments contain the highest reported sediment

**TABLE 2**
**Comparative Data on Total Hg, Methyl Hg, and Percent of Methyl Hg in This Study with Those Reported from the Literature**

| | Location | Contamination Source | Organisms Total Hg Range (Avg) ng/g = ppb | Methyl Hg Range (Avg) ng/g = ppb | Methyl Hg as a Percent of Total Hg | Sediment Total Hg Range (Avg) ng/g = ppb | Raw Water Total Hg Range (Avg) mg/L = ppb | Reference |
|---|---|---|---|---|---|---|---|---|
| | | | Plankton | | | | | |
| **freshwater** | | | | | | | | |
| lake phytoplankton | Onondaga L., NY | Wastewater | — | 32 | 24 | — | 0.0003 | 17 |
| lake phytoplankton | Onondaga L., NY | Wastewater | 1830@ | 220* | 12 | n.d.–30,000 | — | 16 |
| lake phytoplankton | El-Temsah L., Egypt | Wastewater | 876–2,170 (904) | — | — | 1,060–4,360 (3,150) | 0.11–0.39 (0.13) | 14 |
| **lake zooplankton** | Clear Lake, CA | Hg mine | 80–661 | 27–67 | 9–83 | 270–183,000 | 0.005–0.05 | This study and 5 |
| lake zooplankton | Onondaga L., NY | Wastewater | 650@ | 260 | 40 | — | 0.0003 | 17 |
| lake zooplankton | Little Rock Lake, WI | Atmospheric | 45 | — | — | <0.1 | 0.00012 | 1 |
| lake zooplankton | Lakes in Sweden | Atmospheric | — | — | 45–85 | 90–300 | — | 20 |
| lake zooplankton | Davis Creek Reservoir, CA | Gold mining | 830–4,100 | — | — | 500–150,000 | 0.006 | 15 |
| lake zooplankton | El-Temsah L., Egypt | Wastewater | 1,060–3,170 (1,230) | — | — | 1,060–4,360 (3,150) | 0.11–0.39 (0.13) | 14 |
| lake zooplankton | Lakes in Finland | Atmospheric | — | — | 45–100 (80) | — | — | 24 |
| lake zooplankton | Lakes in Ontario, Canada | Atmospheric | — | 19–448 | — | — | — | 30 |

# Mercury in Lower Trophic Levels of Clear Lake Aquatic Ecosystem

| Sample | Location | Source | | | | | | Ref. |
|---|---|---|---|---|---|---|---|---|
| lake zooplankton | Natural Lakes in Quebec, Canada | Atmospheric | ~75–350 (193) | ~55–175 | 15–50 | — | 0.2 | 32,38 |
| lake zooplankton | New reservoirs in Quebec, Canada | Atmospheric | 110–300 (185) | 150–850 | 45–85 | — | 0.2 | 32,38 |
| lake zooplankton | Lakes in Ontario and Quebec, Canada | Atmospheric | 26–377 (108) | — | — | 3–267 (80) | — | 28 |
| stream zooplankton | Mountain streams, TN | Atmospheric | — | — | 23–100 (75) | — | — | 25 |
| total lake plankton | Clear Lake, CA | Hg mine | 69–855 | 10–29 | 3–18 | 270–183,000 | 0.005–0.05 | This study and 5 |
| lake plankton (>10 μm) | Churchill R., Canada | New reservoir | <10–2,110 | — | — | 6,000–14,000 | 0.022–0.036 | 12 |
| lake plankton (>73 μm) | Churchill R., Canada | New reservoir | <10–1,230 | — | — | 6,000–14,000 | 0.022–0.036 | 12 |
| **marine** | | | | | | | | |
| marine phytoplankton | Pelagic ocean | Atmospheric | 410 | — | — | — | — | 13 |
| marine phytoplankton | Kastela Bay, Yugoslavia | Chlor-alkali | — | — | — | 90–1,370 | — | 13 |
| marine zooplankton | Kastela Bay, Yugoslavia | Chlor-alkali | — | — | — | 90–1,370 | — | 13 |
| marine zooplankton | Lavaca Bay, TX | Chlor-alkali | 100–200 | — | — | — | — | 26 |
| marine zooplankton | Pelagic ocean | Atmospheric | 55–388 (191) | — | — | — | — | 36 |
| marine zooplankton | Pelagic ocean | Atmospheric | 39–448 (134) | — | — | — | — | 37 |

**TABLE 2 (continued)**
Comparative Data on Total Hg, Methyl Hg, and Percent of Methyl Hg in This Study with Those Reported from the Literature

| | | | Organisms | | | Sediment | Raw Water | |
| --- | --- | --- | --- | --- | --- | --- | --- | --- |
| | Location | Contamination Source | Total Hg Range (Avg) ng/g = ppb | Methyl Hg Range (Avg) ng/g = ppb | Methyl Hg as a Percent of Total Hg | Total Hg Range (Avg) ng/g = ppb | Total Hg Range (Avg) mg/L = ppb | Reference |
| | | | Benthic Invertebrates | | | | | |
| **freshwater oligochaetes** | Clear Lake, CA | Hg mine | 416–41,671 (3,000) | 5–20 (9) | 0.3–4.8 | 270–183,000 | 0.005–0.05 | This study and 5 |
| **chironomids** | Clear Lake, CA | Hg mine | 194–27,686 (2,000) | 5–62 (20) | 0.6–5.8 | 270–183,000 | 0.005–0.05 | This study and 5 |
| amphipods | Onondaga L., NY | Wastewater | 200–1,300 (397) | 52–338 (103) | ca. 27 | — | 0.0003 | 17 |
| chironomids | Onondaga L., NY | Wastewater | 300–1,800 (790) | 108–486 | ca. 27 | — | 0.0003 | 17 |
| amphipods | Onondaga L., NY | Wastewater | 210–300 | 100–150# | 0.5 | 0–30,000 | — | 16 |
| chironomids | Onondaga L., NY | Wastewater | 350–1,900 | 170–1,000# | 0.5 | 0–30,000 | — | 16 |
| insects (detritivores/grazers) | Duncan Lake, Quebec, Canada | Atmospheric | 120–143 (129) | 14–61 (40) | 10–50 | 36–59 | — | 29 |
| insects (grazers/predators) | Duncan Lake, Quebec, Canada | Atmospheric | 136–256 (189) | 102–124 (111) | 49–75 | 36–59 | — | 29 |
| insects (detritivores/grazers) | LaGrand complex, Quebec, Canada | New reservoir | 139–1,020 (417) | 64–106 (76) | 6.3–76 | 33–275 | — | 29 |
| chironomids | LaGrand complex, Quebec, Canada | New reservoir | 285–1,075 (946) | 64–76 (71) | 6–26 | 33–275 | — | 29 |

| Taxon | Location | Source | | | | | | Ref. |
|---|---|---|---|---|---|---|---|---|
| chironomids | Lakes in Ontario, Canada | Atmospheric | 97–177 (158) | — | — | 28–311 (134) | 0.0015–0.0022 | 33 |
| oligochaetes | Lakes in Ontario, Canada | Atmospheric | 164 | — | — | 28–311 (134) | 0.0015–0.0022 | 33 |
| amphipods | Lakes in Ontario, Canada | Atmospheric | 98–158 (128) | — | — | 28–311 (134) | 0.0015–0.0022 | 33 |
| *Sergentia* (detritivore) | Lakes in Sweden | Atmospheric | 71–4,680 | — | 45–85 | 90–300 | — | 20 |
| *Chironomus* (detritivore) | Lakes in Sweden | Atmospheric | 115–1,350 | — | 45–85 | 90–300 | — | 20 |
| *Cryptochironomus* (predator) | Lakes in Sweden | Atmospheric | 45 | — | 45–85 | 90–300 | — | 20 |
| *Procladius* (predator) | Lakes in Sweden | Atmospheric | 42–173 | — | 45–85 | 90–300 | — | 20 |
| chironomids | Churchill R., Canada | New reservoir | 1,400–1,500* | — | — | 6,000–14,000 | 0.022–0.036 | 12 |
| oligochaetes | Churchill R., Canada | New reservoir | nd–4,900* | — | — | 6,000–14,000 | 0.022–0.036 | 12 |
| annelid | Lakes | Atmospheric | 500–1,500 | 20–180* | 4–36 | — | — | 27 |
| **marine** | | | | | | | | |
| *nereis* (polychaete) | Nissum Broad, Denmark | Chlor-alkali | 225–1,350 | — | — | 100–22,000 | — | 18,19 |
| *arenicola* (polychaete) | Nissum Broad, Denmark | Chlor-alkali | 335–760 | — | — | 100–22,000 | — | 18,19 |
| *maldanes* (polychaete) | Novaya Zemlya, Barents Sea | Atmospheric? | 130–450 | 80–340 | 18–75 | — | 0.002–0.015 | 31,34 |

*Note:* All values presented as dry weight unless specified otherwise. — = data not available; * = value calculated from wet weight data with a 7:1 wet/dry ratio; @ = value derived from methyl Hg and ratio values; # = value derived from total Hg and ratio values; nd = not detected.

inorganic Hg contamination (over 183,000 ng/g), yet methyl Hg concentrations in benthic invertebrates were typically one to two orders of magnitude lower than all other reported values for contaminated and uncontaminated sites alike (Table 2). The fact that methyl Hg concentrations in Clear Lake benthic invertebrates are low despite high total Hg concentrations may be due to several factors that may inhibit either methylation or bioaccumulation. Such factors might include (1) the high level of productivity of Clear Lake, (2) the alkaline nature of the lake, or (3) the high concentration of sulfides in water and sediments.

Among nonvertebrate biota in Clear Lake, inorganic Hg was found to dominate the total Hg burden, accounting for 75 to >99% of the total, as determined from the methyl to total Hg ratios (Figure 4). A trend of increasing proportion of methyl Hg with increasing trophic rank is consistent with many previous reports (e.g., References 16, 21, 22, and 29). Comparing methyl Hg in biota to environmental concentrations, especially in contrast to total Hg, highlights the influence of bioaccumulation. Each biotic group had methyl Hg burdens at the level of their surrounding medium or higher. Plankton had the highest differential, with methyl Hg concentrations five to six orders of magnitude higher than those in water; benthic invertebrates (with methyl Hg levels only slightly higher than that of the sediment) had the lowest.

Unlike the results from Clear Lake, studies of Hg in total plankton (zooplankton and phytoplankton combined) in other aquatic systems have not found direct positive correlations with Hg levels in water.[12,23] Studies of phytoplankton alone seem to have more varied results. Jackson[12] reported no relation between Hg in phytoplankton vs. water, yet he found inverse relationships of Hg in phytoplankton compared to

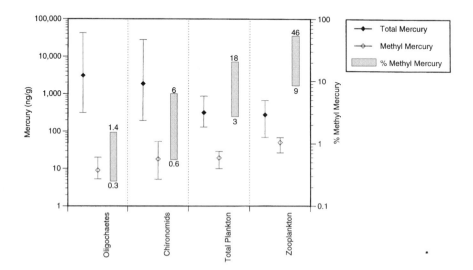

**FIGURE 4**  Means and ranges of total Hg, methyl Hg, and percent methyl Hg in oligochaetes, chironomids, total plankton, and zooplankton in Clear Lake.

environmental concentrations of total Hg, methyl Hg, and Hg methylation rate. Moreover, Hg levels in plankton correlated positively with concentrations of sulfides and suspended organic matter, which are thought to influence the bioavailability of Hg. Jackson[12] concluded that Hg concentrations in plankton reflect the bioavailability of Hg rather than the total concentration of Hg in the environment. Westcott and Kalff[30] showed that zooplankton methyl Hg was best predicted by water color and pH, being positively correlated with water color and negatively correlated with pH.

In other field studies where Hg concentrations in benthic invertebrates were assessed, some found that chironomid Hg body burdens correlated with sediment Hg levels,[16,17] whereas others did not.[12,20] Like Clear Lake, the Lake Onondaga system studied by Becker et al.[16] and Becker and Bigham[17] had fairly high surficial sediment Hg levels (up to 30,000 ng/g), yet rather low chironomid Hg concentrations (350 to 1900 ng/g). Jackson,[12] working in the Churchill River, a system with only moderately contaminated sediments (6000 to 14,000 ng/g), found chironomid Hg burdens to have strongest correlations with sediment total organic carbon (TOC) and iron (Fe), which influence the bioavailability of Hg. In Swedish lakes with sediments of relatively little contamination (90 to 300 ng/g), Parkman and Meili[20] showed that feeding habits had the strongest influence upon chironomid Hg. In these systems, profundal detritivores (including *Chironomus* sp.), as opposed to predators, had the highest concentrations of Hg. They also found correlations between chironomid Hg and pH (negative), water color (positive), and seasonal changes, all of which are factors that may influence Hg bioavailability.

## SUMMARY

In comparison with other studies worldwide, Clear Lake plankton have relatively low total (primarily inorganic) and methyl Hg concentrations. In contrast, Clear Lake benthic invertebrates exhibit some of the highest concentrations of total Hg found at any previously studied aquatic ecosystems, yet some of the lowest comparable methyl Hg values. In general, total Hg in Clear Lake biota exhibited a clear signature of a point source pollutant originating from the Sulphur Bank Mercury Mine, but methyl Hg did not. While there were high methyl Hg levels in biota near the mine, there were also high levels in the far reaches of the Upper Arm, some 15 to 20 km from the mine. These elevated methyl Hg concentrations could be derived from methyl Hg that is (1) produced *in situ* at these distant sites (from relatively low concentrations of inorganic Hg) or (2) produced near the mine (at sites with very high inorganic Hg) and transported via water and/or particulate matter by wind-driven currents to sites distant from the mine.

## ACKNOWLEDGMENTS

We would like to thank the County of Lake for continued support and resources throughout this project. Especially supportive have been the Lake County Board of Supervisors, the Lake Bed Management and Flood Control District, the Air Quality Management District, and the Mosquito and Vector Control District. Dr. Gary Gill with Texas A&M University at Galveston provided valuable input to the design of the study plan and discussion of results. Support from the U.S. EPA Region IX Superfund (68-S2-9005) Program and the U.S. EPA Center for Ecological Health Research (R819658) made this project possible. Thanks to Carolyn d'Almeida and Jeri Simmons of EPA-Superfund and to Karen Morehouse of EPA-ORD. Although the information in this chapter has been funded wholly or in part by the U.S. Environmental Protection Agency, it may not necessarily reflect the views of the agency and no official endorsement should be inferred.

For questions or futher information regarding the work described in this chapter, please contact:

**Thomas H. Suchanek**
Dept. of Wildlife, Fish and Conservation Biology
University of California
Davis, CA 95616
e-mail: thsuchanek@ucdavis.edu

## REFERENCES

1. Watras, C.J. and J.W. Huckabee. 1994. *Mercury Pollution: Integration and Synthesis.* Lewis Publishers, Ann Arbor, MI.
2. Chamberlin, C.E., R. Chaney, B. Finney, M. Hood, P. Lehman, M. McKee and R. Willis. 1990. Abatement and Control Study: Sulphur Bank Mine and Clear Lake. Prepared for the California Regional Water Quality Control Board. Environmental Resources Engineering Department, Humboldt State University, Arcata, CA.
3. Suchanek, T.H., P.J. Richerson, L.A. Woodward, D.G. Slotton, L.J. Holts and C.E.E. Woodmansee. 1993. A survey and evaluation of Hg in sediment, water, plankton, periphyton, benthic invertebrates and fishes within the aquatic ecosystem of Clear Lake, California. Report prepared for EPA Region IX. Ecological Assessment: Sulphur Bank Mercury Mine Superfund Site, Clear Lake, CA.
4. Suchanek, T.H., P.J. Richerson, L.J. Holts, B.A. Lamphere, C. E. Woodmansee, D.G. Slotton, E.J. Harner and L.A. Woodward. 1995. Impacts of mercury on benthic invertebrate populations and communities within the aquatic ecosystem of Clear Lake, California. *Water Air Soil Pollut.* 80:951–960.
5. Suchanek, T.H., L.H. Mullen, B.A. Lamphere, P.J. Richerson, C.E. Woodmansee, D.G. Slotton, E.J. Harner and L.A. Woodward. 1998. Redistribution of mercury from contaminated lake sediments of Clear Lake, California. *Water Air Soil Pollut.* 104(1/2):77–102.
6. Richerson, P.J., T.H. Suchanek and S.J. Why. 1994. The causes and control of algal blooms in Clear Lake. Clean Lakes Diagnostic/Feasibility Study for Clear Lake, California. Final Report to Lake County Flood Control and Water Conservation District, California State Water Resources Control Board and U.S. EPA. ca. 200pp.
7. Ecology and Environment. 1990. Brown and Bryant Site Assessment, November 16, 1990.
8. California Regional Water Quality Control Board (CRWQCB) Central Valley Region. 1986. Summary of Mercury Data Collection at Clear Lake. December 1, 1986.

9. Slotton, D.G., T.H. Suchanek, L.H. Mullen and P.J. Richerson. 1997. Mercury trends in Clear Lake fishes. Proceedings of the First Annual Clear Lake Science and Management Symposium, Lakeport, CA. September 13, 1997. 5pp.
10. Bloom, N.S. and E.A. Crecelius. 1987. Distribution of silver, mercury, lead, copper and cadmium in central Puget Sound sediments. *Mar. Chem.* 21:377–390.
11. Bloom N.S. 1989. Determintation of picogram levels of methymercury by aqueous phase ethylation, followed by cryogenic gas chromatography with cold vapor atomic fluorescence detection. *Can. J. Fish. Aquat. Sci.* 46:1131–1140.
12. Jackson, T.A. 1988. The mercury problem in recently formed reservoirs of Northern Manitoba (Canada): effects of impoundment and other factors on the production of methyl mercury by microorganisms in sediments. *Can. J. Fish. Aquat. Sci.* 45:97–121.
13. Zvonaric, T. and P. Stegnar. 1987. Total mercury, cadmium, copper, zinc and arsenic contents in surface sediments from the coastal region of the central Adriatic. *Acta Adriat.* 28:65–72.
14. Abo-El-Wafa, O. and H.I. Abdel-Shafy. 1987. Concentration of mercury and arsenic in El-Temsah Lake. In S.E. Lindberg and T.C. Hutchinson, eds., *Heavy Metals in the Environment Vol. 2*, New Orleans Conf. Proceedings, pp. 265–267.
15. Slotton, D. 1987. Mercury accumulation in a new reservoir system. In S.E. Lindberg and T.C. Hutchinson, eds., *International Conference on Heavy Metals in the Environment*. September 1987, New Orleans. Symposium Proceedings, pp. 63–65.
16. Becker, D.S., G.N. Bigham and M.H. Murphy. 1993. Distribution of mercury in a lake food web. Poster from 14th Annual Meeting of the Society of Environmental Toxicology and Chemistry, Houston, TX.
17. Becker, D.S. and G.N. Bigham. 1995. Distribution of mercury in the aquatic food web of Onondaga Lake, New York. *Water Air Soil Pollut.* 80:563–571.
18. Kiorbøe, T., F. Mohlenberg and H. U. Riisgard. 1983. Mercury levels in fish, invertebrates and sediment in a recently recorded polluted area (Nissum Broad, Western Limfjord, Denmark). *Mar. Pollut. Bull.* 14:21–24.
19. Andersen, H.B. 1992. The expansion of mercury contamination, five years after discovery. *Mar. Pollut. Bull.* 24:367–369.
20. Parkman, H. and M. Meili. 1993. Mercury in macroinvertebrates from Swedish forest lakes: influence of lake type, habitat, life cycle, and food quality. *Can. J. Fish. Aquat. Sci.* 50:521–534.
21. Kidd, K.A., R.H. Hesslein, R.J.P. Fudge and K.A. Hallard. 1995. The influence of trophic level as measured by $^{15}N$ on mercury concentrations in freshwater organisms. *Water Air Soil Pollut.* 80:1011–1015.
22. Lasorsa, B. and S. Allen-Gil. 1995. The methylmercury to total mercury ratio in selected marine, freshwater, and terrestrial organisms. *Water Air Soil Pollut.* 80:905–913.
23. Rang, S.A. and P.M. Stokes. 1987. Seasonal variation and uptake and loss of cadmium, lead and mercury in Cladophora in the Niagra River. In S.E. Lindberg and T.C. Hutchinson, eds., *Heavy Metals in the Environment Vol. 2*, New Orleans Conf. Proceedings, pp. 259–261.
24. Surma-Aho, K. and J. Paasivirta. 1986. Organic and inorganic mercury in the food chain of some lakes and reservoirs in Finland. *Chemosphere* 15:353–372.
25. Huckabee, J.W., S.A. Janzen, B.G. Blaylock, Y. Talmi and J.J. Beauchamp. 1978. Methylated mercury in brook trout (Salvelinus fontinalis): absence of an in vivo methylating process. *Trans. Am. Fish. Soc.* 107:848–852.
26. Palmer, S.J. 1992. Mercury bioaccumulation in Lavaca Bay, Texas. M.S. Thesis. 139pp.

27. Gardner, W.S., D.R. Kendall, R.R. Odom, H.L. Windom and J.A. Stephens. 1978. The distribution of methyl mercury in a contaminated salt marsh ecosystem. *Environ. Pollut.* 15:243–251.
28. Tremblay, A., M. Lucotte and D. Rowan. 1995. Different factors related to mercury concentration in sediments and zooplankton of 73 Canadian lakes. *Water Air Soil Pollut.* 80:961–970.
29. Tremblay, A., M. Lucotte and I. Rheault. 1996. Methylmercury in a benthic food web of two hydroelectric reservoirs and a natural lake of Northern Quebec (Canada). *Water Air Soil Pollut.* 91:255–269.
30. Westcott, K. and J. Kalff. 1996. Environmental factors affecting methyl mercury accumulation in zooplankton. *Can. J. Fish. Aquat. Sci.* 53:2221–2228.
31. Ali, I.B., C.R. Joiris and L. Holsbeek. 1997. Total and organic mercury in the starfish *Ctenodiscus crispatus* and the polychaete *Maldenes sarsi* from the Barents Sea. *Sci. Total Environ.* 201:189–194.
32. Plourde, Y., M. Lucotte and P. Pichet. 1997. Contribution of suspended particulate matter and zooplankton to MeHg contamination of the food chain in midnorthern Quebec (Canada) reservoirs. *Can. J. Fish. Aquat. Sci.* 54:821–831.
33. Wong, A.H.K., D.J. McQueen, D.D. Williams and E. Demers. 1997. Transfer of mercury from benthic invertebrates to fishes in lakes with contrasting fish community structures. *Can. J. Fish. Aquat. Sci.* 54:1320–1330.
34. VonBurg, R. 1995. Toxicology update. *J. Appl. Toxicol.* 15:483–493.
35. Suchanek, T.H., P.J. Richerson, L.J. Mullen, L.L. Brister, J.C. Becker, A. Maxson and D.G. Slotton. 1997. The role of the Sulphur Bank Mercury Mine site (and associated hydrogeological processes) in the dynamics of mercury transport and bioaccumulation within the Clear Lake aquatic ecosystem. Report prepared for the USEPA, Region IX Superfund Program. 245pp, plus 9 Appendices and 2 Attachments.
36. Williams, P.M. and H.V. Weiss. 1973. Mercury in the marine environment: concentration in sea water and in a pelagic food chain. *J. Fish. Res. Board. Can.* 30:293–295.
37. Knauer, G.A. and J.H. Martin. 1972. Mercury in a marine pelagic food chain. *Limnol. Oceanogr.* 17:868–876.
38. Chaire de Recherche en Environnement. 1995. Sources et devenir du mercure dans les réservoirs hydroélectriques. Rapport annuel 1994-1995. Chaire de recherche en environnement Hydro-Québec/CRSCG/UQAM, Montréal, Québec.

# Section III

## Risk

# 15 Resources at Risk: A Forest Fire-Based Hazard/Risk Assessment

*Timothy A. Burton, Deirdre M. Dether, John R. Erickson, Joseph P. Frost, Lynette Z. Morelan, William R. Rush, John L. Thornton, Cydney A. Weiland, and Leon F. Neuenschwander*

## CONTENTS

Abstract ..................................................................................................................271
Introduction ............................................................................................................272
Methodology ..........................................................................................................273
  Step 1 ................................................................................................................273
  Step 2 ................................................................................................................273
    Forested Vegetation Outside the Historical Range of Variability ...............273
    Fire Ignition ..................................................................................................275
    Wildlife Habitat Persistence ........................................................................275
    Watershed Hazard (Erosion and Sedimentation Potential) .........................277
    Fisheries Condition ......................................................................................277
  Step 3 ................................................................................................................278
  Step 4 ................................................................................................................278
Results ....................................................................................................................279
Discussion ..............................................................................................................280
Acknowledgments ..................................................................................................283
References ..............................................................................................................283

## ABSTRACT

On the 2.6-million-acre Boise National Forest (NF) in southwestern Idaho, wildfires have burned nearly 50% of the ponderosa pine forest over the last nine years. Much of this forest has burned with uncharacteristic intensity. The historic fire regime — one marked by nonlethal surface fires that removed dense understories of saplings or pole-sized trees and increased nutrient availability — has changed. The altered fire regime now results in severe, stand-replacing fires that kill large areas of forest and return them to grass- and shrub-dominated landscapes. Preliminary analysis shows the remaining ponderosa pine on the Boise NF could be fragmented, with only isolated patches remaining, within the next 20 years by severe, stand-replacing wildfire.

In partnership with the University of Idaho, the Boise NF has developed a Geographic Information System (GIS)-based "hazard/risk assessment" model that estimates where the forest ecosystems are most at risk to severe, large wildfires burning in conditions outside the historical range of variability (HRV) and evaluates important resources at risk to these fires. The hazard/risk assessment links five submodels. When the submodels are linked, the assessment estimates where severe, large wildfires burning in conditions outside HRV would severely deplete late-successional habitat needed by old-growth-dependent and other wildlife species, accelerate naturally high levels of erosion and sedimentation, and increase the likelihood that identified fish populations will not persist.

The hazard/risk assessment is most appropriately used to approximate the relative size and extent of the fire-based ecosystem problem on the forest — the result of excluding fire from fire-adapted ponderosa pine ecosystems. It is intended to "nest" between the large-scale analysis undertaken as part of the Upper Columbia River Basin assessment and the site-specific evaluation performed for landscape- and project-level analysis.

**Keywords:** *fire regime, late-successional habitat, sedimentation, historical range of variability, ponderosa pine*

## INTRODUCTION

The Boise NF has an especially acute focus on forest ecosystem health:

**Its ponderosa pine forests are among the endangered and threatened ecosystems in the U.S.**[1]

Historically maintained by frequent, low-intensity fire, the 1.1-million acres of ponderosa pine forests encompassed by the Boise NF have been altered by decades of fire suppression, grazing, and logging which have removed fire-adapted species. In these and other areas throughout the Interior West, ponderosa pine forests are now dominated by dense stands of Douglas-fir and other fire-sensitive species.[1]

When wildfires now occur in ponderosa pine forests with altered fire regimes, they are more intense, severe, and larger than traditionally experienced. The historic, nonlethal surface fires that removed dense understories of saplings or pole-sized trees and increased nutrient availability have been succeeded by stand-replacing fires that return large areas of forest to grass and shrubland.[2]

In the Boise NF, wildfires in ponderosa pine forest have been increasingly large and severe since 1986. Nearly 500,000 acres of NF land (about 50% of the Boise NF's ponderosa pine forest and almost 20% of the land managed by the Boise NF) have burned. Many of these acres have burned with uncharacteristic intensity. Costs to suppress these fires and undertake emergency watershed rehabilitation have exceeded $100 million. In many severely burned areas, soil productivity and aquatic, wildlife, and plant habitat have been critically damaged.[3,4]

Preliminary analysis shows the remaining ponderosa pine forest could be fragmented, with only isolated pockets remaining, within the next 20 years.[5] To respond to this threat to the forest's ponderosa pine ecosystem, a Boise NF interdisciplinary team, working in partnership with the University of Idaho, has developed a GIS-based "hazard/risk assessment."

The assessment estimates *on a relative, forestwide basis* where forest ecosystems are most at risk to severe, large wildfires burning in conditions outside the HRV and evaluates important resources at risk to these fires. The hazard/risk assessment links five submodels — forested vegetation outside HRV, fire ignition, wildlife habitat persistence, watershed hazard (erosion and sedimentation potential), and fisheries condition. When linked, these submodels estimate where severe, large wildfires burning in vegetation conditions outside HRV would alter the composition, structure, and function of an ecosystem by

- Severely depleting late-successional habitat needed by old-growth-dependent and other wildlife species
- Accelerating naturally high levels of erosion and sedimentation
- Increasing the likelihood that identified fish populations will not persist

## METHODOLOGY

In developing the hazard/risk assessment, the team used GIS tools and state-of-the-art computer software designed to process and analyze spatial information.* The assessment was formulated through the following steps.

### STEP 1

Five GIS submodels were first created to evaluate hazards for specific resources. These submodels included forested vegetation outside HRV, fire ignition, wildlife habitat persistence, watershed hazard (erosion and sedimentation potential), and fisheries condition.

### STEP 2

For each of the five submodels, a relative hazard rating, ranging from 1 (lowest) to 5 (highest), was assigned to each subwatershed. (The 378 subwatersheds on the Boise NF are drainages averaging 6000 acres in size.)

The submodels and sample hazard ratings include the following:

### Forested Vegetation Outside the Historical Range of Variability

This submodel locates areas where ponderosa pine is or could be climax or a major seral species and examines the density of the forested vegetation in these areas based

---

* The assessment was written using ARC/INFO Version 7.03 and uses automated machine language (AML) to process data in the GRID, ARCPLOT, ARCEDIT, and TABLES modules. Most of the analysis was performed using rasterized data in the GRID module, ARCPLOT for graphic output, and TABLES for reports. Data were analyzed and displayed on a system that included an IBM RISC-6000 "390" server and AIX 3.2.5 operating system on a Thinwire Ethernet local area network (LAN).

on June 1992 LANDSAT satellite imagery classification developed at a 30-m pixel resolution.

First, LANDSAT imagery was used to determine current forest cover types for the Boise NF. Forest cover type includes an indication of species composition, density, and, in some cases, maturity (immature/mature).

Second, the NF determined that, historically, many forests were dominated by open stands of ponderosa pine, dry Douglas-fir, and dry grand fir habitat types with densities ranging from a few trees per acres to up to 40 or 50 trees per acre. This information was assembled using historical structure information from the Boise Basin,[6] analysis from the Deadwood Landscape Assessment,[7] and documentation of research in similar habitat types in Montana.[8] The historic density would have produced open stands with less than 30% crown closure, a determination which was subsequently correlated to the LANDSAT satellite imagery cover types.

A four-person interdisciplinary team then determined which cover types represented forest vegetation outside HRV. In determining forested vegetation outside HRV, cover types were evaluated to select those

- Where ponderosa pine is or could be climax or a major seral species, using consistency tables to help determine the appropriate successional pathways[9]
- Which consisted of moderate (more than 30% but less than 70%) and dense (greater than 70%) crown closure

Eight cover types were determined to represent forest vegetation outside HRV. Samples include Douglas-fir/ponderosa pine (dense), Douglas-fir/ponderosa pine (moderate), and Douglas-fir/ponderosa pine/grand fir (dense).

The remaining cover types have a higher likelihood of being close to or within HRV and, therefore, are less susceptible to uncharacteristic fires burning in vegetation outside HRV. These cover types include those where ponderosa pine is not or could not be a major seral species, such as Douglas-fir/lodgepole pine/subalpine fir (dense). They also include those cover types which reflect the vegetation density found historically, such as open (less than 30% crown closure) cover types which contain ponderosa pine.

To determine which subwatersheds have substantial amounts of forested vegetation outside HRV and, therefore, which are at moderate to high hazard, subwatersheds were evaluated based on the number of acres in cover types that represent forest vegetation outside HRV relative to the total number of acres in the subwatershed (minus clouds and shadow, which mask actual on-the-ground conditions).

A scale was then used to express relative hazard of forest vegetation outside HRV on a 1 (lowest hazard) to 5 (highest hazard) scale. For example, a subwatershed in which 45% or more of the acreage included forested vegetation outside of HRV was rated as a "5," while one in which 24 to 35% of the acreage included forested vegetation outside of HRV was rated as a "3."

Percentages assigned to each hazard rating were verified by conducting proportional analysis. Through this process, a frequency distribution was generated by plotting the number of 30-m pixels within subwatersheds (Y-axis) against the percentage of forested vegetation outside HRV found in the subwatersheds (X-axis). A curve was visually fit to the frequency distribution. The curve resembled a standard, bell-shaped curve which indicated a normal distribution. Because the ratings were determined on a "1–5" scale, five equal areas under the curve were delineated to determine logical breaks for the hazard ratings.

The forest vegetation submodel includes two important assumptions; namely, that LANDSAT satellite imagery can be appropriately used to assess forest vegetation outside HRV at a forestwide level and that vegetative hazard for a subwatershed can be appropriately assessed without considering the vegetative hazard of adjacent subwatersheds.

## Fire Ignition

The fire ignition submodel evaluates where fires, both lightning and human caused, have historically started based on Boise NF fire records from 1956 to 1994.

The NF's fire ignition database was first sorted by section (640 acres) to determine the number of total ignitions in each section. The total number of ignitions varied from 0 to 14. The number of ignitions per section was overlain with a map of subwatersheds, and a fire ignition score was then assigned to each subwatershed based on the highest number of ignitions in any one section of the subwatershed.

A scale was then used to express relative risk of fire ignition on a 1 (lowest hazard) to 5 (highest hazard) scale. For example, a subwatershed in which more than eight ignitions occurred in any one section was rated as a "5," while one in which four to five ignitions occurred in any one section was rated as a "3."

These ratings were developed based on professional judgement and were later ratified through proportional analysis. A frequency distribution curve for the number of subwatersheds selected in each hazard rating score was developed. The resulting curve was bell shaped, indicating a normal distribution.

The ignition submodel includes several assumptions, including the assumption that a subwatershed is at relatively high risk to fire ignition if any section within that subwatershed is at high risk and the assumption that ignitions will occur at the same frequency as they have over the past several years (1956 to 1994).

## Wildlife Habitat Persistence

Wildlife evolved with fire. However, recent fires — burning larger and more intensely in some habitats than noted historically — represent a concern to wildlife persistence for some species. There is evidence that catastrophic uncertainty poses greater risks to population persistence than demographic or even environmental types of uncertainty.[10] The frequency and severity of extreme fire events, combined with management activities, may be critical to the long-term productivity and persistence of certain wildlife populations.[11]

The wildlife habitat persistence submodel is based in part on the assumption that extensive, contiguous, stand-replacing fires are the primary threat to wildlife persistence.[11] This submodel first assesses the amount and distribution of mid- and late-seral wildlife habitat in a subwatershed and then determines the amount and distribution of this habitat considered outside HRV, and therefore considered at risk to large, uncharacteristic fires.

First, LANDSAT satellite imagery cover types were combined with Digital Elevation Model (DEM) information such as elevation, slope, and aspect to develop a map of habitat types. (Habitat types reflect potential natural vegetation and indicate the successional pathway following a disturbance.) Habitat types with similar successional pathways and disturbance regimes were combined into "habitat type groups."

"Habitat at risk" and "habitat *not* at risk" was then delineated by identifying habitat groups of mid- and late-seral habitat *outside* HRV and *within* HRV, respectively. Habitat at risk includes Douglas-fir and dry grand fir habitat types with moderate to dense canopy closure, while habitat not at risk includes Douglas-fir habitat types with open canopy and lodgepole pine and subalpine fir habitat types with moderate to dense canopy.

Persistence hazard ratings were developed to reflect the likelihood that suitable habitat will *not* persist, with the assumption that the more extensive the vegetation outside HRV, the higher the likelihood that extensive, uncharacteristic wildfire might occur, and that mid- and late-seral habitats would not persist. Hazard ratings were assigned based on the amount, location, and size of mid- and late-seral habitat remaining in the subwatershed after the vegetation was removed through wildfire. The amount and distribution of later seral habitat remaining within a subwatershed are a function of the amount of habitat present initially and the amount potentially eliminated by uncharacteristic fire. Amount, location, and size were all considered because all are key to the habitat requirements of several threatened, endangered, sensitive, and/or management indicator species, including the white-headed woodpecker and fisher.

A scale was then used to express the relative risk that habitat will not persist on a 1 (lowest hazard) to 5 (highest hazard) scale. For example, a subwatershed in which 10% or less of a subwatershed would remain in mid- or late-seral habitat, with no patch at least 350 acres in size, was rated as a "5," while one in which 15% or less would remain in mid- or late-seral habitat, with one patch at least 350 acres in size, was rated as a "3." (Low-elevation subwatersheds which primarily consist of grass, brush, and shrublands were not included in this analysis.)

These ratings were developed based on professional judgement and were later ratified through proportional analysis. As part of this process, the relationships between the percent of mid- or late-seral habitat remaining within a subwatershed and the number of subwatersheds and number of 30-m pixels, respectively, were compared. The resulting curves were very similar, thus verifying the correlation between the relatively small data unit (30-m pixel) and analysis unit (subwatershed).

As part of the proportional analysis, a different set of criteria — using only "percent mid- or late-seral habitat remaining" and ignoring patch size — were used to formulate hazard ratings. It was determined that although this set of criteria produced similar results as the set that included patch size, the patch-size version more accurately reflected on-the-ground conditions.

The wildlife persistence submodel includes several important assumptions, including the assumption that stand-replacing fire is the primary change agent influencing the quantity, quality, and distribution of wildlife habitat across the forest and that areas of high wildlife habitat hazard will eventually burn in a stand-replacing fire.

**Watershed Hazard (Erosion and Sedimentation Potential)**

The watershed hazard submodel is based on the knowledge that there are inherent differences in natural (undisturbed) sedimentation rates from land types (areas with similar soils and landforms and, therefore, similar hazards and capabilities) within a subwatershed and that following wildfire there is the potential for accelerated sedimentation.[12-15] This submodel evaluates potential natural sediment yield, as determined from land types.

The average natural sedimentation rate for each subwatershed was first determined. First, the area encompassed by each land type within a subwatershed was calculated, and the natural sedimentation rate associated with each land type was identified from the Boise NF GIS database. The average natural sedimentation rate for each subwatershed was then calculated by dividing the total natural sediment yield for the watershed (expressed in tons/year) by the area encompassed by the subwatershed (expressed in square miles).

A scale was then used to express the relative watershed hazard (erosion and sedimentation potential) on a 1 (lowest hazard) to 5 (highest hazard) scale. For example, a subwatershed with an average natural sediment yield of 75 or more ton/mi$^2$/year was rated as a "5," while one with an average yield of 36 to 49 ton/mi$^2$/year was rated as a "3."

This scale was developed based on professional judgement and field review of several post-fire areas. A cumulative sedimentation rate analysis was then completed for the subwatersheds. This analysis displayed the sedimentation rates at break points for 20, 40, 60, and 80% of the total number of subwatersheds (corresponding with the same number of proportional hazard rating breaks within the 1 to 5 rating scheme). Although this analysis showed slightly different breaks, the rating scale developed through professional judgement was used because the Boise NF hydrologist believed the professional values more accurately reflected the relatively high natural soil erosion and sedimentation rates found on the Boise NF, much of which is underlain by the erosive, granitic soils of the Idaho Batholith. In addition, the differences between the two rating scales are quite minor, given the relative, broad-scale, "coarse filter" nature of the submodel.

The watershed hazard submodel includes important assumptions, including the assumption that erosion and sedimentation rates will increase following wildfire relative to their natural rates.

**Fisheries Condition**

The fisheries condition submodel selects spring/summer chinook salmon and bull trout as indicator species because in Idaho chinook have been listed as "endangered" and bull trout as "warranted but deferred" under the Endangered Species Act of 1973. The submodel uses a scheme to prioritize watersheds for species protection, along with identified population strength and fragmentation factors.[16] The fisheries condition submodel assumes that large wildfires burning in conditions outside HRV would lead to environmental disturbances (floods, etc.) that decrease the likelihood of persistence for those fish populations low in abundance (chinook salmon) or important to regional populations (local bull trout populations).

Ratings for each of three components — species, relative population strength, and isolation — were assigned to each subwatershed, based in part on sampling information located in the Boise NF's Aquatic Survey Database. These components were used to identify the strongest chinook salmon and bull trout populations, as well as nearby weakened populations with the greatest chance for recovery. The three components were then averaged to calculate an overall hazard rating for each subwatershed.

In general, for chinook salmon, moderate and high hazard subwatersheds (rated 3 or higher on the 1 to 5 scale) are those where spawning and rearing habitat for chinook salmon exists. For bull trout, moderate and high hazard subwatersheds are those where, within strong regional populations, there is risk that local populations will not persist. Those populations are relatively lower in abundance, smaller in areal extent, isolated from other populations, and therefore less likely to recover from uncharacteristic fire.

Percentages assigned to each hazard rating were verified by conducting proportional analysis. A frequency distribution curve was generated for the number of subwatersheds selected in each hazard rating class. This curve shows a relatively flat, but bell-shaped distribution. The data appear to be distributed in four general clusters; no subwatersheds were rated in the 0.00 to 0.99 category. The first (ratings between 1.00 and 1.99) include subwatersheds where bull trout and chinook salmon have not been found or are not known to migrate. Data in the second cluster (2.00 to 2.99) represent subwatersheds where bull trout are known or expected to migrate and overwinter. Data in the third cluster (3.00 to 3.99) are represented mostly by subwatersheds where bull trout production is relatively strong (strong focal habitats). Subwatersheds in the fourth cluster (4.00 to 5.00) are represented by production of endangered chinook salmon and/or weak bull trout production (weak focal and adjunct habitats).

The fisheries condition submodel contains several assumptions, including the assumption that a population's sensitivity or resilience to large, severe disturbance depends on its overall condition, as defined by relative size, isolation potential, and availability of all life-history forms.

## STEP 3

An overall "high risk" rating was assigned to a subwatershed if it received moderate ("3") or higher hazard ratings from ALL FIVE submodels. A break point of "3" was selected because it approximated a midpoint on the scale, and the team believed it was more important to *include* some areas potentially at lesser risk than to *exclude* areas potentially at greater risk.

## STEP 4

A watershed was rated as "high risk" if at least *one* subwatershed within it received an overall high risk rating. (The 82 watersheds on the Boise NF are larger drainages, about 30,000 acres in size, which consist of several subwatersheds.) This assignment reflects the Boise NF's observation that the recent uncharacteristic wildfires are burning across vast landscapes and entire watersheds.

## RESULTS

The hazard/risk assessment was designed in part to answer two questions:

1. **Where are forest ecosystems most at risk to severe, large wildfires burning outside HRV?** — Based on current information and analysis, the forest ecosystems most at risk to severe, large wildfires burning outside HRV include large areas of moderate and dense forest where ponderosa pine is or was a major seral species *and* where moderate to high numbers of fires have occurred. By linking the fire ignition submodel (which can identify those subwatersheds with moderate to high levels of fire ignition) with the forested vegetation outside the HRV submodel (which can identify those subwatersheds with moderate to high hazard for forested vegetation outside HRV), the assessment estimates that up to 152 subwatersheds (total of 1,196,781 acres) are those most at risk to severe, large wildfires burning in vegetation conditions outside HRV (Figure 1).
2. **What important resources are at risk to these severe wildfires?** — To determine what important resources are at risk to these fires, the hazard/risk assessment estimated where severe, large wildfires burning in vegetation conditions outside HRV would affect specific wildlife, watershed, and fisheries resources. By linking all five submodels included in the assessment, analysis indicates that in 20 watersheds (total of 610,389 acres) all of these important resources could be affected by severe, large wildfires burning in vegetation conditions outside HRV (Figure 2).

Comparative information is illustrated in Table 1:

### TABLE 1
### Forest Most at Risk from Uncharacteristic Wildfire

|  | Acres | % of Forest[a] |
|---|---|---|
| Most at risk to large, uncharacteristic fire | 1,196,781[b] | 40 |
| Important resources affected by fire | 610,389[c] | 20 |

[a] Percentage relative to Boise NF *encompassed* area of 3,000,000 acres, as captured in 1992 LANDSAT satellite imagery. Includes about 350,000 intermingled state, other federal, and private land. Net Boise NF is about 2,650,000 administered acres.

[b] Figure represents *all* acres within 152 subwatersheds, including some grass and shrublands, subalpine fir, etc.

[c] Figure represents *all* acres within 20 watersheds, including some grass and shrublands, subalpine fir, etc.

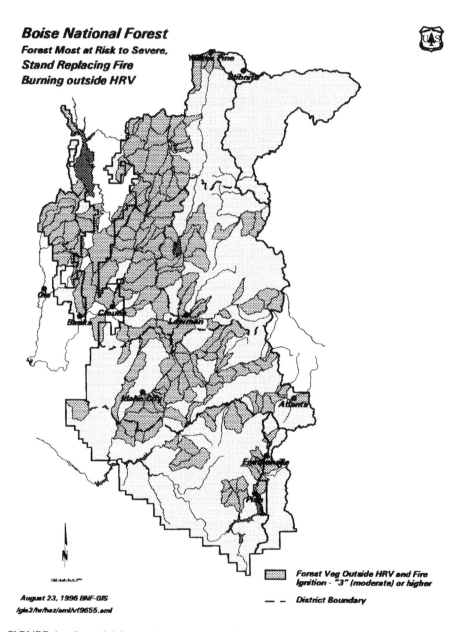

**FIGURE 1**  Potential for uncharacteristic wildfire.

## DISCUSSION

The hazard/risk assessment is designed to evaluate the *relative* size and extent of the Boise NF's challenge in managing sustainable, resilient, and resistant ponderosa pine ecosystems. It also tells land managers where to "go look closer" — where to begin evaluating site-specific conditions at a finer scale; where to begin determining

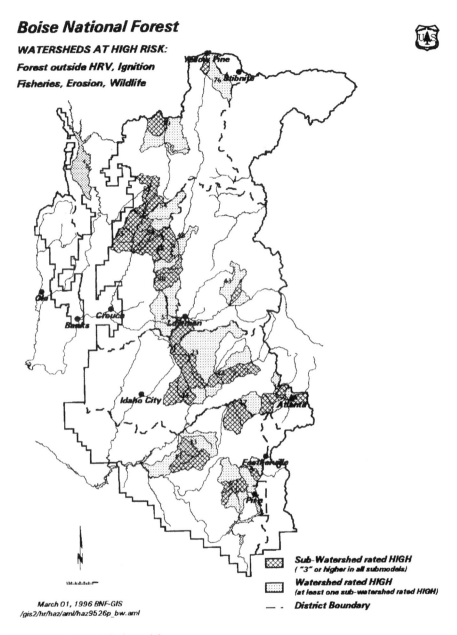

**FIGURE 2** Watersheds at risk.

a "desired future condition" for a landscape at risk; and, finally, where and what specific projects might be designed and undertaken, if needed, to begin restoring sustainable ecosystem conditions across the landscape.

The assessment is intended to "nest" between the large-scale analysis undertaken as part of the Upper Columbia River Basin assessment and the more site-specific evaluation performed for watershed- and landscape- or project-level analyses. The assessment is compatible with the U.S. Forest Service National Hierarchical Framework of Ecological Units. The Boise NF lies in Section M332A (Idaho Batholith) of Province M332 (Middle Rocky Mountain Steppe-Coniferous Forest-Alpine Meadow).[17] Habitat types developed as a basis for the wildlife persistence model were in turn developed based on "section" information established by the Upper Columbia River Basin assessment. Thus, information from the hazard/risk assessment can be aggregated to ecological "sections" at a larger scale.

Because the assessment was developed to analyze conditions on a forestwide basis, it should not be used for more site-specific watershed- and landscape- or project-level work without further evaluation and refinement.

The hazard/risk assessment represents an important addition to the "analysis toolbox" available to today's land managers. It recognizes the potential for large, severe wildfires burning with altered fire regimes to damage important resources and to substantially interrupt successional pathways historically not experienced on the large scale we see today. Because it focuses on potential effects to fisheries populations, late-successional wildlife habitat, and sedimentation, the hazard/risk assessment highlights the consequences of severe, stand-replacing fires burning outside historical patterns to disturb the dynamics of an entire ecosystem.

Given the potential loss of ponderosa pine-dominated forests on the Boise NF in the next 20 years, the hazard/risk assessment can be a primary tool for prioritizing areas most at risk for further evaluation. The model's structure is particularly well suited to examine situations like this in which time and resources for assessment and resolution are limited, because the assessment uses selected criteria to progressively narrow the area of consideration to one which is "do-able."

The assessment's use of GIS as the "modeling medium" is particularly appropriate in examining landscape conditions because GIS can analyze large amounts of data and sophisticated relationships across extensive areas. Since GIS is a widely used, state-of-the-art analysis tool, it lends itself especially well to sharing information among resource specialists from different agencies and organizations. It also facilitates expansion of the hazard/risk assessment to incorporate different ownerships and boundaries (if desired), since the challenges to ecosystem health cross jurisdictional boundaries and affect resources and resource users at many scales.

Forest scientists recognize that to restore the resistance and resilience of ecosystems with altered fire regimes, land managers must use several tools, including fire and timber harvest.[18,19] The Boise NF will need to conduct low-intensity fire under prescribed conditions, to begin restoring fire-dependent ecosystems, as well as to remove ground fuels and recycle nutrients; and thinning to remove less fire-resistant trees such as Douglas-fir and grand fir, while leaving the larger, fire-resistant ponderosa pine. (In today's altered landscapes, thinning is needed to remove trees from dense areas where prescribed fire alone could result in a lethal, stand-replacing wildfire.) By identifying the areas most at risk, the hazard/risk assessment takes land managers "to the ground" to look closer, with the possible outcome that some of these restoration treatments may be prescribed. If so, the hazard/risk assessment may then support the adjustment in the Boise NF's management course needed to incorporate different types of timber harvest and more extensive use of prescribed fire than traditionally undertaken.

## ACKNOWLEDGMENTS

The authors thank members of the Boise National Forest Leadership Team for their support and guidance, especially Dave Rittenhouse, Cathy Barbouletos, Jim Lancaster (retired), Rich Christensen, and Laurie Tippin. The authors also thank Ingra Draper, Cameo Flood, Bob Giles, Nadra Angerman, Diane McConnaughey, Bert Strom, Diana Wall, Julie Weatherby, and Bob Wing for their assistance in methodology and data analysis and display. Special thanks to Ralph Thier for reviewing an earlier version of this work.

For questions or futher information regarding the work described in this chapter, please contact:

> **Cydney A. Weiland**
> Boise National Forest
> 1249 S. Vinnell Way
> Boise, ID 83709
> e-mail: cweiland/r4_boise@fs.fed.us

## REFERENCES

1. Noss, R.F., E.T. LaRoe, III, and J.M. Scott. 1995. Endangered ecosystems of the United States: a preliminary assessment of loss and degradation. Biological Report 28, National Biological Service, U.S. Department of the Interior. pp. 12, Appendix A.
2. Crane, M.F. and W.C. Fischer. 1986. Fire Ecology of the Forest Habitat Types of Central Idaho. General Technical Report INT-218. USDA Forest Service, Intermountain Research Station, Ogden, UT.
3. USDA Forest Service, Boise NF. 1992. Foothills Wildfire Timber Recovery Project. Environmental Assessment. pp. 3, 29–31.
4. USDA Forest Service, Boise NF. 1995. Boise River Wildfire Recovery. Final Environmental Impact Statement. pp. III-38–43, III-51–52, III-75–77, III-79, III-80–83.
5. Neuenschwander, L. 1995. Unpublished data presented to Boise NF District interdisciplinary teams; February–April, 1995.
6. Sloan, J. 1994. Historical Density and Stand Structure of an Old Growth Forest in the Boise Basin of Central Idaho. Pre-publication draft. USDA Forest Service, Intermountain Forest and Range Experiment Station, Ogden, UT.
7. USDA Forest Service, Boise NF. 1994. Deadwood Landscape Assessment.
8. Arno, S.F., J.H. Scott, and M.G. Hartwell. 1995. Age-Class Structure of Old Growth Ponderosa Pine/Douglas-Fir Stands and Its Relationship to Fire History. Research Paper INT-RP-481. USDA Forest Service, Intermountain Forest and Range Experiment Station, Ogden, UT.
9. Steele, R., R.D. Pfister, R.A. Ryder, and J.A. Kittams. 1981. Forest Habitat Types of Central Idaho. General Technical Report INT-114. USDA Forest Service, Intermountain Forest and Range Experiment Station, Ogden, UT.

10. Shaffer, M.L. 1987. Minimum viable populations: coping with uncertainty. In *Viable Populations for Conservation*. M.E. Soule, ed. Cambridge Press. pp. 69–86.
11. Erickson, J.R. and D.E. Toweill. 1994. Forest health and wildlife habitat management on the Boise National Forest, Idaho. In *Journal of Sustainable Forestry* ("Assessing Forest Ecosystem Health in the Inland West") 2:1–2; R.N. Sampson and D.L. Adams, eds. Haworth Press, Binghamton, NY.
12. Megahan, W.F. and D.C. Molitor. 1975. Erosional effects of wildfire and logging in Idaho. In *Watershed Management Symposium*, August 11–13, 1975, Logan, UT, ASCE Irrigation and Drainage Division. pp. 423–442.
13. Helvey, J.D. 1980. Effects of a north central Washington wildfire on runoff and sediment production. *Water Resource Bulletin* 16(4): 627–634.
14. Schultz, R.L., J. Cauhorn, and C. Montagne. 1986. Quantification of erosion from a fire and rainfall event in the Big Belt Range of the Northern Rocky Mountains. Montana Agriculture Experiment Station, Montana State University, in cooperation with the USDA Forest Service, Helena National Forest.
15. Trondel, C.A. and G.S. Bevenger. 1994. Effects of fire on streamflow and sediment transport, Shoshone National Forest, Wyoming. In Proceedings of Greater Yellowstone Fire Symposium.
16. Rieman, B.E. and J.D. McIntyre. 1993. Demographic and Habitat Requirements for Conservation of Bull Trout. General Technical Report INT-302. USDA Forest Service, Intermountain Forest and Range Experiment Station, Ogden, UT.
17. McNab, W.H. and P.E. Avers. 1994. Ecological Subregions of the United States: Section Descriptions. WO-WSA-5. USDA Forest Service, Washington, D.C.
18. Agee, J.K. 1995. Forest fire history and ecology of the Intermountain West. In *Inner Voice*, March–April, 1995. pp. 6–7.
19. Mutch, R.W. 1995. Prescribed fire and the double standard. In *Inner Voice*, March–April, 1995. pp. 8–9.

# 16 Uncovering Mechanisms of Interannual Variability from Short Ecological Time Series

*Alan D. Jassby*

## CONTENTS

Abstract ..................................................................................................285
Introduction ...........................................................................................286
Isolating Modes of Variability ............................................................287
Identifying Causal Factors ..................................................................295
Parameterizing Relationships .............................................................299
Choosing Among Models ...................................................................301
Concluding Remarks ...........................................................................303
Acknowledgments ................................................................................304
References .............................................................................................304

## ABSTRACT

The processes underlying year-to-year variability in almost any ecological variable are typically numerous and nonlinear. Because of the relatively short length of most instrumental records — commonly several decades or less — the data do not readily divulge these mechanisms. Here, we outline a simple practical procedure for analyzing such records. The procedure has evolved during the course of data analyses for several lake and estuarine sites in California. It consists of (1) decomposing spatial-temporal series into individual modes of variability using a special application of rotated principal component analysis, (2) exploring the qualitative relations between individual variability modes and plausible causal factors using general additive models, and (3) parameterizing the relations for hypothesis testing. The parameterization is seen primarily as the final step in identifying dominant mechanisms for inclusion in a predictive model, whether numerical or statistical, not as a predictive tool in itself. Nonetheless, it might be useful at times to arrive at a simple predictive statistical tool by (4) choosing from among plausible parameterizations identified in Item 3 the one that minimizes the prediction error. These steps are briefly illustrated by specific applications from Lake Tahoe, Castle Lake, and the northern San Francisco Estuary.

**Keywords:** *data analysis, estuaries, interannual variability, lakes, statistical models*

## INTRODUCTION

Long-term ecological investigations often include at least one ecological variable that serves as an "integrative" index of ecological conditions. Inland water studies, for example, usually include either chlorophyll *a* as a measure of primary producer biomass or carbon-14 uptake rates as a measure of primary productivity. These indices are thought to be integrative in the sense that they are responsive to many ecological processes. Chlorophyll *a* levels, for example, can reflect both the availability of nutrients as well as the structure of the food web. Time series for these integrative variables are therefore also time series for an important dimension of ecological health or, more neutrally, ecological status. Understanding the causal factors behind the interannual and decadal variability of these time series is a central issue in learning how to anticipate and, where appropriate, modify ecological change.

The very nature of these variables, as responsive to multiple ecological interactions, is also a barrier to understanding their temporal variability. Because they participate in many causal pathways, their interannual and longer-term variability can be a complex combination of processes. The challenge of unraveling this complexity and distinguishing the most important mechanisms underlying interannual variability — and usually there are only a few — is made all the more difficult by the relatively short lengths of many of these time series. Long-term studies date mainly from the 1960s or later, which means that essentially only 30 to 40 observations are available for examining year-to-year changes.

Here, we present a series of simple steps that has proven to be well suited to the analysis of such series. The central feature of the procedure is a special application of principal component analysis for decomposing time series with a higher than annual frequency into seasonal "modes" of variability, each of which is characterized by its own time series. The technique is related to a spatial regionalization approach commonly used in meteorology and oceanography[1] that has a natural extension to the seasonal domain.[2] Probably the earliest application is a meteorological one to a monthly time series of Central England temperatures.[3] The first ecological application appears to be that of Smith and Hayden[4] to a weekly time series of snow goose population abundance. Jassby et al.[5] extended the application to the combined spatial-seasonal domain. The method has the effect of deseasonalizing the original series, resulting in a small number of annualized series that are then available for further investigation. Although this subsequent investigation can take many forms, we describe one specific approach that uses exploratory graphical techniques to arrive at a set of simple parameterized models. These models are then compared for their predictive capabilities. We rely on several studies of lake and estuarine sites in California for examples.

## ISOLATING MODES OF VARIABILITY

A formal treatment of principal component analysis (PCA), as well as advice and caveats regarding its use, can be found in many general, multivariate statistics texts[6] and fewer, more specialized texts.[1,7] Our object here is to gather together and recount the various issues that need to be considered by ecologists applying PCA to interannual variability. These considerations have evolved largely from the practical experience of researchers studying geophysical fluids. We focus on a simple geometric description of PCA. Suppose $\mathbf{x}(t) = [x_1(t), x_2(t),...,x_{12}(t)]^T$ is the vector of 12 monthly means of a variable $x$ for year $t$, and $\overline{\mathbf{x}} = [\overline{x}_1, \overline{x}_2,..., \overline{x}_{12}]^T$ is the corresponding vector of long-term means for each month. In the state-space representation of these data, each yearly anomaly (i.e., deviation from the long-term mean) $\mathbf{x}(t) - \overline{\mathbf{x}}$ occupies a point in 12-dimensional euclidean space. The $i$th month is represented by the $i$th axis, and the projection of year $t$ on the $i$th axis is simply $x_i(t) - \overline{x}_i$. A PCA of the 12 monthly variables can be described as a rotation of the axes so that the anomalies are expressed in a new and special coordinate system:

$$\mathbf{x}(t) - \overline{\mathbf{x}} = a_1(t)\mathbf{e}_1 + a_2(t)\mathbf{e}_2 + ... + a_{12}(t)\mathbf{e}_{12} \tag{1}$$

where $\mathbf{e}_i = [e_{1,i},...,e_{12,i}]^T$ is the $i$th new axis, and $a_i(t)$ is the $i$th coordinate for year $t$ in the new coordinate system. Because we have simply rotated the original axes about their origin, they are still orthogonal and of unit length. The new coordinate system selected by PCA, however, has a special additional property: the first $m$ axes, for any $m = 1, 2, ..., 11$, account for more of the data's variability than any other m axes. For example, the projections of the observations on the first axis have a larger variance than for any other possible line, and similarly for the projections on the plane formed by the first two axes.

Note that this approach implicitly assumes a fixed seasonal pattern and treats all variability as annual deviations from this fixed pattern. In some cases, it may be more accurate to allow for a long-term evolution in the seasonal pattern. One approach is *Seasonal and Trend Decomposition using Loess* (STL), a filtering procedure for seasonal time series that has as its primary goal the estimation of a changing seasonal component.[8] STL involves a sequence of smoothing operations based on locally weighted regression (loess). PCA can then be applied to the residual series after removal of the identified seasonal component. Another concern lies with the stationary properties of the data. Demeaning the data from their monthly means, as we do here, may not always be an innocuous procedure when the data are nonstationary. In the latter case, it would be worthwhile to consider alternatives such as cointegration analysis that have proven particularly powerful in the analysis of long-term relationships in economic data.[9]

Mathematically, $\mathbf{e}_i$ is the $i$th eigenvector of the covariance matrix of the original variables $x_1(t), x_2(t),...,x_{12}(t)$, and the variance accounted for by $\mathbf{e}_i$ is the $i$th eigenvalue $\lambda_i$. In some PCA applications, the eigenvectors and eigenvalues of the correlation rather than covariance matrix are extracted. The correlation matrix generally is more appropriate when the original variables are disparate in their units or scale of measurement. For our goal, however, namely, identifying the major sources of

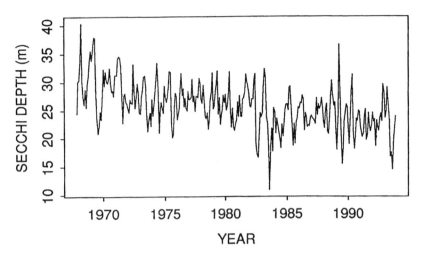

**FIGURE 1** Monthly means of Secchi depth at the Index Station of Lake Tahoe.

interannual variability, the variables are all derived from the same time series and the covariance matrix is the appropriate object.[10] In meteorological and oceanographic applications, the $e_i$ are often known as *empirical orthogonal function* (EOF) coefficients and the $a_i$ as amplitudes. In other contexts, particularly the social sciences, the eigenvectors are scaled by the square root of the corresponding eigenvalue and referred to as *principal component loadings*; the corresponding $a_i$ are called *scores*. We will also refer to the $e_i$ as modes, in the sense of form or pattern of variability.

As an example, consider the time series of Secchi depth for Lake Tahoe, collected at a fixed station (the "Index Station") approximately every 12 days since 1967 (Figure 1).[11] Secchi depth is simply the depth at which a 25-cm white disk disappears from the view of a surface observer and is a measure of transparency to the human eye. It integrates the combined effects of inorganic particulate material and phytoplankton and is therefore responsive to climate, erosional processes in the watershed, and biological processes in the lake itself. The data were aggregated by calculating monthly trapezoidal means. In general, years should be split at a time when the serial correlation is low;[3] otherwise, there is the danger of assigning the manifestation of a single mechanism to two consecutive years, ultimately confusing identification of the underlying mechanisms. For Secchi depth, serial correlation is at a local minimum for September through October and again for March through April. We chose to split years at the end of September. Our 12 variables are therefore the months of the water year, from October through September of the following year.

The first four modes (eigenvectors), the variance (eigenvalue) associated with each mode, and their corresponding amplitude time series are illustrated in Figure 2. Note that more than 85% of the total variance is associated with these four modes. We will return to the shapes of these modes later.

By choosing $m < 12$, we can collapse the dimensionality of the data from 12 to $m$ dimensions. The properties of the new coordinate system ensure that these m axes

# Uncovering Mechanisms of Interannual Variability

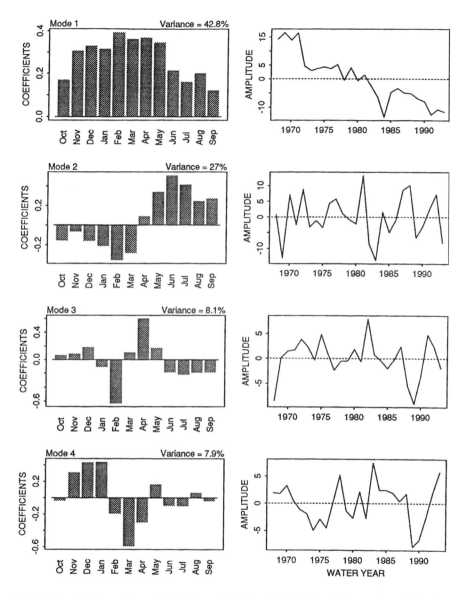

**FIGURE 2** Modes and amplitudes determined by a PCA of the 12 time series of monthly mean Secchi depths.

retain as much of the data's variability or information as possible. The smaller the value of $m$, the fewer the variables we need to examine, but the less information retained. How do we choose the value of $m$? Jolliffe[7] describes the available selection procedures, of which there are a bewildering number. One of the early researchers on this problem, H. F. Kaiser, remarked that "it is really an easy problem in one sense; I use to make a selection rule up every morning before breakfast."[1]

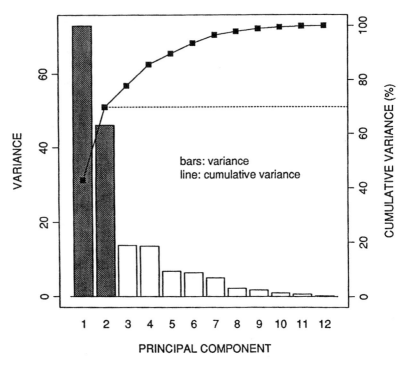

**FIGURE 3** Scree graph of the variances determined by a PCA of the 12 time series of monthly mean Secchi depths.

Based on practical experience and simulation studies, two types of procedures appear to stand out above the others.[1] The first is the *scree test*, popularized by Cattell,[12] and its relatives. A plot of the variance (eigenvalue) associated with each axis (eigenvector) vs. the index of the axis is known as a *scree graph*, because of its resemblance to the accumulation of coarse rock debris (i.e., scree) resting against the base of a cliff (Figure 3). In the scree test, all axes up to and including the first major inflection point in the slope of the scree graph are considered to be worth retaining. In the case of Secchi depth, for example, the scree test suggests a three-dimensional subspace.

The second test is a Monte Carlo procedure developed by Overland and Preisendorfer[10] known as *Rule N*. Rule N consists of applying PCA to a n by p data matrix formed by random samples from a standard normal distribution, where n is the number of observations (in our application: years) and p is the number of variables (in our application: months) for the observed data set. The fraction of the total variance explained by each axis is compared to the value found with the observed data set. The procedure is repeated many times (at least 1000). If the variance for the m-th axis of the real data set exceeds those for the random data sets at least 95% of the time, then the m-th axis is deemed worth keeping: it explains more of the data set's variability than one would have expected from chance alone. Efron and Tibshirani[13] discuss a similar bootstrap approach.

One of the problems with Rule N is that it steps out of the purely mathematical domain where no model is assumed to the statistical domain where the data are assumed to have certain properties. The observed variables may, in fact, have more structure than a normal variate. When these variables are time series of monthly means, they may exhibit serial correlation reflecting some kind of persistence from one year to the next. If this is in fact the case, Rule N should be modified so that the random variables are generated with the same kind of serial correlation structure.[5] An additional problem with Rule N in its basic form is that the test for any axis beyond the first is not conditional on the results for the preceding axes. To address this difficulty, Stauffer et al.[14] proposed comparing fractions of the variance *remaining* after accounting for preceding axes, rather than fractions of the total variance itself.

In the case of Secchi depth, Rule N indicates that only the first two axes need to be retained; the scree test suggests three. This disagreement among criteria is not uncommon, and a compromise value of $m$ must often be chosen. Given the exploratory nature of this approach, several values of $m$ could be investigated. For illustrative purposes, we will assume that $m = 2$ in what follows. These first two axes account for a total of 69.8% of the original variance (Figure 3).

The combination of PCA and a subsequent examination of the variances associated with each axis have enabled us to choose a subspace in which to further examine the data. We have effectively reduced the dimensionality of the original data set from 12 to 2, while retaining 70% of the variability. Do the retained axes have any further use in understanding the causal factors underlying year-to-year variability? Typically, the coefficients associated with the first axis are all of the same sign and magnitude (Figure 2). For the second axis, the first half of the coefficients contrast with the second half. Subsequent axes exhibit increasingly complex coefficient patterns, resembling a series of harmonics. These predictable shapes are called *Buell patterns*[15] and are largely an artifact of the PCA itself, not a property of the data. PCA is designed to define subspaces preserving as much of the data's variability as possible, not to provide physically meaningful transformations of the original variables.[7] Unfortunately, some researchers expend much effort trying to find meaning in these axes.

An additional, physically meaningful set of axes can be found, however. Once a subspace has been defined through PCA and a choice for $m$, the axes defining the subspace can be rotated, this time to facilitate insight into the variability. Various rotation methods exist, using different strategies to arrive at what is called *simple structure*.[16] An effective way to visualize simple structure rotation is to consider each variable $x_j - \bar{x}$ in euclidean $m$-space with coordinates $e_{j1}, e_{j2}, \ldots, e_{jm}$ (the "factor diagram").[17] The variables will often form subgroups or clusters in this space. To achieve simple structure, the axes of this space are rotated so that they line up as best as possible with the clusters. In the resulting coordinate system, only one of the coefficients $e'_{j1}, e'_{j2}, \ldots, e'_{jm}$ associated with each variable $x_j - \bar{x}$ will have a large value and the other coefficients will be small, at least to the extent that simple structure can be achieved. Simulations show that a simple structure rotation will reveal mechanisms generating the variability even when these remain disguised by the original PCA.[15]

Two types of rotation are used. *Orthogonal* rotations retain the orthogonality of the original axes, while *oblique* rotations do not. Oblique rotations are more flexible and can locate cluster positions more accurately. They are preferred when the goal is identifying underlying mechanisms. Further, the modes produced by oblique rotation are less sensitive to overestimates of the subspace dimension $m$.[18] Simulation studies and other considerations suggest that, of those methods that are readily available, the *varimax* method should be used for orthogonal rotation and the *promax* method for oblique rotation.[15] For certain rotation algorithms, more meaningful results are sometimes obtained with the principal component loadings rather than the eigenvectors.

We subjected the first two axes of our example to both a varimax and a promax rotation. The two rotations differed little in this case (Figure 4). Compared to the original axes, however, two new and very different modes emerge. Note how the variability is more equally distributed between the two rotated modes than between the original ones. Also, there are fewer coefficients of intermediate value, April being a notable exception. The second mode coincides with the annual period of strong thermal stratification when a temperature discontinuity at approximately 20 m separates a well-mixed upper layer from a cooler, deeper layer of the lake. In contrast, the first mode coincides with a period when the discontinuity slowly weakens and deepens, and the deeper waters are gradually mixed into surface waters. April is a transition month in which strong stratification often develops. Together these modes suggest that most of the interannual variability is due to two independent mechanisms, one operating during spring and summer thermal stratification and the other controlling variability during the rest of the year. Thus, the analysis gives us strong clues for and constraints on the underlying mechanisms.

With the Secchi depth time series, we binned each year's data by time (months) in order to obtain our multidimensional representation of that year. When the data have a spatial as well as a temporal dimension, they can be binned by spatial subregions as well as by time periods. Consider Castle Lake, for example, where primary productivity measurements are made at a fixed station from the surface to 30+ m throughout the summer.[5] Each year can be represented as a collection of 24 variables by binning the data by month (June through September) and by depth interval (5-m intervals from 0 to 30 m) (Figure 5A).

Sometimes it is useful to define the spatial extent of bins with other than fixed geographic coordinates. Consider the northern San Francisco Estuary, for example, where chlorophyll *a* is measured at 24 fixed stations at 1 m depth approximately every month.[19] This area lies along the continuum between freshwater river habitat and the ocean and therefore exhibits a strong horizontal salinity gradient. Many important phenomena in the estuary, such as spatial maxima in turbidity and chlorophyll, happen at particular places in the salinity gradient. The data could be binned by time and latitude-longitude, instead of by time and depth as in the case of Castle Lake, but, because the salinity gradient responds to fluctuating freshwater flow, locations in the salinity gradient do not correspond to fixed geographic locations. An effective alternative is to redimensionalize the problem, binning the data by month and by salinity (specific conductance) interval. Due to certain constraints on

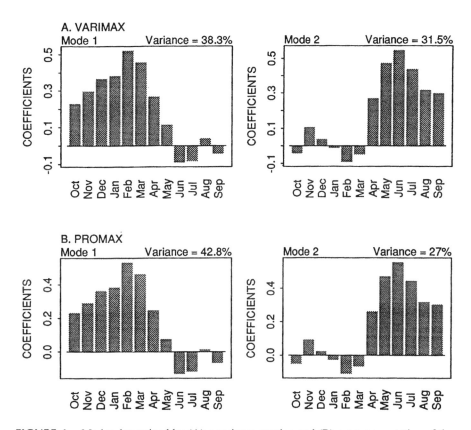

**FIGURE 4** Modes determined by (A) a varimax rotation and (B) a promax rotation of the first two modes in Figure 2.

the data, we ended up binning by 4 time and 4 conductance intervals, a total of 16 bins (Figure 5B).[19]

Thus, the technique possesses enough flexibility to accomodate data with various kinds of spatial and temporal characteristics. By including a spatial dimension, the resulting (rotated) modes are often easier to interpret. The choice of bins is a critical step that must reflect an ecological understanding as well as familiarity with PCA: the fewer the bins, the more reliable the modes and amplitudes, but the more difficult the physical interpretation. One has to choose bins that reflect how fast conditions evolve along both the spatial and temporal dimension, but that still respect the number of observations (years). In practice, good results have been obtained when the number of bins is approximately the same size as the number of observations, although a smaller number of bins is preferable if the intra-annual spatial or temporal variability is low enough. Bins should be of equal size within the spatial and also within the temporal dimension.[20]

A related issue is how to aggregate the data within bins. For the Lake Tahoe Secchi disk and the Castle Lake primary productivity data, we aggregated using the trapezoidal mean. For the San Francisco Estuary chlorophyll data, we aggregated

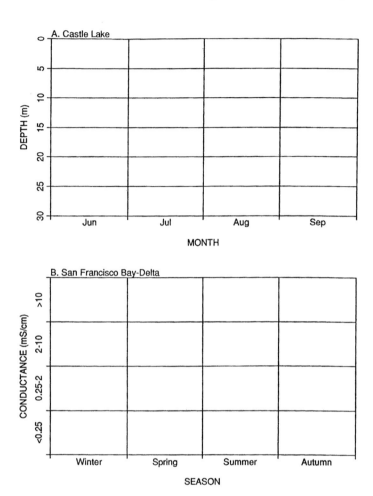

**FIGURE 5** Space-time binning schemes for (A) Castle Lake and (B) the northern San Francisco Estuary.

within bins by using the median. By using the median, we effectively remove the influence of extreme but isolated phenomena that would otherwise require preliminary identification and removal. One striking example is the large but rare forest fires that strongly affect Lake Tahoe primary productivity.[21] An alternative is to use robust estimates of the covariance matrix.[22]

The use of PCA has a controversial past in ecology and other disciplines with similar analytical problems. One of the problems, as described previously, has been a failure to rotate the subset of $m$ axes in order to achieve simple structure. In that case, one often ends up trying to interpret artifactual patterns. A second major difficulty has been that the variables analyzed together are usually completely different and may even have different units. A physical interpretation for the linear combinations of these variables that emerge from PCA is therefore often tortuous and unsatisfying. The PCA application presented here is not subject to these difficulties. Each variable (bin) represents an aggregation in space and time of the same measured field quantity, so that the principal components themselves have a simple interpretation as modes of space-time variability and the amplitudes are simply the strengths of each mode from one year to the next. This particular application of PCA seems to have a high success rate at decomposing relatively short time series into constituent modes of variability.

## IDENTIFYING CAUSAL FACTORS

The next step is to explore relations between the individual modes of variability and other ecological variables suspected to be causal factors. The shapes of the individual modes — the patterns of coefficients or loadings — can often be identified with well-known seasonal phenomena, which then guides us in choosing which variables to explore. In the case of Castle Lake primary productivity, for example, two modes were found, one representing interannual variability of deeper-water spring algal blooms and the second representing variability of late-summer, mixed-layer populations.[5] Prior data from Castle Lake, as well as the general body of limnological understanding, led to an exploration of climatic effects on the first mode and cascading trophic interactions on the second.

Once a (rotated) mode has been characterized, the key to subsequent exploration is the corresponding amplitude time series. It describes the year-to-year strength of the mode. One can use the amplitude time series itself or a closely related time series that may be more physically meaningful. In the case of San Francisco Estuary chlorophyll, for example, the first mode was identified largely with the bin representing the summer season and 2 to 10 mS $cm^{-1}$ conductance range (Figure 5B).[19] As the second mode had only a small coefficient for this bin, one can conclude that the first mode essentially represented interannual chlorophyll variability in this bin. By using the time series for median chlorophyll in this bin instead of the amplitude time series for the first mode, we were able to obtain a longer time series and one that had a more familiar interpretation. In the same vein, one might want to use seasonally averaged Secchi depth in both the unstratified and stratified seasons at Lake Tahoe as replacement series for the first and second amplitudes, respectively (Figure 4). The price one pays for a more familiar variable, however, is the probable introduction of more noise into the series. It is no longer a "pure" representation of a single variability mode, which may make its relationships with other variables less clear.

Regardless of which series we choose, how do we explore the relationships with other variables while imposing a mimimum of preconceptions on these relationships? Although each series represents a single mode of variability, any number of variables may function as causal factors. Therefore, we need a method that is inherently multivariate. In principle, the most flexible approach would be to use one of several nonparametric regression procedures to approximate the relationship, including the use of autoregressive terms if these were warranted. One example is the *local regression* model in which the regression surface is approximated locally by a weighted linear or quadratic polynomial.[23] Another is the *tree-based model*, which is basically a multidimensional step function approximation to a response surface.[24] Both of these methods permit interaction among the predictor variables. In practice, though, these unrestricted methods probably require too much data if our time series are only a few decades long. At the other extreme, generalized linear models[25] are too restrictive as an exploratory tool. Although they permit nonlinearities, extensive examination of residual and partial residual plots are required to identify them, and manual iteration of model specification and diagnostic checking is required.

An intermediate approach is to permit variables to enter in a nonlinear manner, but to constrain them to be additive. This class of models is sometimes known as general additive models. They all have the feature of identifying nonlinearities directly by incorporation into the model itself. Through the use of certain smoothing techniques, the variables are modeled nonparametrically and so the data themselves suggest the form of the nonlinearities. Several approaches to general additive modeling exist, including alternating conditional expectation (ACE),[26] additivity and variance stabilization (AVAS),[27] and generalized additive modeling (GAM).[28] Some aquatic ecological applications include studies of phytoplankton[29] and macrobenthos[30] using ACE, phytoplankton[19] and northern anchovy[31] using AVAS, and striped bass[32] and walleye pollock[33] using GAM.

The GAM procedure is best introduced as an extension of generalized linear models, which are in turn the natural extension of classic linear models. If $Y$ is the response variable, the $Z_i$ are predictor variables, and $E(Y) = \mu$, a generalized linear model takes the following form:

$$g(\mu) = \alpha + \sum_{i=1}^{p} \beta_i Z_i \tag{2}$$

where g is a *link* function describing how the expected value of $Y$ depends on the linear combination of predictors, and $\alpha$ and the $\beta_i$ are constants; g can be any monotonic differentiable function. The dependence of the variance of $Y$ on the mean $\mu$ is specified independently of the link function by a variance function $V(\mu)$, i.e., the variance is not assumed to be homogeneous. Once a link and variance function have been chosen through some sort of exploratory data analysis, the parameters $\alpha$ and $\beta_i$ are estimated by maximum likelihood, using an iteratively reweighted least-squares algorithm. Nonlinear biological phenomena are often expressed as classical linear models by transforming the response (e.g., with a log transform), but the transformation often leads to unnatural scales, does not necessarily result in homogeneity of variance, and may not even be defined for certain response values. Generalized linear models avoid these problems, yet remain almost as tractable as classical linear models with regard to summary statistics and hypothesis testing.

For the GAM procedure, Equation 2 is still appropriate, but the $Z_i$ are replaced by $f_i(Z_i)$, where the $f_i$ are some kind of scatterplot smoother:

$$g(\mu) = \alpha + \sum_{i=1}^{p} \beta_i f_i(Z_i) \tag{3}$$

The $\beta_i$ and $f_i$ are chosen through an iterative smoothing process. The algorithm cycles through each variable $i$ in turn and smooths the partial residuals obtained by subtracting from $g(\mu)$ current estimates for all additive terms $j \neq i$, resulting in updated estimates of $\beta_i$ and $f_i$. Two choices of smoother are available for nonparametric models: *loess smoothers*[23] and *smoothing splines*.[34] Loess smoothers in this context are the one-dimensional versions of the locally weighted polynomial models mentioned earlier. Smoothing splines are cubic splines with knots at the unique values of the predictor variables. In either case, the degree of smoothness must be specified. In principle, this can be done automatically through some kind of cross-validation procedure, but in the case of GAM the computational cost is high. Loess has an advantage in that the smoothness parameter, known as the *span*, is intuitive; it simply indicates how many nearest neighbors to use for computing the weighted regression estimate of the smoothed value: the bigger the span, the smoother the fit. The smoothness parameter for smoothing splines does not have such an accessible visual interpretation.

As an example, consider how year-to-year variability of the striped bass (*Morone saxatilis*) survival index for the San Francisco Estuary depends on hydrological conditions (Figure 6). The index is a measure of the proportion of eggs that develop and survive to become young-of-the-year bass.[32] We examined how survival was affected by the two main features of hydrological conditions in the estuary: (1) diversion of flow in the upper estuary into water projects for agriculture and other uses and (2) the remaining undiverted flow of water through the estuary. Diversion was expressed as a percentage of total water inflow to the estuary. The flow through the estuary was indexed by the position of the 2 ppt near-bottom salinity position, known as $X_2$ and expressed as distance along the axis of the estuary from its mouth: the higher the flow through the estuary, the smaller (more seaward) is $X_2$. $X_2$ has come into use as a surrogate for flow because of its relative ease of measurement, as well as its physical and biological significance.[32] Based on life history and behavior of striped bass, we used the April through July averages for both $X_2$ and diversion. Because of the small number of data points (22 years), no additional predictors were considered.

Separate plots of survival index against each of $X_2$ and diversion suggest negative relationships with a lot of scatter in both cases (Figure 7A). A naive conclusion would be to attempt a multiple linear regression. More cautiously, we could begin an iterative exploration of partial residual plots, while still remaining within the context of linear models. Most efficiently, we could fit a generalized additive model. Generalized additive models do require some preconceptions in that we need to choose a link function, a variance specification, and the degree of smoothing. Given the evidence for increasing variance with increasing survival index in Figure 7A, natural choices for the link and variance are $g = \ln$ and $V = \mu$. We choose the smoothest possible loess smoother by setting the span equal to 1, which means that all points are used in calculating the smooth at each individual point.

The fitted GAM suggests that survival index has a nonlinear dependency on $X_2$, but a more or less linear dependency on diversion (Figure 7B). The $X_2$ effect is unimodal with a peak between 70 and 80 km, while diversion has a monotonic negative effect. Note that the additive predictors have been plotted in Figure 7B so

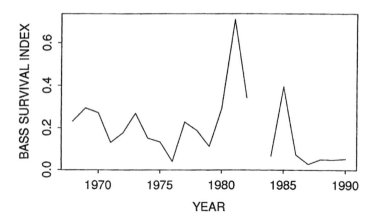

**FIGURE 6**  Annual estimates of striped bass survival index for the San Francisco Estuary. Because of extremely high flows in 1983, accurate estimates could not be obtained.

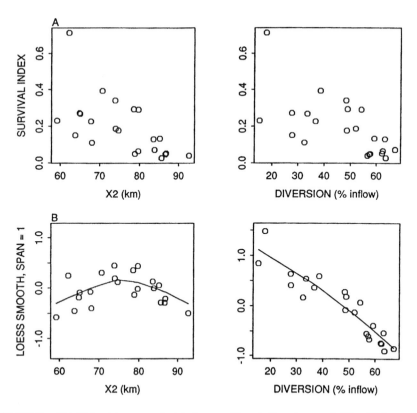

**FIGURE 7**  (A) Striped bass survival index vs. $X_2$ and diversion, and (B) additive terms $\beta_i f_i(Z_i)$ (Equation 3) for a generalized additive model of the striped bass survival index, where the $f_i$ are loess smooths with the span equal to 1. The points are partial residuals.

**TABLE 1**
**Linear Model for (ln Transformed) Striped Bass Survival**

| Term | Coefficient | SE | t | P |
|---|---|---|---|---|
| Intercept | −37.8 | 15.3 | −2.47 | 0.024 |
| $X_2$ | 1.01 | 0.40 | 2.51 | 0.022 |
| $X_2^2$ | −0.00643 | 0.00241 | −2.67 | 0.016 |
| Diversion | −0.0726 | 0.0328 | −2.22 | 0.040 |

that the vertical scales are comparable. The difference in vertical extent for the two predictors shows that diversion had the bigger effect on survival over this period.

Confirmatory analyses are available for generalized additive models. In particular, terms can be tested for nonlinearity, and new observations can be predicted with corresponding confidence intervals. These tests are approximate and, at least in our experience with small data sets, sometimes misleading. For example, a chi-squared test for the nonlinearity of the $X_2$ term in Figure 7B[35] was not significant, which contrasts with the more exact results described later. Here, though, we emphasize the use of these models for diagnostic purposes, specifically to suggest parametric transformations for terms in the model. Once a form has been established for each effect, the individual effects can be parameterized and their significance tested exactly in more conventional ways.

## PARAMETERIZING RELATIONSHIPS

So far we have identified independent modes of variability and, for each mode, determined possible causal factors and the general form of their effect: linear or, if nonlinear, the approximate shape. The third step is to describe these effects in the form of a classical (or generalized) linear model for exact (or asymptotically exact) testing of significance using the general additive model as a guide. In the case of striped bass survival, for example, we can incorporate the unimodal response to $X_2$ with a quadratic term:

$$\ln(\text{survival}) = \alpha + \beta_1 X_2 + \beta_2 X_2^2 + \beta_3 \cdot \text{diversion} \quad (4)$$

The values for the coefficients, as well as their significance, are summarized in Table 1. The results support the incorporation of both $X_2$ and diversion, as well as the nonlinear form for $X_2$. These results make sound ecological sense and confirm a prior more qualitative understanding of the forces affecting striped bass survival.[32]

In order to gain more confidence about conclusions from the significance tests, the models should be assessed for their correspondence with underlying assumptions, as well as their possible dependence on only one or a few influential years. Diagnosis of classical linear models is a well-developed field that is treated in many texts[36,37] and need not be considered here. For the sake of completeness, we provide what

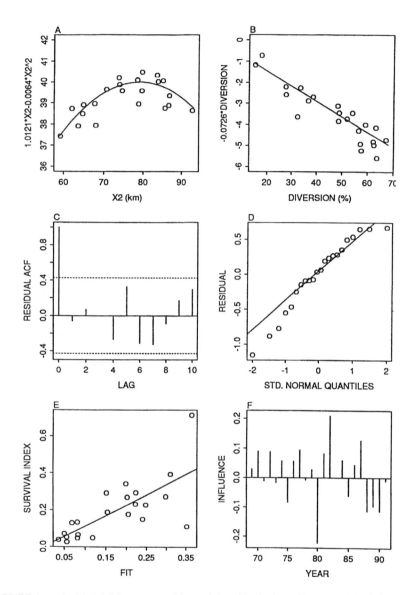

**FIGURE 8** (A, B) Additive terms with partial residuals for a linear model of the striped bass survival index; (C) autocorrelation function of the residuals; (D) normal quantile-quantile plot of the residuals; (E) survival index vs. (back-transformed) fitted values; and (F) influence of individual observations on fitted values.

might be considered a minimum set of diagnostic plots for the model of Equation 4 (Figure 8). These include (A, B) a plot of the additive terms with associated partial residuals; (C) a plot of the autocorrelation function of the residuals showing the absence of serial correlation; (D) a normal quantile-quantile plot of the residuals suggesting that the data have a longer left-hand tail than the normal distribution; (E) a plot of actual vs. (back-transformed) fitted values showing no unusual behavior of the residuals that could suggest misspecification of the model; and (F) an influence plot illustrating the influence of each observation on the fitted values. Given the strong influence of 1980 and 1982, a further step would be to investigate the model with either and both of these points removed to test the stability of the model specification.

## CHOOSING AMONG MODELS

Our focus in the previous step was to confirm the importance of specific causal variables and the manner of their effect. Often, more than one model specification is capable of accomplishing this end, in the sense of incorporating the shapes from the general additive model, having significant coefficients, and passing various diagnostic tests. If we are interested in the best possible prediction, then we no longer have the same flexibility in choosing a model. The question arises: Which of the models capable of describing the existing data gives the best prediction for new data? More formally, we can ask which model minimizes the prediction error (*PE*):

$$PE = E\{y - \hat{y}(\mathbf{x})\}^2 \tag{5}$$

where $y$ is the response, $\hat{y}$ is the predicted value given the vector-valued predictor variable $\mathbf{x}$, and the expectation is over new realizations of the joint distribution of $y$ and $\mathbf{x}$.

Procedures for estimating prediction error fall into two classes: analytical expressions and various kinds of resampling procedures.[13] The analytical forms start with the average squared residual (*ASR*):

$$ASR = \frac{1}{n}\sum_{i=1}^{n}(y_i - \hat{y}_i)^2 \tag{6}$$

where $n$ is the number of observations. The *ASR* is similar to the *PE*, but with the $\hat{y}_i$ being the fitted values using existing data. The *ASR* is obviously an optimistic estimate of *PE*, as it uses the same data for testing the model as were used for fitting the model. The difference $PE - ASR$ is, in fact, known as the *optimism*. More accurate analytical measures of *PE* incorporate an estimate of the optimism by "penalizing" *ASR* according to the number of parameters to be estimated, aside from the intercept. Although including more predictors inevitably leads to a decrease in ASR, the values for their associated parameters have a higher uncertainty; it is the balance between these two trends that determines how well future values can be predicted. A simple example of such a measure is the adjusted ASR:

$$\text{adjusted } ASR = \frac{1}{(n-2p)}\sum_{i=1}^{n}(y_i - \hat{y}_i)^2 = ASR + \frac{2p}{n-2p}ASR \tag{7}$$

where p is the number of parameters. Other measures for *PE* incorporate estimates of the residual variance $\hat{\sigma}^2$. One example is a form of the $C_p$ statistic suggested by Mallows.[38]

$$C_p = ASR + \frac{2p\hat{\sigma}^2}{n} \tag{8}$$

The unbiased estimate of residual variance $\hat{\sigma}^2$ is given by

$$\hat{\sigma}^2 = \frac{1}{(n-p)} \sum_{i=1}^{n} (y_i - \hat{y}_i)^2 \tag{9}$$

When comparing several models, the largest one can be used to provide a value for $\hat{\sigma}^2$.

The other class of procedures for estimating *PE* involves various types of resampling, including cross validation and bootstrapping. Among these, the 0.632 bootstrap estimator[39] appears to perform the best. In this method, a given model is refitted many times using bootstrap samples (random samples with replacement) of the original observations. For each model, the observations not included in the corresponding bootstrap sample are compared with model predictions. Specifically, in the case of classical linear models, an average error rate is first estimated by

$$\hat{\varepsilon}_0 = \frac{1}{n} \sum_{i=1}^{n} \frac{1}{B_i} \sum_{b \in C_i} (y_i - \hat{y}_{i,b})^2 \tag{10}$$

where  $\hat{y}_{i,b}$ = the estimate of $y_i$ from the model based on the *b*th bootstrap sample
$C_i$ = the set of indices for bootstrap samples not containing the *i*th observation
$B_i$ = the number of such samples

Theoretical arguments indicate that the prediction error is then given by

$$PE_{0.632} = ASR + 0.632 \cdot (\hat{\varepsilon}_0 - ASR) \tag{11}$$

Note that in all of these cases (Equations 7, 8, and 11) the estimate of prediction error is broken down into the sum of *ASR* and an estimate of the optimism.

To illustrate, we compared these three *PE* estimates for different models of striped bass survival, using all possible subsets of predictor variables in Equation 4 (Table 2). $PE_{0.632}$ was calculated with 1000 bootstrap samples (using the bootpred S-PLUS function of Efron and Tibshirani[13] available from the statistics archive at Carnegie-Mellon University). In all three cases, Equation 4 is the model with the smallest *PE*. For every subset of variables, $PE_{0.632}$ gives the most conservative

## TABLE 2
### Estimates of Prediction Error for Different Subsets of Predictor Variables in a Linear Model of (ln Transformed) Striped Bass Survival

| Variables | Adjusted ASR | $C_p$ | $PE_{0.632}$ |
|---|---|---|---|
| $X_2$, $X_2^2$, diversion | 0.35 | 0.33 | 0.37 |
| $X_2$ | 0.39 | 0.38 | 0.41 |
| $X_2^2$ | 0.37 | 0.37 | 0.42 |
| $X_2$, $X_2^2$ | 0.39 | 0.37 | 0.43 |
| $X_2^2$, diversion | 0.42 | 0.39 | 0.44 |
| $X_2$, diversion | 0.43 | 0.40 | 0.47 |
| Diversion | 0.44 | 0.42 | 0.48 |

estimate of *PE*, followed by the adjusted *ASR*. The estimates differ slightly among their rankings of the different models. The adjusted *ASR* and $C_p$ suggest that $X_2$ alone should be ranked third or fourth rather than second. Although differences among rankings are minor in this case, different results clearly can be obtained with different estimators. As no definitive statement can be made yet about the relative performance of these estimators for small numbers of observations, several probably should be calculated and compared. From a practical point of view, though, the resampling estimators, among which $PE_{0.632}$ appears to be the best, have several advantages. The number of parameters *p* does not need to be known, and the residual variance $\hat{\sigma}^2$ does not need to be estimated. Therefore, they are more generally applicable and more robust than analytical measures such as Equations 7 and 8.[13]

## CONCLUDING REMARKS

The four-step program outlined in this chapter can be summarized as follows:

1. Aggregrate the data for each year into space-time bins to form a collection of annualized time series; apply rotated principal component analysis to this collection; and interpret the resulting sets of component loadings as spatial-seasonal modes of interannual variability. PCA accomplishes several important tasks. It reduces the dimensionality of data collected throughout the year at multiple locations, isolates patterns of spatial-temporal coherence that presumably are independent modes underlying variability in the data set, and filters out noise by identifying significant modes.
2. Calculate the amplitude time series associated with each of the modes identified in Item 1; construct some type of generalized additive model relating each amplitude time series to the set of possible forcing factors; and plot the additive predictors in order to determine the shape of each partial dependency on individual factors.

3. Parameterize these shapes in the form of classical or generalized linear models for confirmatory testing of the relationships; and diagnose the models for inconsistency with assumptions or a dependence on unusual observations.
4. Decide among plausible parameterizations by estimating the prediction error for each one using several different methods.

This program, in whole or in part, has been applied successfully to several lake and estuarine sites where the instrumental record is two to three decades long. With time series of this length, typically two significant modes of variability and two causal factors per mode can be identified. The two modes usually account for about two thirds of the variability, and the subsequent statistical models account for more than half the variability in each mode. In terms of prediction, the remaining uncertainty is large and perhaps restrictive. In terms of investigating the underlying mechanisms, however, the approach is clearly valuable and can be used to identify processes for inclusion, regardless of the modeling approach.

## ACKNOWLEDGMENTS

I thank Charles Goldman of the University of California at Davis for the Secchi depth time series and Chuck Armor of the California Department of Fish and Game for the striped bass survival index data. The California Department of Water Resources provided hydrological data. Two anonymous reviewers offered helpful suggestions which were incorporated into the manuscript. This work was supported by the U.S. EPA Center for Ecological Health Research at the University of California at Davis (R819658). Although the information in this chapter has been funded in part by the U.S. EPA, it does not necessarily reflect the views of the agency and no official endorsement should be inferred.

For questions or further information regarding the work described in this chapter, please contact:

> **Alan Jassby**
> Dept. of Environmental Science and Policy
> University of California
> Davis, CA 95616
> e-mail: adjassby@ucdavis.edu

## REFERENCES

1. Preisendorfer, R.W. 1988. *Principal component analysis in meteorology and oceanography* (C.D. Mobley, compiler). Elsevier, New York.
2. Jassby, A.D. and T.M. Powell. 1990. Detecting changes in ecological time series. *Ecology* 71:2044–52.

3. Craddock, J.M. 1965. A meteorological application of principal component analysis. *Statistician* 15:143–56.
4. Smith, T.J.I. and B.P. Hayden. 1984. Snow goose migration phenology is related to extratropical storm climate. *Int. J. Biometeorol.* 28:225–33.
5. Jassby, A.D., T.M. Powell and C.R. Goldman. 1990. Interannual fluctuations in primary production: direct physical effects and the trophic cascade at Castle Lake, California. *Limnol. Oceanogr.* 35:1021–38.
6. Chatfield, C. and A.J. Collins. 1980. *Introduction to multivariate analysis.* Chapman and Hall, New York.
7. Jolliffe, I.T. 1986. *Principal component analysis.* Springer-Verlag, New York.
8. Cleveland, R.B., W.S. Cleveland, J.E. McRae and I. Terpenning. 1990. STL: a seasonal-trend decomposition procedure based on loess. *J. Official Stat.* 6:3–73.
9. Engle, R.F. and C.W.J. Granger, eds. 1991. *Long-run economic relationships.* Oxford University Press, New York.
10. Overland, J.E. and R.W. Preisendorfer. 1982. A significance test for principal components applied to a cyclone climatology. *Mon. Weather Rev.* 110:1–4.
11. Goldman, C.R. 1988. Primary productivity, nutrients, and transparency during the early onset of eutrophication in ultra-oligotrophic Lake Tahoe, California-Nevada. *Limnol. Oceanogr.* 33:1321–33.
12. Cattell, R.B. 1966. The scree test for the number of factors. *J. Multivar. Behav. Res.* 1:245–76.
13. Efron, B. and R.J. Tibshirani. 1993. *An introduction to the bootstrap.* Chapman and Hall, New York.
14. Stauffer, D.F., E.O. Garton and R.K. Steinhorst. 1985. A comparison of principal components from real and random data. *Ecology* 66:1693–8.
15. Richman, M.B. 1986. Rotation of principal components. *J. Climatol.* 6:293–335.
16. Thurstone, L.L. 1947. *Multiple factor analysis.* University of Chicago Press, Chicago, IL.
17. Afifi, A.A. and V. Clark. 1990. *Computer-aided multivariate analysis.* 2nd ed. Van Nostrand Reinhold, New York.
18. Dingman, H.F., C.R. Miller and R.K. Eyman. 1964. A comparison between two analytical rotation solutions where the number of factors is indeterminate. *Behav. Sci.* 9:76–80.
19. Jassby, A.D. and T.M. Powell. 1994. Hydrodynamic influences on interannual chlorophyll variability in an estuary: upper San Francisco Bay-Delta (California, U.S.A.). *Estuarine Coastal Shelf Sci.* 39:595–618.
20. Karl, T.R., A.J. Koscielny and H.F. Diaz. 1982. Potential errors in the application of principal component (eigenvector) analysis to geophysical data. *J. Appl. Meteorol.* 21:1183–6.
21. Jassby, A.D., C.R. Goldman and T.M. Powell. 1992. Trend, seasonality, cycle and irregular fluctuations in primary productivity at Lake Tahoe, California-Nevada, U.S.A. *Hydrobiology* 246:195–203.
22. Devlin, S.J., R. Gnanadesikan and J.R. Kettenring. 1981. Robust estimation of dispersion matrices and principal components. *J. Am. Stat. Assoc.* 76:354–62.
23. Cleveland, W.S. and S.J. Devlin. 1988. Locally-weighted regression: an approach to regression analysis by local fitting. *J. Am. Stat. Assoc.* 83:596–610.
24. Breiman, L., J.H. Friedman, R. Olshen and C.J. Stone. 1984. *Classification and regression trees.* Wadsworth International Group, Belmont, CA.
25. McCullagh, P. and J.A. Nelder. 1989. *Generalized linear models.* 2nd ed. Chapman and Hall, London.

26. Breiman, L. and J.H. Friedman. 1985. Estimating optimal transformations for multiple regression and correlation (with discussion). *J. Am. Stat. Assoc.* 80:580–619.
27. Tibshirani, R. 1988. Estimating transformations for regression via additivity and variance stabilization. *J. Am. Stat. Assoc.* 83:394–405.
28. Hastie, T. and R. Tibshirani. 1990. *Generalized additive models.* Chapman and Hall, London.
29. Millet, B. and P. Cecchi. 1992. Wind-induced hydrodynamic control of the phytoplankton biomass in a lagoon ecosystem. *Limnol. Oceanogr.* 37:140–6.
30. Millet, B. and O. Guelorget. 1994. Spatial and seasonal variability in the relationships between benthic communities and physical environment in a lagoon ecosystem. *Mar. Ecol. Prog. Ser.* 108:161–74.
31. Cury, P., C. Roy, R. Mendelssohn, A. Bakun, D.M. Husby and R.H. Parrish. 1995. Moderate is better: exploring nonlinear climatic effects on the Californian northern anchovy *(Engraulis mordax). Can. Special Pub. Fish. Aquat. Sci.* 121:417–24.
32. Jassby, A.D., W.J. Kimmerer, S.G. Monismith, C. Armor, J.E. Cloern, T.M. Powell, J.R. Schubel and T.J. Vendlinski. 1995. Isohaline position as a habitat indicator for estuarine populations. *Ecol. Appl.* 5:272–89.
33. Swartzman, G., E. Silverman and N. Williamson. 1995. Relating trends in walleye pollock (Theragra chalcogramma) abundance in the Bering Sea to environmental factors. *Can. J. Fish. Aquat. Sci.* 52:369–80.
34. Wahba, G. 1990. Spline functions for observational data. CBMS-NSF Regional Conference Series. SIAM, Philadelphia, PA.
35. Hastie, T.J. 1992. Generalized additive models. In J.M. Chambers and T.J. Hastie, eds., *Statistical models.* S. Wadsworth & Brooks, Pacific Grove, CA.
36. Belsley, D.A., E. Kuh and R.E. Welsch. 1980. *Regression diagnostics.* John Wiley & Sons, New York.
37. Cook, R.D. and S. Weisberg. 1982. *Residuals and influence in regression.* Chapman and Hall, New York.
38. Mallows, C. 1973. Some comments on $C_p$. *Technometrics* 15:661–75.
39. Efron, B. 1983. Estimating the error rate of a prediction rule: improvements on cross-validation. *J. Am. Stat. Assoc.* 78:316–31.

# 17 Developing Realistic Air Pollution Exposure/Dose Criteria for Ecological Risk Assessments

*Allen S. Lefohn*

## CONTENTS

Abstract .................................................................................................................. 307
Introduction ........................................................................................................... 308
Hazard Identification ............................................................................................ 308
Ambient Exposure Characterization .................................................................... 309
Dose/Exposure-Response Assessment ................................................................. 314
An Example of Applying Exposure and Response Information ........................ 315
Future Research Directions .................................................................................. 316
Conclusion ............................................................................................................ 316
References ............................................................................................................. 317

## ABSTRACT

There is a need for flexible problem-solving approaches that can link ecological measurements and data with the decision-making needs of environmental managers. Increasingly, ecological risk assessment is being suggested as a way to address this wide array of ecological problems. This chapter discusses the ambient exposure characterization component associated with the analysis phase of risk assessment methodology. Using surface ozone ($O_3$) as an example, specific guidance is provided on future research directions that are needed to assist scientists and policymakers in improving the quality of data that are available for quantifying this phase of the risk analysis. Future vegetation research efforts should focus on applying real-world $O_3$ hourly data that mimic actual geographic locations. Future research involving attempts to mathematically predict $O_3$ uptake by vegetation with hourly average concentration information must consider the sensitivity of the plant at time of exposure, as well as the limitation associated with the micrometeorological models employed. Results from actual soil moisture, light conditions, temperature, and humidity interactions with plant uptake must be reflected in future modeling efforts.

The implementation of research efforts in this area will provide results which can be used in future ecological risk assessments that will potentially be associated more with dose-related statistics rather than exposure-related metrics.

**Keywords:** *criteria, dose, exposure, ozone, risk assessment, W126*

## INTRODUCTION

The ecological problems facing environmental scientists and decisionmakers are numerous and varied. There is a need for flexible problem-solving approaches that can link ecological measurements and data with the decision-making needs of environmental managers. Increasingly, ecological risk assessment is being suggested as a way to address this wide array of ecological problems.

Ecological risk assessment evaluates the likelihood that adverse ecological effects may occur or are occurring as a result of exposure to one or more stressors.[1] Ecological risk assessment includes three primary phases: problem formulation, analysis, and risk characterization.[2] Within problem formulation, important areas include identifying goals and assessment endpoints, preparing a conceptual model, and developing an analysis plan. The analysis phase involves evaluating exposure to stressors and ecological effects. In risk characterization, key elements are (1) estimating risk through integration of exposure and stressor-response profiles, (2) describing risk by discussing lines of evidence and determining ecological adversity, and (3) preparing a report.

The analysis phase, which follows problem formulation, includes two principal activities: characterization exposure and ecological effects. Using surface ozone ($O_3$) as an example, this chapter focuses on the exposure characterization activity of the analysis phase. In addition, it provides specific guidance on future research directions that are needed to assist scientists and policymakers in improving the quality of data available for quantifying the exposure characterization phase of the risk analysis.

## HAZARD IDENTIFICATION

One of the first steps in a risk assessment is the identification of a specific form of the pollutant or suite of pollutants that may be either emitted within or transported into a geographic area, as well as the concentration, timing, and length of exposure that is experienced. In many cases, for assessing possible impacts of air pollutants, it is not difficult to identify the pollutants of interest. Usually a specific action is proposed where a source is known to emit specific sets of pollutants, such as dioxin, furan, chlorine, or heavy metals. For other situations, one may be concerned about the deposition of sulfate and nitrate, where the original source of emissions may be hundreds of miles from the area of interest. The latter situation is more complex and requires specific attention to better understand the relationship between cumulative deposition and resulting effects. The former example is more straightforward in that the distribution of concentration as a function of time and space can be predicted using chemical and transport models.

An additional important aspect of pollutant identification is the possible additive relationship that may exist when individual pollutants interact with the ecosystem. It is important to carefully identify the pollutants of concern, as well as to quantify the co-occurrence, sequential, and complex-sequential patterns of exposure that occur during multiple pollutant exposures.[3] For example, in the 1970s, there was concern that vegetation exposed to both $O_3$ and sulfur dioxide ($SO_2$) suffered greater impacts when the pollutants co-occurred than when exposed to either pollutant individually. However, Lefohn and Tingey[3] and Lefohn et al.[4] characterized the temporal relationship between $O_3$ and $SO_2$ for all sites in the U.S. that monitored both pollutants. The authors reported that the co-occurrence of the two air pollutants (i.e., at simultaneous occurrences of hourly average concentrations $\geq 0.05$ ppm) was fairly rare. The authors reported further that experiments described in the literature that had used constant concentrations of each pollutant simultaneously applied were not useful for the standard-setting process because the exposures did not mimic those that occurred in ambient air. For developing more accurate predictions for assessing effects that depend upon linking experimental results with ambient conditions, it is important that specific attention be given to air quality characterization to better understand the co-occurrence exposure patterns that are experienced at the area of interest.

## AMBIENT EXPOSURE CHARACTERIZATION

After identifying the important pollutants and their ambient concentrations, the biological components of the ecosystem that may be at risk need to be identified. As part of the risk analysis, for assessing possible environmental effects, one needs to know if there are specific times of the day or seasonal periods when the biological components may be most sensitive to the emitted or transported pollutants. As part of the analysis phase, the following questions should be addressed.

- For the pollutant(s) of concern, does an exposure-dose relationship exist between the pollutant and the biological components?
- If dose (i.e., the amount of pollutant that gets into the plant) is difficult to measure, does a good substitute for dose exist in the form of an exposure index metric?
- Do all concentrations have an equal potential for possible biological impact?
- Is there a specific threshold that exists when relating exposure with response?

It is important to characterize the intensity, frequency, and duration of actual or predicted exposures to adequately relate exposure with biological response. One of the most important challenges has been the identification of exposure indices or metrics that mathematically relate exposures with biological effects. As part of revising the U.S. National Ambient Air Quality Standard, the U.S. EPA[5] summarized the history of the development of exposure indices that have been used as substitutes for dose in relating $O_3$ exposure with agricultural and tree growth reduction.

For almost 70 years,[6] air pollution specialists have explored alternative mathematical approaches for summarizing ambient air quality information in biologically meaningful forms that can serve as a surrogate for dose for both human health and vegetation effects purposes. Although understanding the effects of specific doses is important when attempting to link exposures with the potential for air pollutant effects, little is known about the concentration and, in some cases, the form of the pollutant that enters the organism.[5] At present, not enough is known to allow us to quantify, with a great deal of certainty, the links between exposure and dosage. Hence, scientists have explored the relationships between various exposure characteristics and plant response. Although it would be helpful if one could use actual $O_3$ dose measurements to predict cause-and-effect relationships, the current effort to quantify air pollution dose and relate it to vegetation effects is limited; most scientists use $O_3$ exposure measurements as a surrogate for dose for predicting vegetation effects.

The search for an exposure index that relates well with plant response has been the subject of intensive discussion in the research community.[5,7-12] Both the magnitude of a pollutant's concentration and the length of exposure are important considerations when attempting to develop a realistic exposure index. However, evidence exists in the literature to indicate that the magnitude of vegetation responses to air pollution is more closely associated with the level of concentration than the length of exposure. For $O_3$, the short-term, high concentration exposures are identified as being more important than long-term, low concentration exposures (see Reference 5). Lefohn and Benedict[13] hypothesized in their work that the higher concentrations should be given greater weight than the lower values. However, it was not until Musselman et al.[14,15] and Hogsett et al.[16] performed the experiments that the hypothesis was confirmed. Because the plants used in their experiments were well watered, conditions were optimized for pollutant uptake, and, therefore, dose was more closely related to exposure. The results reported by these authors[13-16] indicated that for dose considerations the higher concentrations should be weighted more than the lower values because they caused a greater negative response. The implication of the hypothesis raised by Lefohn and Benedict[13] and confirmed by the works of Musselman et al.[14,15] and Hogsett et al.[16] is that

> For a given specific time and place, the potential for any $O_3$ hourly average concentrations to impact vegetation is relative to the value of any other hourly average concentration, whether the concern be exposure or dose.

There have been several papers published that question the importance of higher concentrations in relation to the midlevel concentrations. Tonneijck and Bugter,[17] Krupa et al.,[18-20] Grünhage and Jäger,[21,22] Tonneijck,[23] Legge et al.,[24] and Grünhage et al.[25] have suggested that midlevel hourly average concentrations of $O_3$ are more important than higher hourly average concentrations in affecting vegetation.

In view of the discussion that had emerged in the scientific literature regarding the importance of high concentrations vs. midlevel and lower values, the Canadian Vegetation Objective Working Group (VOWG) thoroughly evaluated the work described by Krupa et al.[19,20] and Legge et al.[24] The Canadian VOWG[26] concluded that there was little support for using an exposure index that focused on the midlevel vs. the higher concentrations. The findings were based on the following:

- Cumulative exposure indices that focused on the higher hourly average concentration performed considerably better in exposure-response models than the index proposed by Krupa et al.[20] which focused on the midlevel values.
- Inaccurate use by Krupa et al.[19,20] of some of the NCLAN data was identified.
- The exposure index that focused on the midlevel concentrations predicted greater losses to vegetation at remote ambient $O_3$ monitoring sites in Canada (where losses were not documented) than those that occurred at sites which experienced much higher $O_3$ exposures and where documented $O_3$ effects on vegetation have occurred. The exposure index predicted much greater vegetation losses at remote northern areas of Ontario (i.e., Experimental Lakes Area) where crop effects have not been documented. Similarly, high losses were predicted for remote areas in Cormack, Newfoundland and Vegreville, Alberta.

In continuing its review, the VOWG noted that Legge et al.[24] have pointed out that although midrange concentrations are important, if high concentrations were to occur during the time of day when plants were most sensitive, then the higher concentrations would also be important. However, based on its observation that the exposure index that focused on the midlevel concentrations predicted greater losses at remote ambient $O_3$ monitoring sites than those sites which experienced much higher $O_3$ exposures where effects had been observed, the VOWG concluded that the results reported by Legge et al.[24] were difficult to rationalize. The VOWG concluded that there was sufficient evidence that cumulative exposure indices that weight the higher hourly average concentrations more than the midlevel concentrations should be used for developing exposure-response relationships.[26]

The U.S. EPA[5] has evaluated the results reported by Tonneijck and Bugter,[17] Krupa et al.,[18-20] and Tonneijck.[23] The agency noted that the conclusions should be interpreted with caution and that peak-weighted cumulative exposure indices were appropriate for developing exposure-response relationships to predict $O_3$ vegetation effects.[5,27,28]

Grünhage and Jäger[21,22] and Grünhage et al.[25] investigated $O_3$ flux to grasslands in Germany and developed models that attempted to take into consideration the timing of exposure to vegetation and atmospheric conductivity. Based on their modeling results, the authors predicted that the greatest uptake by vegetation occurred during the period when high hourly average $O_3$ concentrations were not present (e.g., 0900 to 1559 h). Their models predicted that this period was the most important because atmospheric resistance was lowest and $O_3$ fluxes were highest.

Grünhage and Jäger[22] and Grünhage et al.[25] applied their models to data from several of the U.S. EPA Clean Air Status and Trends Network (CASTNet) sites and reported that the midlevel concentrations appear to have a greater potential for impacting vegetation than the higher concentrations. The CASTNet sites are located far from areas where $O_3$ precursors are generated, and, thus, hourly average concentrations ≥0.10 ppm will occur late in the afternoon. Musselman et al.[12] reported that high hourly average concentrations (e.g., ≥0.10 ppm) occur frequently at many rural agricultural and forested sites in the U.S. In addition, Musselman et al.,[12] using data from the U.S. EPA Aerometric Information Retrieval System (AIRS) database, which includes rural data from CASTNet as well as from other rural locations in the U.S., reported that high hourly average concentrations occurred during the 0900 to 1559 h daylight window.

Based on the results reported by Musselman et al.,[12] it appears that modeling results described by Grünhage and co-workers using U.S. data are affected by several important assumptions (e.g., atmospheric conductivity and vegetation conductivity) that may not be applicable to rural agricultural and forested sites in the U.S. Therefore, before applying models that can be used to predict pollutant uptake universally, it is very important to quantify pollutant uptake for different vegetation species grown at different rural locations. In conclusion, due to the limitations of the assumptions built into the previous models, at this time, the evidence, as reviewed by the U.S. EPA[5] and the Canadian VOWG,[26] continues to support the importance of the higher hourly average concentrations in relation to the midlevel and lower values.

Several different types of $O_3$ exposure indices have been proposed as surrogates for dose. A 6-h, long-term seasonal mean $O_3$ exposure parameter was used by Heagle et al.[29] Also, Heagle et al.[30] reported the use of a 7-h, experimental-period mean. The 7-h (0900 to 1559 h) mean, calculated over an experimental period, was adopted as the statistic of choice by the U.S. EPA National Crop Loss Assessment Network (NCLAN) program.[31] Toward the end of the program, NCLAN redesigned its experimental protocol and applied proportional additions of $O_3$ to its crops for 12-h periods.

As an alternative to the seasonal mean exposure metric, hypothesizing that the higher $O_3$ concentrations were more important in eliciting adverse effects on agricultural crops than the lower values, Lefohn and Benedict[13] introduced an exposure parameter that accumulated all hourly concentrations equal to and above a threshold level (i.e., 0.10 ppm). The exposure parameter was similar to that used by Oshima,[32] in which the difference between the value above 0.10 and 0.10 ppm was summed.

In the late 1980s, the focus of attention turned from the use of long-term seasonal means to cumulative indices (i.e., exposure parameters that sum the products of concentrations multiplied by time over an exposure period). The cumulative index parameters proposed by Oshima[32] and Lefohn and Benedict[13] were similar. Both parameters gave equal weight to the higher hourly concentrations, but ignored the concentrations below a subjectively defined minimum threshold (e.g., 0.10 ppm).

The use of an artificial threshold concentration as a cutoff point eliminates any possible contribution of the lower concentrations to vegetation effects. Recognizing the disadvantage, Lefohn and Runeckles[7] suggested a modification to the Lefohn and Benedict[13] exposure index by weighting individual hourly mean concentrations of $O_3$ and summing over time. Lefohn and Runeckles[7] proposed a sigmoidal weighting function that was used in developing a cumulative, integrated exposure index.

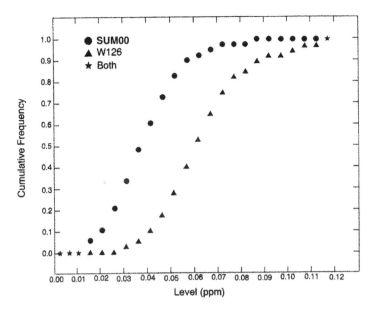

**FIGURE 1** A comparison between the resulting cumulative frequencies for the exposure parameters, sum of all hourly average concentrations (SUM00), and the sigmoidally weighted, integrated exposure index (W126).

The sigmoidal weighting function (referred to as the W126 index) was multiplied by each of the hourly mean concentrations; thus, the lower, less biologically effective concentrations were included in the integrated exposure summation.

The U.S. EPA[5] summarized the results of retrospective studies performed using NCLAN data to test the hypothesis that cumulative indices that describe $O_3$ exposures may adequately serve as a dose surrogate for describing exposure/dose-response relationships for agricultural crops. The results reported by Lefohn et al.[33] and Lee et al.[34,35] demonstrated that cumulative indices could be used in relating $O_3$ exposure to vegetation effects. In many cases, the cumulative indices perform better than seasonal mean indices.

Although some cumulative indices offer the advantage of focusing on the higher hourly average concentrations, not all cumulative indices (i.e., the summation of all hourly concentrations, SUM00) achieve this goal.[10] Lefohn et al.[10] showed that the SUM00 cumulative exposure index focused too much attention on the lower hourly average concentrations instead of the higher values. Figure 1 shows a comparison between the resulting cumulative frequencies for the two exposure parameters; (1) sum of all hourly average concentrations (SUM00) and (2) the sigmoidally weighted, integrated exposure index (W126). The figure shows that the SUM00 approaches its maximum relative value (i.e., 1.0 at the top of the figure) faster than the W126. This observation implies that the maximum value of the SUM00 is more associated with the lower hourly average concentrations than the W126. Thus, the figure shows that the SUM00 exposure index gives greater weight to the lower values instead of the biologically more effective higher values.

In conclusion, data in the published literature support the observation that (1) cumulative seasonal exposure and (2) the higher concentrations are important features of exposure for both crops and trees.[5,36] The selection of an appropriate exposure time of interest should take into account the cumulative impact from repeated peaks over an entire growing season, while including the mid-level concentrations. At this time, for risk characterization purposes, exposure indices that weight the hourly $O_3$ concentrations differentially appear to be the best candidates for relating exposure with predicted plant response. Peak-weighted, cumulative indices appear to have major advantages over both the mean (e.g., 7-h seasonal mean) index and the index that cumulates all hourly average concentrations (i.e., SUM00).[5]

## DOSE/EXPOSURE-RESPONSE ASSESSMENT

Characterization of the relationship between exposure and dose and the incidence and severity of the adverse effect is associated with the dose/exposure-response assessment. As a first step, one needs to identify the biological species and its sensitivity that may be at risk due to the air pollutants emitted or transported into the area. In many cases, existing monitoring data allow for the characterization of air pollutant concentrations over time. Given this information, one needs to identify the (1) concentration levels and their temporal characteristics and (2) the dose-response relationship with the biological components of concern. In many cases, the dose-response relationship has not been quantified; however, for some vegetation, experimental exposure-response relationships between air pollutants and biological components have been developed. Using ambient air quality data, it is possible to identify when the maximum concentration and duration (i.e., exposure) occur (e.g., at night for 2 h during the month of July). Unfortunately, there are only a few species where sensitivities are known and documented. Therefore, little is known about whether the biological species of interest within the ecosystem is more sensitive during a specific part of the day or a specific part of the growth season. If the maximum exposure and the sensitivity of the biological organism do not overlap, the pollutant may not be of concern to the particular biological organism.

The U.S. EPA[36] proposed that a 12-h diurnal window, which includes the daylight hours from 8:00 a.m. to 8:00 p.m. (0800 to 1959 h), is an appropriate basis for a possible national air quality standard designed to protect a wide range of vegetation growing across the U.S. The agency pointed out that most plants are believed to open their stomata during the day and close them at night. However, the U.S. EPA[5] noted that the inherent variability in stomatal opening makes identification of a set time period for $O_3$ exposure problematic.

It is important to note that in the U.S. the day length during summer months at all locations is greater than 12 h. In some locations, such as Montana, the day length is greater than 16 h. Several authors (e.g., see Reference 37) have noted that in many rural areas in the U.S. peak concentrations occur after the 12-h window. The inherent variability in stomatal opening makes it difficult to limit a set time period during the day to focus on accumulating $O_3$ exposure. Some experimental results indicate that plants are sensitive to $O_3$ at night. Because in rural areas in the Southeast U.S. peak concentrations are known to occur after the 12-h window, effects may occur if the stomates are open in the evening.

## AN EXAMPLE OF APPLYING EXPOSURE AND RESPONSE INFORMATION

It is important that experimental research be performed to identify those exposure regimes that elicit an adverse effect on the organism of interest. Applying the varying exposure regimes, it is possible to predict the potential effects of a specific air pollutant on a component of the biological system. However, as noted later, the important components of the exposure regime should be included with the dose surrogates (i.e., exposure index) that are used to relate dose/exposure and vegetation response.

The U.S. EPA has depended heavily on experimental data derived from the NCLAN. In these studies, crops were grown under typical farm conditions and exposed in open-top chambers to ambient and above-ambient $O_3$ concentrations. The modified ambient treatments in the experimental chambers contained numerous high peaks (hourly $O_3$ concentrations above 0.10 ppm), occurring more frequently than in typical ambient air quality distributions.[12,38] In developing exposure-response models, the agency has depended almost solely on regression models using a cumulative-weighted index. In many of the experiments, the NCLAN exposure regimes exhibited numerous occurrences of high hourly average concentrations that produced 10% or greater growth loss. The frequency of these high concentrations has raised questions as to whether a cumulative exposure index alone is adequate for quantitatively predicted losses.

In addressing the frequent occurrence of high hourly average $O_3$ concentrations, Lefohn et al.[39] described an approach to identify areas that may be at risk in the Southern Appalachian Mountains. Results from seedling and tree experiments operated in open-top chambers were used to characterize $O_3$ exposure regimes that resulted in growth loss under controlled conditions. Many experiments experienced numerous occurrences of hourly average concentrations ≥0.10 ppm. Available $O_3$ monitoring data were characterized for the states of Alabama, Georgia, South Carolina, North Carolina, West Virginia, Tennessee, Kentucky, and Virginia using the W126 biologically based, cumulative exposure index. As a part of the analysis, both the occurrences of hourly average $O_3$ concentrations ≥0.10 ppm and the soil moisture conditions in the geographic area were considered. The focus on the higher hourly average concentrations occurred because of the frequent number of high values documented in the vegetation experiments. Combining exposure information with moisture availability and experimental exposure-response data, the extreme northern and southern portions of the Southern Appalachian area were identified as having the greatest potential for possible vegetation effects. The authors noted that the study was based mostly on results from individual tree seedlings grown in chambers and pots. The investigators recommended that additional research was needed to identify what differences in effects might be observed if exposures were similar to those experienced in forests. Furthermore, it was recommended that future investigations verify the location and presence of specific vegetation species and amounts and whether actual growth losses occurred in those areas of concern identified in their study.

## FUTURE RESEARCH DIRECTIONS

As we have seen, ecological risk assessment includes three primary phases: problem formulation, analysis, and risk characterization. The analysis phase involved evaluating exposure characterization and ecological effects. The important elements are estimating risk through integration of exposure and biological-response profiles. In the late 1970s and early 1980s, vegetation researchers attempted to apply ambient-type $O_3$ exposures to vegetation. Unfortunately, the fumigation protocol used in the NCLAN experiments resulted in occurrences of numerous peak hourly average concentrations in many of the experimental treatments near the 10% growth reduction level. However, based on uncertainties discussed earlier in this chapter, future vegetation research efforts should focus on applying real-world $O_3$ hourly data that mimic actual geographic locations.

Ozone flux is affected by (1) the sensitivity of the plant at the time of day and growth stage when plants are most sensitive, (2) atmospheric conductivity, and (3) the number of hourly average concentrations in the high, mid, and low range that define the exposure regime (i.e., the distribution). Any attempt to mathematically predict $O_3$ uptake by vegetation with hourly average concentration information must consider the sensitivity of the plant at time of exposure. The development of $O_3$ flux uptake models described by Grünhage and co-workers represents an important attempt to establish a meaningful relationship between $O_3$ exposure and vegetation effects. The authors have included some of the factors that are important when using ambient air quality data to predict vegetation effects. However, soil moisture, light conditions, temperature, and humidity are also important considerations which affect vegetation sensitivity at time of exposure and cannot necessarily be generalized without recording actual uptake measurements. These factors, as well as others, may cause inconsistent results when attempting to relate atmospheric exposure potential or cumulative exposure indices with vegetation effects. For developing better terrestrial risk management models, future vegetation research should be directed at better understanding the mechanisms that are responsible for $O_3$ uptake (i.e., dose) by the plant and under what conditions uptake is most optimal. It may be possible to integrate these results so that future ecological risk assessment considerations might be associated more with dose-related statistics than with exposure-related metrics.

## CONCLUSION

Ecological risk assessment is suggested as a way to address a wide array of ecological problems. Ecological risk assessment includes three primary phases: problem formulation, analysis, and risk characterization. The analysis phase involves evaluating exposure to stressors and ecological effects. As part of the analysis phase, the following questions need to be addressed. (1) For the pollutant(s) of concern, does an exposure-dose relationship exist between the pollutant and the biological components? (2) If dose (i.e., the amount of pollutant that gets into the plant) is difficult to measure, does a good substitute for dose exist in the form of an exposure index metric? (3) Do all concentrations have an equal potential for biological impact? (4) Is there a specific threshold that exists when relating exposure with response?

Although understanding the effects of specific doses is important when attempting to link exposures with the potential for air pollutant effects, little is known about the concentration and, in some cases, the form of the pollutant that enters the organism. At present, not enough is known to allow us to quantify the links between exposure and dosage. Hence, scientists have explored the relationships between various exposure characteristics and plant response. Although it would be helpful if one could use actual $O_3$ dose measurements to predict cause-and-effect relationships, the current effort to relate air pollutant dose to vegetation effects is limited mostly to relating $O_3$ exposure measurements to vegetation effects.

Future vegetation research efforts should focus on applying real-world $O_3$ hourly data that mimic actual geographic locations. Future research involving attempts to mathematically predict $O_3$ uptake by vegetation with hourly average concentration information must consider the sensitivity of the plant at time of exposure, as well as assumptions and limitations built into the models. Actual soil moisture interactions with plant uptake must also be reflected in future modeling efforts. In addition, light conditions, temperature, and humidity are important considerations which affect vegetation sensitivity at time of exposure and cannot necessarily be generalized without taking actual uptake measurements. The implementation of research efforts in this area will hopefully provide results which can be used in future ecological risk assessments that will potentially be associated more with dose-related statistics than exposure-related metrics.

For questions or further information regarding the work described in this chapter, please contact:

**Allen Lefohn**
A.S.L. & Associates
111 No. Last Chance Gulch, Suite 4A
Helena, MT 59601
e-mail: asl_associates@compuserve.com

## REFERENCES

1. U.S. Environmental Protection Agency. 1992. Framework for Ecological Risk Assessment. Washington, DC: Risk Assessment Forum, U.S. Environmental Protection Agency. EPA/630/R-92/001.
2. U.S. Environmental Protection Agency. 1996. Proposed Guidelines for Ecological Risk Assessment. *Fed. Reg.* 61 (175):47553–47631.

3. Lefohn, A.S. and D.T. Tingey. 1984. The co-occurrence of potentially phytotoxic concentrations of various gaseous air pollutants. *Atmos. Environ.* 18:2521–2526.
4. Lefohn, A.S., C.E. Davis, C.K. Jones, D.T. Tingey and W.E. Hogsett. 1987. Co-occurrence patterns of gaseous air pollutant pairs at different minimum concentrations in the United States. *Atmos. Environ.* 21:2435–2444.
5. U.S. Environmental Protection Agency. 1996. Air Quality Criteria for Ozone and Related Photochemical Oxidants. Office of Research and Development, Washington, D.C. EPA/600/P-93/004a-cF.
6. O'Gara, P.J. 1922. Abstract of paper: Sulphur dioxide and fume problems and their solutions. *Ind. Eng. Chem.* 14:744.
7. Lefohn, A.S. and V.C. Runeckles. 1987. Establishing a standard to protect vegetation — ozone exposure/dose considerations. *Atmos. Environ.* 21:561–568.
8. Hogsett, W.E., D.T. Tingey and E.H. Lee. 1988. Exposure indices: concepts for development and evaluation of their use. In W.W. Heck, O.C. Taylor and D.T. Tingey, eds., *Assessment of Crop Loss from Air Pollutants*. Elsevier Applied Science Publishing, London, pp. 107–138.
9. Tingey, D.T., W.E. Hogsett and E.H. Lee. 1989. Analysis of crop loss for alternative ozone exposure indices. In T. Schneider, S.D. Lee, G.J.R. Wolters and L.D. Grant, eds., *Atmospheric Ozone Research and Its Policy Implications*. Elsevier Science Publishers B.V., Amsterdam, The Netherlands, pp. 219–225.
10. Lefohn, A.S., V.C. Runeckles, S.V. Krupa and D.S. Shadwick. 1989. Important considerations for establishing a secondary ozone standard to protect vegetation. *JAPCA* 39:1039–1045.
11. Lefohn, A.S. (ed.). 1992. *Surface-Level Ozone Exposures and Their Effects on Vegetation*. Lewis Publishers, Chelsea, MI, 366 pp.
12. Musselman, R.C., P.M. McCool and A.S. Lefohn. 1994. Ozone descriptors for an air quality standard to protect vegetation. *J. Air & Waste Manage. Assoc.* 44:1383–1390.
13. Lefohn, A.S. and H.M. Benedict. 1982. Development of a mathematical index that describes ozone concentration, frequency, and duration. *Atmos. Environ.* 16:2529–2532.
14. Musselman, R.C., R.J. Oshima and R.E. Gallavan. 1983. Significance of pollutant concentration distribution in the response of 'red kidney' beans to ozone. *J. Am. Soc. Hortic. Sci.* 108:347–351.
15. Musselman, R.C., A.J. Huerta, P.M. McCool and R.J. Oshima. 1986. Response of beans to simulated ambient and uniform ozone distribution with equal peak concentrations. *J. Am Soc. Hortic. Sci.* 111:470–473.
16. Hogsett, W.E., D.T. Tingey and S.R. Holman. 1985. A programmable exposure control system for determination of the effects of pollutant exposure regimes on plant growth. *Atmos. Environ.* 19:1135–1145.
17. Tonneijck, A.E.G. and R.J.F. Bugter. 1991. Biological monitoring of ozone effects on indicator plants in the Netherlands: initial research on exposure-response functions. *VDI Ber.* 901:613–624.
18. Krupa, S.V., W.J. Manning and M. Nosal. 1993. Use of tobacco cultivars as biological indicators of ambient ozone pollution: an analysis of exposure-response relationships. *Environ. Pollut.* 81:137–146.
19. Krupa, S.V., M. Nosal and A.H. Legge. 1994. Ambient ozone and crop loss: establishing a cause-effect relationship. *Environ. Pollut.* 83:269–276.
20. Krupa, S.V., L. Grünhage, H.-J. Jäger, M. Nosal, W.J. Manning, A.H. Legge and K. Hanewald. 1995. Ambient ozone ($O_3$) and adverse crop response: a unified view of cause and effect. *Environ. Pollut.* 87:119–126.

21. Grünhage, L. and H.-J. Jäger. 1994. Influence of the atmospheric conductivity on the ozone exposure of plants under ambient conditions: considerations for establishing ozone standards to protect vegetation. *Environ. Pollut.* 85:125–129.
22. Grünhage, L. and H.-J. Jäger. 1996. Critical levels for ozone, ozone exposure potentials of the atmosphere or critical absorbed doses for ozone: a general discussion. In L. Kärenlampi and L. Skärby, eds., Critical Levels for Ozone in Europe: Testing and Finalizing the Concepts. UN-ECE Workshop Report. Univ. of Kuopio, Dept. of Ecology and Environmental Science, pp. 151–168.
23. Tonneijck, A.E.G. 1994. *Use of several plant species as indicators of ambient ozone: Exposure-response relationships.* In J. Fuhrer and B. Achermann, eds., Critical Levels for Ozone — A UN-ECE Workshop Report. Proceedings of the UN-ECE Workshop on Critical Levels for Ozone, Bern, Switzerland, November 1–4, 1993. Published by the Swiss Federal Research Station for Agricultural Chemistry and Environmental Hygiene CH-3097 Liebefeld-Bern, Switzerland, pp. 288–292.
24. Legge, A.H., L. Grünhage, M. Nosal, H.-J. Jäger and S.V. Krupa. 1995. Ambient ozone and adverse crop response: an evaluation of North American and European data as they relate to exposure indices and critical levels. *Angew. Bot.* 69:192–205.
25. Grünhage, L., H.-J. Jäger, H.-D. Haenel, K. Hanewald and S. Krupa. 1997. PLATIN (plant-atmosphere interaction) II: Co-occurrence of high ambient ozone concentrations and factors limiting plant absorbed dose. *Environ. Pollut.* 98:51–60.
26. Canadian Vegetation Objective Working Group. 1997. Canadian 1996 NOx/VOC Science Assessment. Report of the Vegetation Objective Working Group. Science Assessment and Policy Integration Division, Atmospheric Environment Service, Environment Canada, Toronto, Ontario. ISBN-1-896997-12-0.
27. U.S. Environmental Protection Agency. 1996. Review of National Ambient Air Quality Standards for Ozone-Assessment of Scientific and Technical Information. Office of Air Quality Planning and Standards, Research Triangle Park, NC. EPA-452/R-96-007.
28. U.S. Environmental Protection Agency. 1997. National Ambient Air Quality Standards for Ozone: Final Rule. *Fed. Reg.* 62 (138):38855–38896. July 18, 1997.
29. Heagle, A.S., D.E. Body and G.E. Neely. 1974. Injury and yield response of soybean to chronic doses of ozone and sulfur dioxide in the field. *Phytopathology* 64:132–136.
30. Heagle, A.S., S. Spencer and M.B. Letchworth. 1979. Yield response of winter wheat to chronic doses of ozone. *Can. J. Bot.* 57:1999–2005.
31. Heck, W.W., O.C. Taylor, R.M. Adams, G.E. Bingham, J.E. Miller, E.M. Preston and L.H. Weinstein. 1982. Assessment of crop loss from ozone. *JAPCA* 32:353–361.
32. Oshima, R.J. 1975. Development of a system for evaluating and reporting economic crop losses caused by air pollution in California. III. Ozone dosage — crop loss conversion function — alfalfa, sweet Corn. IIIA. Procedures for production, ozone effects on alfalfa, sweet corn and evaluation of these systems. California Air Resources Board, Sacramento, CA.
33. Lefohn, A.S., J.A. Laurence and R.J. Kohut. 1988. A comparison of indices that describe the relationship between exposure to ozone and reduction in the yield of agricultural crops. *Atmos. Environ.* 22:1229–1240.
34. Lee, E.H., D.T. Tingey and W.E. Hogsett. 1988. Evaluation of ozone exposure indices in exposure-response modeling. *Environ. Pollut.* 53:43–62.
35. Lee, E.H., D.T. Tingey and W.E. Hogsett. 1989. Interrelation of Experimental Exposure and Ambient Air Quality Data for Comparison of Ozone Exposure Indices and Estimating Agricultural Losses. U.S. Environmental Protection Agency, Corvallis Environmental Research Laboratory, Corvallis, OR. Contract No. 68-C8-0006.

36. U.S. Environmental Protection Agency. 1996. Review of National Ambient Air Quality Standards for Ozone-Assessment of Scientific and Technical Information. Office of Air Quality Planning and Standards, Research Triangle Park, NC. EPA-452/R-96-007.
37. Lefohn, A.S. and V.A. Mohnen. 1986. The characterization of ozone, sulfur dioxide, and nitrogen dioxide for selected monitoring sites in the Federal Republic of Germany. *JAPCA* 36:1329–1337.
38. Lefohn, A.S. and J.K. Foley. 1992. NCLAN results and their application to the standard-setting process: protecting vegetation from surface ozone exposures. *J. Air Waste Manag. Assoc.* 42:1046–1052.
39. Lefohn, A.S., W. Jackson, D.S. Shadwick and H.P. Knudsen. 1997. Effect of surface ozone exposures on vegetation grown in the Southern Appalachian Mountains: identification of possible areas of concern. *Atmos. Environ.* 31:1695–1708.

# 18 Survey Methodologies for the Study of Ecosystem Restoration and Management: The Importance of Q-Methodology

*John T. Woolley, Michael V. McGinnis, and William S. Herms*

## CONTENTS

Abstract .................................................................................................................. 321
Values and Science in Ecological Restoration ..................................................... 322
Two Survey Methodologies ................................................................................. 323
Q-Methodology ..................................................................................................... 323
A Brief Summary of the Findings ....................................................................... 327
Conclusions ........................................................................................................... 330
Acknowledgments ................................................................................................ 331
References ............................................................................................................. 331

## ABSTRACT

Ecological restoration involves human attempts to return physical sites to some state preceding some human disturbances. There are more and more restoration efforts underway. However, scientists have shown it is rarely possible to know what ecosystems looked like or how they functioned. Restoration policymaking, therefore, should be understood as a transcientific enterprise which involves the intermingling of facts and values. In an examination of the place of values and science in river restoration, we describe two survey techniques, Q-methodology and attitudinal surveys. We find that Q-methodology is an important complement to traditional survey techniques and argue that Q techniques should be used to identify the range of values that may influence the restoration planning process.

**Keywords:** *restoration, discourse, Q-methodology, ecosystem, values*

> *When we copy nature not only do we get things we don't want, but some critical parts are always missing, parts that we can't copy, the best parts of all. These are the signature processes of nature: self-repair, regeneration, and rebirth. Nature holds on to them tightly; they are not just copyright, they are copyproof.*[1]

## VALUES AND SCIENCE IN ECOLOGICAL RESTORATION

One question for restorationists is how far "back" should they go to restore "nature"? This is a complicated question because of the transcientific character of the science of ecology. Ecologists cannot tell us what is natural. In their study of the role of ecology as a method in the development of conservation strategies, Shrader-Frechette and McCoy (pp.102–103)[2] write:

> [E]cologists cannot always specify what is "natural" … Knowing that one is acting in accord with nature is often defined as a condition in existence before the activities of humans who perturbed the system.… The definition is flawed, however, both because it excludes humans, a key part of nature, and because there are probably no fully natural environments or ecosystem anywhere. Because natural systems continually change, it is difficult to specify a situation at one particular time, rather than another time, as natural. We are unable to define natural in a way free of categorical values. [our emphasis]

Therefore, ecological restoration will entail the intermingling of scientific facts and sociocultural values. In addition, ecological restoration will likely take place in a historical context that includes a number of participants (i.e., restorationists) who often hold diverse values, beliefs, perceptions, and ethical orientations.[3-6]

The practice of restoration is predicated on the development of institutional capacity to deal with a multitude of discourses that may be found in a particular context and place.[4] Despite an extensive and growing literature on the theory and practice of restoration, there has not been an investigation of how restorationists, including philosophers, scientists, or theorists, connect restoration values to restoration practice. Development of a more practical mode of restoration theory requires an examination of the actual discourses of restorationists.

This chapter has two objectives. First, we describe Q-methodology as an alternative to the more conventional attitudinal survey. We draw from the social and cognitive sciences to outline how Q-methodology can compliment ordinary survey techniques to investigate the place of values and science in ecological restoration.

Second, we briefly describe the results of a Q-study of river restorationists conducted in 1997. The individuals participating in the study as subjects were involved in one of three river restoration projects, the Yakima River in Washington and the Upper Sacramento and Santa Ynez Rivers in California. These cases involve a wide array of watersheds in terms of economic activities, urbanization, and topography.

## TWO SURVEY METHODOLOGIES

Research on identifying the "environmental" values, beliefs, attitudes, and perceptions of the general public, environmentalists, policymakers, and scientists tends to employ various survey instruments.[7-11] A review of this extensive literature can be found elsewhere.[7-10] Generally, this research strives to make the role of values, perceptions, and beliefs more explicit in modeling environmental concern.[11] Two fundamental social paradigms in modern societies have been identified — the industrial world view (technocentrism) and an emerging ecologically oriented world view (ecocentrism).[10,11] Mail and telephone survey questionnaires are used to gather information on the kinds of factors or conditions that contribute to these two world views.[10] These surveys do not necessarily predict human behavior, but can provide an idea of public perception and world view at a particular point of time. As O'Riordan (p. 8)[11] writes, "Although public opinion polls are sometimes flawed (with responses heavily influenced by the perceived responses of the questions and particular prompts from the questioner), if undertaken consistently and professionally, they offer real insights into public attitudes toward important issues."

In general, previous research has consistently shown widespread support for environmental protection among the general public,[11] but has yet to address issues and concerns related to the values for or against ecological restoration. Our survey of this literature shows that there exists no study that uses data taken from attitudinal surveys (the R-technique) to illuminate the relationship between the values associated with ecological restoration. In addition, there remains a weak link made between values, perceptions, and beliefs and the structure of policy preferences (especially with respect to restoration policymaking).[12,13] There are studies of public support for protection and restoration of salmon to the Pacific Northwest,[14] but not general ecological restoration.

We propose the adoption of an alternative to R-technique to examine the place of values in the politics of ecological restoration. This alternative is the so-called Q-study of the discourses of restorationists involved in river restoration. Q-methodology is an "intensive, intentional alternative" to conventional opinion surveys and has been used to investigate an array of issues and concerns.[15,16] Q-methodology has been used to develop profiles of debates and the applied meanings of various concepts in a wide variety of situations.[17] It has been shown to be an especially useful methodology for those interested in studying the values held by individuals because, as described further later, it emphasizes the individual respondent's subjective orientation and the structure of held values and beliefs. R-technique, by contrast, typically involves much more involvement by the researcher in defining *a priori* the structure of the ideas and concepts (or general traits) with which respondents are given an opportunity to express agreement or disagreement, item by item.[18]

## Q-METHODOLOGY

In contrast to R-technique, Q-methodology is not defined by the attempt to draw a sufficiently large, representative sample from some underlying population. Rather,

**TABLE 1**
**Design Framework for Q-Sample Composition**

| Position on Restoration | World View | Types of Argument |
|---|---|---|
| A. Pro-restoration | (a) Ecocentric | 1. Definitive |
| B. Anti-restoration | (b) Technocentric | 2. Designative |
|  |  | 3. Evaluative |
|  |  | 4. Advocative |

*Note:* Q-sample = (Position on Restoration) (World View) (Type of Argument) = 16 (m). N = 48 statements.

Q-methodology demonstrates, using an appropriately diverse set of subjects (respondents), how people impose an order on and organize a representative set of value statements. Q-methodology requires that the respondents be selected to provide a reasonable representation of points of view in the relevant discourse. That is, the subjects should roughly reflect the range of views in the larger population, but need not reproduce their relative frequency. The goal of Q-methodology is to "simulate" rather than to "represent" the subjective structure of beliefs and values held by individuals and to compare these individuals with one another.

In the parlance of Q-methodology, the entirety of expressed values and preferences with respect to a subject constitute the "communication concourse." A concourse encompasses a population of "stimulus items" (i.e., subjectively communicable statements of opinion) bearing on the relevant issue(s). In practice, procedures are developed to assure that the concourse incorporates a very wide array of combinations of possible value positions. Each Q-study engages subjects by asking them to "sort" a representative sample of statements drawn from the relevant communication concourse.

Our design for creating a broad sample of value statements was developed following the strategy outlined in Dryzek and Berejikian.[18] We started by gathering 144 actual statements that reflected a range of sentiments and positions concerned with restoration. These statements were gathered from a range of journals, newspapers, technical reports, government documents, advertisements, and from informant interviews with restorationists.

These statements were then sorted according to a 2 × 2 × 4 factorial matrix. Again, the matrix design builds on prior work by Dryzek and Berejikian[18] and is informed by our discussions with informants. The first dichotomous factor distinguished pro- vs. anti-restoration sentiments. The second dichotomous factor distinguished "ecocentric" vs. "technocentric" views of restoration (consistent with References 10 and 11). In this matrix design, then, there were two kinds of anti-restoration sentiments: ecocentric and technocentric. We further classified these four groups of statements in accordance to Toulmin's[19] well-known characterization of claims used in argumentation: Definitive, concerning the meaning of terms; Designative, concerning questions of fact; Evaluative, concerning the worth of something that does or could exist; and Advocative, concerning something that should or should not exist.[18] The design framework is found in Table 1, which provides for 16 categories of statements.

The anti- and pro-restoration positions have been characterized in some detail by McGinnis.[20] Some theorists argue that restoration is ethically suspect because natural areas and species are irreplaceable, and there is a view that a restored nature is a "fake" nature insofar as it is not a product of a historical natural process. Other anti-restoration arguments are possible, including that restoration is unnecessary or that restoration is technically infeasible. Still other scholars defend restoration and propose that it is ethically and scientifically justified.[5,20-22]

Drawing on the literature on "environmental" world views, each statement was categorized as either ecocentric or technocentric (consistent with References 10 and 11) and either pro-restoration or anti-restoration (consistent with Reference 20) and was classified as definitive, designative, evaluative, or advocative. Working with this classification scheme, we sorted our 144 statements and then reduced them to a sample of 48, which were selected to assure that each cell in the factor matrix was represented in the sample by three statements. These procedures follow Dryzek and Berejikian[18] and Brown.[16] The 48 statements gathered represented our Q-sample and are listed in Table 2. The items are selected so as to represent all sorts of alternative sentiments roughly equivalently, not proportionately to their actual occurrence in any real-world discourse.

Each of our 26 subjects were participants in river and watershed restoration. The participants included scientists; industry representatives; private property owners; environmentalists; and representatives of federal, state, and local resource agencies. Participants included 13 females and 13 males. Among the identified respondents, there were two from federal resource agencies, four state resource agency representatives, two local government representatives, six environmental (river and salmon advocacy) group representatives, seven private industry representatives (including major property owners), and five academics (including two environmental philosophers and biological or ecological scientists) (N = 26). Subjects obviously can have more than one "identity." One respondent, for example, is a male who works as a scientist for a federal resource agency.

Each subject was asked to sort the 48 statements in a way that reflects his/her attitudes about the item itself and his/her judgment of the relationship among all other items. Within the constraints of a roughly "normal" distribution which is imposed on the sort, Q-sorting is entirely self-referential (items are ranked relative to one another in terms of a subject's personal understanding and subjective preference). The respondents to Q-sort have a broader range of "distances" to deal with than in typical public opinion items (which usually involve five- or seven-item scales), and they score each items in relationship to other items rather than in isolation (as traditionally is the case in public opinion polls). Hence, the subject, rather than the researcher, imposes a frame of reference on the interpretation of particular concepts or their relationships.

Respondents were asked to sort statements as follows. The respondents were directed to select two items for the most extreme categories (closest and farthest from the subject), three for the next extreme, and so on, until all statements had been placed in 1 of 11 categories. The most important statements, the ones closest

## TABLE 2
## Factor Analysis of Restoration Q-Sorts

Orthogonal Rotation; Prerotation Method: Varimax

### Rotated Factor Pattern

| Subject | Factor 1 | Factor 2 | Factor 3 | Factor 4 |
|---|---|---|---|---|
| R5  | 82* | −7  | 23  | −1  |
| R23 | 81* | 7   | 2   | 12  |
| R19 | 79* | 2   | 17  | 32  |
| R3  | 77* | 30  | 24  | 4   |
| R12 | 75* | 16  | 19  | 5   |
| R7  | 74* | 26  | 1   | 41* |
| R22 | 72* | 19  | 15  | −15 |
| R2  | 71* | 7   | 30  | 21  |
| R15 | 71* | 7   | 45* | 21  |
| R13 | 70* | −1  | 44* | −7  |
| R1  | 70* | 3   | 12  | 13  |
| R24 | 65* | 34  | −12 | −28 |
| R4  | 63* | 18  | 43* | 6   |
| R6  | 60* | 16  | 44* | 31  |
| R8  | 55* | 36  | 9   | 40  |
| R17 | 53* | 22  | 25  | 39  |
| R16 | −22 | 80* | −13 | 0   |
| R25 | 33  | 74* | 3   | −7  |
| R14 | 2   | 73* | −8  | 28  |
| R9  | 16  | 66* | 19  | 37  |
| R26 | 11  | 58* | −5  | 3   |
| R10 | 48* | 54* | −6  | 33  |
| R20 | 50* | 51* | 26  | 43* |
| R11 | 16  | −22 | 73* | −18 |
| R18 | 46* | −2  | 71* | 13  |
| R21 | 3   | 20  | −11 | 79* |

### Variance Explained by Each Factor

| Factor 1 | Factor 2 | Factor 3 | Factor 4 |
|---|---|---|---|
| 8.840769 | 3.717894 | 2.341429 | 2.100152 |

*Note*: Printed values are multiplied by 100 and rounded to the nearest integer. Values greater than 0.404 (significant) have been flagged by an *.

to the subject's own point of view, reflect not only strong agreement, but stronger agreement than with items in the next closest pile. Items in the middle piles are relatively unimportant even if the respondent agrees or disagrees with the statement. The final result is a bell-shaped distribution of items with a large number of items (in this study 20 out of 48) in the three central categories.

## A BRIEF SUMMARY OF THE FINDINGS

We analyzed the 26 sorts following the procedures outlined in Brown.[16] First, the sorts were statistically analyzed using factor analysis to identify the groupings among respondents. If there was strong agreement among participants, i.e., if participants tended to arrange the statements in very similar ways, there would be only one factor. If only one factor existed, the discourse of restoration would be a monistic and unified one.

Our analysis extracted four factors (or groups of respondents) using principal components analysis and then rotated the factors by the varimax method (an orthogonal rotation simplifying the column structure). The total of four factors was arrived at following the conventional rule of thumb to retain factors with eigenvalues greater than 1 and was also consistent with the results of a scree plot. These four factors (or groups) are identified in Table 3.

The factors were then analyzed in order to identify the sorts (respondents) that could be taken to most clearly define the underlying factors. The idealized sorts were based on actual sorts that had high factor loadings on the factor of interest if that particular factor accounted for a very high proportion of their communality score. Thus, the first idealized sort was based on three respondents whose loadings on Factor 1 were 0.82, 0.80, and 0.72, respectively, representing 0.92, 0.97, and 0.94 of their commonalties, respectively. The second idealized sort was based on respondents whose loadings on Factor 2 were 0.80 and 0.73, respectively, representing 0.91 and 0.86 of their commonalties. The third and fourth idealized sorts were represented by the actual sorts of a single respondent.

Following Brown,[16] the idealized sorts are calculated as follows. First, the scores assigned to each statement by the respondents in the relevant group is weighted by w/wmax, where wmax is the largest score w calculated, following Spearman (1927), as $f/(1 - f^2)$, with f being the initial factor loading. Thus, if the respondent with the largest factor loading had a loading of 0.82, as in our first factor, the Spearman expression would be $0.82/(1 - 0.82^2)$ or 2.50. The formula for weighting would result in a weight of 1 for this respondent and successively smaller weights (0.94, 0.56) for the other respondents. Thus weighted, the scores for particular statements are summed across respondents for each statement. The resulting scores are normalized (subtracting the mean, dividing by the standard deviation) and sorted from high to low, which provides the reconstructed idealized Q-sort underlying Factor 1.

These sorts were averaged together to produce "idealized" sorts that reflect the most distinctive elements of each factor, as shown in Table 3. Most of our respondents load quite strongly on either Factor 1 or Factor 2, although several loaded strongly on Factors 1 and 3 and two respondents loaded almost equally on Factors 1 and 2. Factor 3 has only two respondents loading very strongly on it; only one of those loaded almost exclusively on Factor 3. The fourth factor has only a single respondent loading very strongly on it, although the factor loadings of two others exceed "statistical significance." In conventional analysis of survey responses, the last two factors would probably be tossed out as either being idiosyncratic or as not reflecting any underlying dimension more complex than the particular response item. In

## TABLE 3
### Idealized Factors Sorted According to Pro-Restorationist Factor

| Statement Number | Q-Statement | Pro-Restoration | Anti-Restoration | "Nature" Philosophers | "3rd Path" Optimist |
|---|---|---|---|---|---|
| 15 | There is plenty of wilderness out there, and society does not need to be wasting its precious money "restoring" our land. | –5 | 0 | –2 | –1 |
| 21 | Environmentalists have overstated the need to restore the environment. | –5 | 5 | –2 | 4 |
| 40 | The environment cannot be restored no matter how technologically proficient environmental engineers become. | –4 | –4 | 3 | –4 |
| 2 | Since restoration is impossible by definition, we should quit worrying about myths of nature, and start worrying about human welfare. | –4 | –1 | –1 | 0 |
| 3 | Ecological restoration is used as another justification for the continued damage to natural systems. | –4 | –1 | –1 | –2 |
| 10 | Frankly, …We can fix what's really ugly, but we don't have to cater to every whim of the environmental extremists. | –3 | 1 | –3 | 2 |
| 26 | Dams and salmon can coexist-exist without restoring habitat. | –3 | 1 | –3 | 1 |
| 5 | The problem with restoration efforts is that many restorationists simply do not know when to stop. | –3 | 2 | 0 | 0 |
| 43 | People can manage, manipulate, and repair the environment just as a mechanic can repair a machine. | –3 | 3 | –1 | 0 |
| 37 | Naturalness cannot be restored. | –2 | –5 | 4 | –2 |
| 17 | Deliberate genetic manipulations by scientists merely repeat what has previously taken place in nature. | –2 | –1 | –4 | 3 |
| 44 | One promise of restoration enterprises is the creation of new worlds. | –2 | –1 | –4 | 1 |
| 36 | People should put community needs ahead of restoration needs. | –2 | 0 | –5 | 1 |
| 11 | A formidable barrier to restoration is the inevitable conflict between environmental protection and property rights. | –2 | 4 | 2 | –5 |
| 25 | A restored nature is a false nature; it is an artificial human creation, not the product of a historical natural process. | –1 | –3 | 3 | –3 |
| 42 | Genetically engineered organisms are no different from those animals, plants, and microorganisms that currently exist. | –1 | 0 | –4 | 1 |
| 8 | Restoration is based on the claim that any loss in value of an area is only temporary, and the re-creation of something of equal value. | –1 | 0 | –3 | 1 |
| 34 | Restoration actions should not be taken unless their long-term consequences are clear. | –1 | 1 | 1 | 0 |
| 39 | The river is no longer a natural river, and the river cannot be restored to its natural free-flowing state. | –1 | 1 | 3 | –1 |
| 12 | I have an abiding faith in science and technology to solve complex ecological problems and restore our environment. | –1 | 3 | –5 | –4 |
| 32 | Private property rights should be sacrificed to protect the integrity of ecosystems and foster restoration efforts. | 0 | –4 | 1 | –4 |
| 33 | Human restoration efforts are, by definition, unnatural. | 0 | –3 | 2 | –3 |
| 47 | Artificial habitat cannot replace lost wetlands and riparian zones. | 0 | –3 | 3 | 2 |
| 24 | Eliminating the millions of tons of pesticides and fertilizers spread on the ground ought to be the first step in environmental restoration. | 0 | –2 | 0 | –3 |

## TABLE 3 (CONTINUED)
## Idealized Factors Sorted According to Pro-Restorationist Factor

| Statement Number | Q-Statement | Pro-Restoration | Anti-Restoration | "Nature" Philosophers | "3rd Path" Optimist |
|---|---|---|---|---|---|
| 28 | In restoration work, technology creates at least as many problems as it resolves. | 0 | –2 | 2 | 3 |
| 46 | There are no technologically proven ways to restore nature. | 0 | –2 | 4 | –1 |
| 9 | Since we cannot restore nature, we should focus instead on developing ways to use nature wisely. | 0 | 2 | –2 | 0 |
| 18 | Restoration concerns are as much philosophical as they are technical. | 0 | 4 | 1 | 5 |
| 16 | Wilderness cannot be created. | 1 | –3 | 5 | 0 |
| 38 | Restorationists work with the scientific community and others to address information needs. | 1 | 0 | 0 | –1 |
| 20 | In restoration, small scale technology is preferred to large scale. | 1 | 1 | 0 | –1 |
| 19 | Restoration does not depend on science or technology, but on politics. | 1 | 2 | 2 | 2 |
| 41 | Restoration policies should be made at the local level to reflect local concerns. | 1 | 3 | 0 | 3 |
| 30 | Citizens should have the major say in deciding important restoration issues. | 1 | 5 | –1 | 2 |
| 27 | Failure to attempt restoration denies ecological reality and represents a truly unethical choice. | 2 | –4 | –1 | 0 |
| 23 | Restoration requires adopting an ecological consciousness rooted in species equality, appropriate technology, recycling, and bioregions. | 2 | –2 | 1 | –3 |
| 29 | I advocate a restoration policy whether or not a restored ecosystem is a full-fledged natural entity. | 2 | 0 | –2 | 1 |
| 13 | Restoration strategies and implementation actions must be based upon the best available science. | 2 | 2 | –1 | 0 |
| 7 | Current restoration policies consist of a multitude of sometimes conflicting regulations, enforced by various agencies without an overall coherent plan. | 2 | 3 | 1 | 2 |
| 1 | Hatcheries do not restore wild salmon, they threaten them. | 3 | –2 | 4 | 3 |
| 48 | Ecological restoration must involve social and community restoration. | 3 | 0 | 0 | 5 |
| 6 | I am in favor of preservation policy over and above a restoration policy. | 3 | 1 | 5 | –2 |
| 4 | We need to recognize the uncertainty of the science involved in restoration. | 3 | 4 | 0 | 4 |
| 31 | Ecological restoration, ideally, involves "returning a site to some previous state, with the species richness and diversity. . . before human settlement. | 4 | –1 | 0 | –5 |
| 14 | Preservation and restoration are inseparable. | 4 | 0 | 2 | –2 |
| 35 | Mankind can do a great deal to assist nature in the healing process. | 4 | 2 | –3 | –2 |
| 22 | A restored river is not a luxury, it is a necessity. | 5 | –5 | 1 | –1 |
| 45 | Restoration is good economics as well as good science. | 5 | –1 | –2 | 4 |

Q-methodology, however, distinctively different sorts reflecting only a single respondent may be of great interest if there is some special characteristic of the respondent or if the logic of the sort is especially interesting. For example, if the Q-sort was produced by a well-known proponent of a specific point of view in an ongoing debate, that Q-sort might hold particular interest.

Our purpose here is to characterize Q-methodology as an alternative to R-technique, rather than to provide a comprehensive summary of the findings from this research. For a detailed, descriptive overview of the four discourses, we refer the reader to Woolley and McGinnis.[17] In general, the first idealized discourse, which we have labeled "pro-restoration" in Table 3, regards restoration as factually necessary, not part of some value tradeoff, and, although slightly less emphatically, ethically mandated. The discourse is defined by the four extreme statements: "A restored river is not a luxury, it is a necessity" (+5); "Restoration is good economics as well as good science" (+5); "There is plenty of wilderness out there, and society does not need to be wasting its precious money restoring our land" (-5); and "Environmentalists have overstated the need to restore the environment" (-5).

The second discourse, which we have labeled as "local-control" is perhaps even more explicitly prescriptive, placing far greater emphasis on private property rights and local control of restoration projects. Respondents sharing this discourse apparently view the claims of both restorationists and scientists with skepticism: "Environmentalists have overstated the need to restore the environment" (+5); "We need to recognize the uncertainty of the science involved in restoration" (+4) and emphasize that restoration necessarily involves trade-offs with other important cultural values ("A formidable barrier to restoration is the inevitable conflict between environmental protection and property rights" (+4)), a position the pro-restorationists reject.

We also identified two additional discourses which we have labeled the "nature philosophers" and the "compromise path." The "nature philosophers" express deep distrust of science and technology and of "community needs" as a standard for policy. The deepest feelings are for wilderness and preservation: "Wilderness cannot be created" (+5); "I am in favor of preservation policy over and above a restoration policy" (+5); and "Naturalness cannot be restored" (+4). In this discourse, a fundamental view is that restoration is literally impossible and its products are "false."

## CONCLUSIONS

We used Q-methodology in order to characterize more precisely the various discourses of restoration. These diverse discourses should be taken into account in restoration planning. Q-technique can be a useful way to identify the diverse discourses that may exist in a particular restoration context. To begin to understand the diverse values, beliefs, perceptions, and discourses that may be associated with a particular restoration effort, we have offered Q-methodology as one alternative technique to the more conventional R-technique. Identifying the diverse discourses of restoration, bringing them into the conversation, and talking about them will turn out to be an important part of the restoration process.[4] When this is done, restorationists will find it easier to find common ground with a sizable constituency of participants and supporters. When it is not done, the development of a clear and well-defined consensus among participants will be put off. In such cases the restoration effort is likely to be limited or compromised not by the limitations of scientific or technological knowledge, but by beliefs, perceptions, and values (i.e., discourses) and the inevitable conflict among them.

## ACKNOWLEDGMENTS

The authors gratefully acknowledge the financial support of the National Science Foundation's Ethics and Values Studies Program under award SBR95-11599. The authors thank Rachelle Hollander (at NSF), comments on a previous draft of this chapter from Bill Jordan (at the Society for Ecological Restoration), John Dryzek (Professor, University of Australia, Melbourne), and an anonymous reviewer. We also especially thank participants who generously contributed time to complete Q-sorts. Views and conclusions are those of the authors and are not necessarily those of the NSF or the University of California.

For questions or further information regarding the work described in this chapter, please contact:

> **Michael Vincent McGinnis**
> 7602 Hollister Ave. #202
> Goleta, CA 93117
> e-mail: mcginnis@lifesci.lscf.ucsb.edu

## REFERENCES

1. Ehrenfeld, D. 1998. Death of a Palm. *Orion* 17:9–10.
2. Shrader-Frechette, K.S. and E.D. McCoy. 1994. *Method in Ecology: Strategies for Conservation*. London, Cambridge University Press.
3. Hoerr, W. 1993. The Concept of Naturalness in Environmental Discourse. *Natural Areas Journal* 13:29–32.
4. McGinnis, M.V. and J.T. Woolley. 1997. The Discourses of Restoration. *Restoration and Management Notes* 15(1):74–77.
5. Cowell, M.C. 1993. Ecological Restoration and Environmental Ethics. *Environmental Ethics* 15:19–32.
6. Higgs, E.S. and A. Light. 1996. The Politics of Ecological Restoration. *Environmental Ethics* 18:227–247.
7. Pepper., D. 1985. *The Historical Roots of Environmentalism*. London: Routledge.
8. Jones, R.E. and R. Dunlap. 1992. The Social Bases of Environmental Concern: Have They Changed Over Time? *Rural Sociology* 57:28–47.
9. Dunlap, R.E. and K.D. Van Liere. 1984. Commitment to the Dominant Social Paradigm and Concern for Environmental Quality. *Social Science Quarterly* 65:1013–1028.
10. Olsen, M.E., D.G. Lodwick, and R.E. Dunlap. 1992. *Viewing the World Ecologically*. Boulder, CO: Westview Press.

11. O'Riordan, T. 1995. Frameworks for Choice: Core Beliefs and the Environment. *Environment* 37(8):4–9, 25–29.
12. Norton, B.G. and B. Hannon. 1997. Environmental Values: A Place-Based Theory. *Environmental Ethics* 19:227–245.
13. Bullis, C.A. and J.J. Kennedy. 1991. Value Conflicts and Policy Interpretation: Changes in the Case of Fisheries and Wildlife Managers in Multiple Use Agencies. *Policy Studies Journal* 19:542–552.
14. Smith, C.L. and B.S. Steel. 1997. Values in the Valuing of Salmon. In D. Stander, P. Bisson, and R. Nainan, eds. *Pacific Salmon and Their Ecosystems*. New York: Chapman and Hall.
15. Stephenson, W. 1953. *The Study of Behavior: Q Technique and Its Methodology*. Chicago, IL: University of Chicago Press.
16. Brown, S.R. 1986. Q Technique and Method: Principles and Procedures. In W.D. Berry and M.S. Lewis-Beck, eds. *New Tools for Social Scientists*. Newbury Park, CA: Sage.
17. Woolley, J.T. and M.V. McGinnis. In press. Conflicting Discourses of Restoration. Society of Natural Resources.
18. Dryzek, J.S. and J. Berejikian. 1993. Reconstructive Democratic Theory. *American Political Science Review* 87(1):48–61.
19. Toulmin, S. 1958. *The Uses of Argument*. London: Cambridge University Press.
20. McGinnis, M.V. 1996. Deep Ecology and the Foundations of Restoration. *Inquiry* 39:203–217.
21. Merchant, C. 1986. Restoration and Reunion with Nature. *Restoration and Management Notes* 4(2):68–70.
22. Jordan, W.R., III. 1989. Restoring the Restorationist. *Restoration and Management Notes* 7:55.
23. Kim, J.O. and C.W. Mueller. 1986. *Factor Analysis: Statistical Methods and Practical Issues*. Beverly Hills, CA: Sage.

# 19 The California Water Quality Assessment Spatial Database: A Preliminary Look at Sierra Nevada Riverine Water Quality

*Anitra L. Pawley, Joshua H. Viers, Isaac Oshima, and James F. Quinn*

## CONTENTS

Abstract .................................................................................................................. 333
Introduction ........................................................................................................... 334
  The Water Quality Assessment ....................................................................... 335
  The Importance of Geographic Information Systems (GIS) ........................... 336
  The River Reach File ....................................................................................... 336
Methods ................................................................................................................. 337
  California Water Quality Assessment Data and Reach File
  Structural Constraints ...................................................................................... 338
Results and Conclusions ....................................................................................... 339
  Pollutants That Degrade Water Quality .......................................................... 339
  Predominant Sources of Water Quality Pollution .......................................... 342
  Are Restoration Efforts Linked to Water Quality Impairment? ..................... 343
Future Directions .................................................................................................. 346
Acknowledgments ................................................................................................. 347
References ............................................................................................................. 348

## ABSTRACT

To date, the Water Quality Assessment (WQA) mandated by national legislation (the Clean Water Act) is the most comprehensive dataset summarizing water quality on a national and statewide level. The assessment is the principal means by which the U.S. Environmental Protection Agency (EPA), Congress, and the public evaluate water quality, the progress made in maintaining and restoring water quality, and the extent of remaining problems. The procedures for developing the assessment database and the methods for evaluating and reporting assessment results are being improved to facilitate better reporting of water quality conditions. Here we present the results of our work with the WQA in California (CWQA) for Sierra Nevada rivers and streams. By attaching assessment database records to the EPA River Reach File, we have developed a more usable form of the statewide assessment. The CWQA spatial database is a Geographic Information System (GIS) which will enhance resource managers' ability to visually depict water quality conditions associated with specific rivers and to analyze these conditions to evaluate possible sources of impairment. An analysis of the resultant CWQA for the Sierra Nevada demonstrates the strength and weaknesses of the spatial dataset in its current format. In addition, a preliminary regional comparison of water quality problems and restoration project locations and issues illustrates how the CWQA spatial database can be incorporated into regional analyses that support watershed management decisions.

**Keywords:** *water quality, geographic information systems, watershed management, restoration, regional planning*

## INTRODUCTION

The recognition that the ecological health of rivers and streams are affected by the activities in the watershed has lead to many publications that extol the virtues of watershed management. Though commonly accepted by aquatic scientists as the best method to address water-based environmental problems, watershed management is not easily defined. Perhaps watershed management can best be understood by presenting an example of its perceived purpose for water quality assessment. The Clean Water Action Plan calls for state environmental agencies and state conservationists to jointly convene a process to create "Unified Watershed Assessments" by October 1998 (EPA, 1998). These watershed assessments will support watershed management objectives that encourage comprehensive solutions to water quality problems, promote program efficiency and cost effectiveness by integrating existing programs, and encourage local citizens to get involved in decision making. Because a GIS facilitates data analysis in a geographic context, it is an essential tool for analyzing water quality and performing the regional analyses that support watershed management.

Though Sierra Nevada ecosystems may appear pristine, these systems have been heavily impacted by the human presence. The Sierra Nevada Ecosystem Project (SNEP) report, submitted to Congress in 1996, assessed the historical, physical, biological, ecological, social, and institutional conditions in the Sierra Nevada (Sierra Nevada Ecosystem Project, 1996). The SNEP project team determined that aquatic systems and their associated riparian habitats were the most altered habitats of the Sierra Nevada (Sierra Nevada Ecosystem Project, 1996). Of 67 types of aquatic habitat characterized, almost two thirds of the habitat types were found to be declining in quality and abundance (Sierra Nevada Ecosystem Project, 1996). Despite these findings and the emphasis of the SNEP technical team on GIS techniques to analyze resource conditions, water quality conditions were not directly evaluated in a geographic context (Kattelman, 1996).

The importance of evaluating and summarizing the water quality of Sierra Nevada waterbodies is supported not only by the need to improve regional water quality conditions, but also due to its influence on a large proportion of the state's water resources. The region can be divided into hundreds of small watersheds, which in turn are subdivisions of larger watersheds (Moyle and Randall, 1996). The western Sierra Nevada contains many of the critical and sensitive headwaters that feed California's large riverine systems, the Sacramento and San Joaquin Rivers that meet just south of Sacramento to form the San Francisco Bay-Delta. This large watershed is the focus of a restoration planning effort called the CalFed Bay-Delta Ecosystem Restoration Program, a joint state-federal restoration effort (CALFED Bay-Delta Program, 1997). One of the four central themes of this large-scale restoration program is to restore water quality. A GIS database that summarizes water quality will facilitate the problem identification process essential for restoring this large-scale riverine ecosystem.

The development of the 1994 CWQA spatial database is the first step in a process that will enable California water quality conditions to be evaluated in a spatial context. The spatial format allows researchers and resource managers to compare existing GIS datasets with CWQA attributes such as beneficial uses, impairment status, types of pollutants, and the sources of water quality pollution. Here we provide a brief overview of the development of the 1994 CWQA and discuss the problems associated with creating a spatial water quality summary from tabular data not intended for a GIS format. To illustrate the resultant databases' utility, we provide an overview of water quality in Sierra Nevada rivers and tributaries. In addition, we provide an illustration of the types of analyses the CWQA spatial database can support by superimposing our results with the California Watershed Projects Inventory database. Our goal is to provide a regional overview of whether restoration and conservation efforts are being performed in appropriate locations on appropriate issues to address the water quality problems identified in the CWQA. The California Watershed Projects Inventory, a California-based GIS that documents restoration and conservation project locations and their associated attributes (Pawley and Quinn, 1997).

## THE WATER QUALITY ASSESSMENT

Since 1975, Section 305(b) of the Federal Water Pollution Act (Clean Water Act) has required states to submit a report on the quality of their waters to the EPA every two years. The EPA compiles the data from the state reports, summarizes them, and transmits the summaries to Congress along with an analysis of the status of water quality nationwide (Clifford et al., 1996). The state assessments are based on the extent to which each state's waterbodies meet state water quality standards, as measured against each state's designated beneficial uses. An essential element of the 305(b) process is the development of a list of impaired waterbodies, termed the 303(d) list, which enables the EPA and the California State Water Resources Control Board to set priorities for programs specified by the Clean Water Act to address water quality problems. In cases where beneficial uses are impaired, the states list the sources (e.g., municipal point source, agriculture, combined sewer overflows) and causes (e.g., nutrients, pesticides, metals) of these problems.

States generally do not assess all of their waters each biennium (Clifford et al. 1996). California reviews its assessed rivers every two years and frequently adds assessed areas based on the state's perception of its greatest water quality problems. To this extent, the CWQAs may be somewhat skewed toward waters with the most pollution and may, if viewed as representative of overall water quality, overstate pollution problems.

## THE IMPORTANCE OF GEOGRAPHIC INFORMATION SYSTEMS (GIS)

A GIS is capable of assembling, storing, manipulating, and displaying geographically referenced information, i.e., data identified according to their locations. Spatial data is generally referenced to polygon, vector (lines), or point locations. By linking the water quality assessment data to a linear system of rivers represented in a vector (line) format, we can ensure improved data representation through computer-generated maps and charts that enhance our ability to analyze and improve the CWQA. The CWQA as a GIS facilitates the comparison of CWQA features with other spatially explicit datasets, including land use, water quality, ecological monitoring data, and restoration project locations. Resource managers can combine the products of this effort with other California-based GIS such as the California Watershed Projects Inventory (URL = ice.ucdavis.edu), the Sierra Nevada Ecosystem Project (URL = ceres.ca.gov/snep/data.html), and the Natural Heritage Program (URL = www.dfg.ca.gov/Nddb/nddb.html) to perform watershed analyses aimed at pinpointing resource problems.

## THE RIVER REACH FILE

The EPA Reach Files is a set of hydrographic databases of the surface waters of the U.S. designed to provide a consistent national framework for managing watershed data (Dewald et al., 1997). The files consist of a collection of surface water features (e.g., streams, lakes, reservoirs, and estuaries), each of which is described by a set of attributes and a spatially referenced vector representation. The associated attribute files contain names for major rivers and many other streams and lakes. A unique identifier called the "Reach Code" is assigned to each surface water feature in the Reach File. Using the Reach Code, all types of attributes may be associated with specific locations represented by the vectors or polygons that denote surface water features. The Reach Code can provide the common link for water quality related data from diverse sources.

Reach File 3 is the product of a series of refinements of this vector-based file describing surface water hydrography. The first Reach File, referred to as RF1, was conceived in the 1970s and completed on a national level in 1982. RF1 was created from scanned 1:500,000 scale National Oceanic and Atmospheric Administration (NOAA) maps and supports broad-based national applications. During 1987 and 1988, the Feature File of the U.S. Geological Survey's (USGS) Geographic Names Information System (GNIS) was used to add one new level of streams to RF1, and the resultant product was called RF2. The development of the latest Reach File, RF3, began in 1989 and has produced, to date, a preliminary version known as

RF3-Alpha (Dewald et al., 1997). The RF3 production process involves the overlay of the RF2 file, the GNIS II Feature File, and the USGS 1988 1:100,000-scale Digital Line Graph Version 3 (DLG3) hydrography (Dewald et al., 1997) to improve the accuracy and spatial resolution of the surface water features.

The Reach File is still undergoing revision. In California, the Teale Data Center (URL = http://www.gislab.teale.ca.gov), the California Department of Fish and Game, and the University of California, Davis have been involved in Reach File correction. There are approximately 220,000 records represented in the California hydrography data layer. The hydrography layer consists of flowing waters, standing waters, and both natural and manmade wetlands. The coverages contain two separate feature types: polygons (areas) and vectors (a.k.a. arcs) (lines). Polygon features have attribute codes that identify waterbodies such as lakes, wide river segments, or swamps. Line features have attribute codes that represent streams or shorelines. Edits to the original linework were made during the data conversion by the Teale Data Center from the original DLG3 format to the ARC/INFO$^{(tm)}$ GIS format. Changes included line movements (due to the map edge-matching process), minor corrections of attribute coding, and the closing of open polygons. Vector linear features representing stream flow direction were also added. The connectivity of each reach to its adjacent (upstream and downstream) reaches is important for it will ultimately ensure its use as a functional tool for hydrologic modeling (Horn and Grayman, 1993). The California Department of Fish and Game and the University of California, Davis are correcting and updating RF1, RF2, and in some cases RF3 to reflect errors in naming river reaches and checking the flow direction for all streams (Karen Beardsley and Paul Veisze, personal communication). Because the Reach File is undergoing modification, the water quality assessment data was linked in a fashion that would accommodate these ongoing changes.

## METHODS

In California, there are nine Regional Water Quality Control Boards that work under the direction of the California State Water Resources Control Board and coordinate their water quality assessment efforts with EPA-Region IX. Each Regional Water Quality Control Board evaluates rivers within their boundaries and submits their evaluations through a database to the State Water Resources Control Board every two to four years. From these sources, a statewide tabular database and descriptive summary are produced. The statewide database used for this study, the 1994 CWQA, evaluated more than 3000 waterbodies, including nearly 1300 rivers, and their status in California. The attributes of the tabular database are extensive and include information on beneficial uses, immediate causes of impairment (pollutants and stressors), sources of pollution, and, in a limited number of cases, the percent of impairment (relating to linear river miles) for each waterbody.

In California, there is a growing demand for geographically referenced, water quality assessment data for use in interagency data integration on an ecosystem and watershed scale. Water quality specialists at Region IX of the EPA needed a spatial version of the CWQA database to enhance interagency cooperation and encourage improved input and verification of the CWQA. In the latter part of 1995, the EPA contracted with the University of California, Davis to work cooperatively with EPA-Region IX and the California State Water Resources Control Board. Our goal was to verify the names of the rivers and streams identified in the CWQA and provide a means to attach these records to the EPA RF3 for spatial representation. As noted above, RF3 is a georeferenced set of polygons and linear features (vectors) that provides a common mechanism within the EPA and other agencies for identifying surface water features and for relating diverse water resource databases (Horn and Grayman 1993). By checking each CWQA waterbody name in the tabular data for a corresponding entry in the Reach File, and correcting the RF3 river name where necessary, we provided a means to link the waterbodies (streams, lakes, estuaries) in the CWQA to the river names (termed "Pnames") of the Reach File using the ESRI Software, ArcView 2.1. This process allowed us to generate graphical output in the form of maps and bar charts that summarize water quality for the state and for selected areas within the state.

## CALIFORNIA WATER QUALITY ASSESSMENT DATA AND REACH FILE STRUCTURAL CONSTRAINTS

The problems encountered in linking the 1994 CWQA tabular data to RF3 illustrate the difficulty of combining databases from two independent sources. The tabular CWQA database was not intended for use as an attribute table for a GIS. The CWQA designations were vague, and the resultant product of the linkage was somewhat coarse in resolution with constraints for adequately representing the tabular data. The records included hydrologic unit codes (State Water Resource Control Board hydrologic units), a geographic designation for watersheds, in addition to river names to help identify the location of the rivers assessed. These references were used to determine the general location of the assessed waterbody; however, the hydrologic units are a coarse designation, and, in many cases, the records referred to a large river without reference to the specific section of the river that was impaired. Large rivers were described by upper and lower segments; however, this is also a coarse designation representing many linear miles. Although the authors of the database attempted to provide more specificity by listing the number of river miles impaired, frequently there was no way to connect the designation with the portion of river assessed. For example, for the Upper Sacramento River (from Red Bluff to the Delta), the CWQA specified that 0 miles of the river had a status of good, 155 miles medium, and 30 miles poor; however, it was unclear to which portion of the Upper Sacramento River each rating referred. Consequently, because the CWQA records were referenced to the RF3 linear features (arcs) by river name, the procedure to develop the CWQA spatial database inadvertently provided a finer level of spatial resolution where smaller rivers and tributaries were rated.

In addition, the coarseness of the data caused some problems for regional analyses. Given that rivers extend across regional boundaries, we had to devise a means to exclude river ratings that were outside of the boundary chosen, even though the waterbody identification code in the tabular data did not vary across regional boundaries. We were able to remove most of the erroneous ratings from our summary statistics for the Sierra Nevada by removing the records that referred to the lower portion of rivers, upon verification that "Lower (*name of river*)" was outside of the Sierra Nevada Bioregions. In a more spatially explicit database that was cross listed for numerous geographic descriptors, this approximation would not have been necessary.

The Reach File in its current format also created problems for CWQA spatial database development. Once the CWQA/RF3 linkage process was started, we noted that RF3, though generally quite useful, was not adequate in its present form to accommodate all of the records contained in the CWQA. Corrections in river names were needed to facilitate the linkage of many of the streams that were misrepresented in the Reach File. Because RF3 coverage for much of the state did not contain many open water, wetland, coastline, and groundwater designations, we employed wetland, groundwater, and open waterbody coverages to link records that did not conform to the River Reach File structure. To document Sierra Nevada riverine water quality, we confined this analysis to the Sierra Nevada rivers and streams data that we were able to link using the surface water features available in the EPA River Reach File. This process resulted in the linkage of 246 river and stream ratings to streams within the boundaries defined by the northern and southern Sierra Nevada Bioregions (geographic units adopted by the California Environmental Resource Evaluation System, 1996).

## RESULTS AND CONCLUSIONS

The CWQA spatial database indicates that a fairly large proportion of the rivers and streams systems evaluated during 1994 have been negatively impacted. Of the total number of streams and rivers rated as impaired in California (303d list), nearly half (55 out of 120 rated streams and rivers) are located in the Sierra Nevada Bioregion (see Figure 1). Of the 246 Sierra Nevada streams, rivers, and river segments evaluated, over one third were considered to be of poor quality. These findings support the importance of evaluating the causes and sources of water quality pollution in the Sierra Nevada.

### POLLUTANTS THAT DEGRADE WATER QUALITY

Where possible, the Regional Water Quality Control Board identifies the *pollutants and stressors* that degrade water quality, such as sediment, nutrients, and chemical contaminants (i.e., dioxin and metals). For Sierra Nevada rivers and streams, heavy metal pollution and trace metals (arsenic, manganese, beryllium, molybdenum, etc.) were the most frequently cited pollutants (Figure 2a). Siltation, other inorganic compounds, and mercury were next in importance. Heavy metals and trace elements appear even more important when these factors are considered by river miles (Figure 2b); however, the actual linear miles should be interpreted loosely due to the coarse-grained nature of the dataset. In addition, the analysis by river miles brings pesticides into focus as a dominant pollutant.

# Impaired Waterbodies for the Sierra Nevada (EPA 303d list)

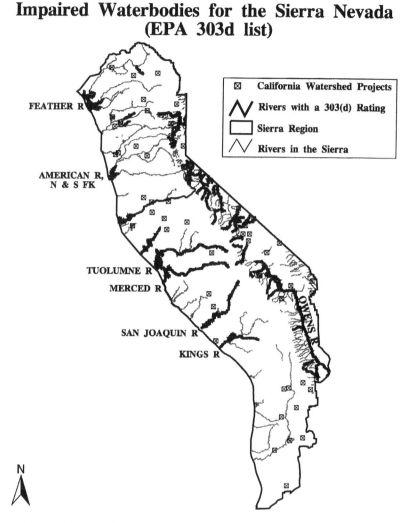

**FIGURE 1** EPA's list of impaired streams and rivers in the Sierra Nevada (the 303d list) and CWPI restoration project locations are depicted for the Sierra Nevada Bioregion.

The spatial database allows a geographic overview of these problems. For all Sierra Nevada streams evaluated, we selected to display the six most prevalent pollutants/stressors calculated by river miles. Figure 2c illustrates the rivers and streams that were rated and their association with these six pollutants/stressors. As expected, eastern and southern Sierra Nevada streams (i.e., the Owens River and East Fork of the Carson River) are heavily impacted by natural trace elements that have become more problematic with reduced flows. Heavy metal and mercury problems occur throughout the Sierra Nevada, which most likely reflects the historic legacy of mining in California. Siltation is not as important over all the river miles included in the analysis; however, it is frequently cited in association with many of the smaller

# California Water Quality Assessment Database

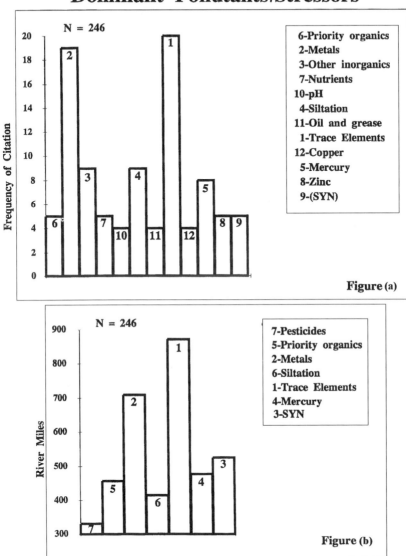

**FIGURE 2a,b** (a) This bar chart summarizes the number of citations recorded for streams and rivers within the Sierra Nevada Bioregion that are affected by specific pollutants or stressors indicated by the 1994 California Water Quality Assessment. (b) This chart depicts the number of river miles affected by specific pollutants/stressors.

rivers (see the frequency of its occurrence in Figure 2c). These results contrast with the overall statewide assessment in which siltation and nutrients are more frequently cited as major pollutants (1994 CWQA, unpublished data). Mining in the Sierra Nevada has exacted a large toll on aquatic systems and is difficult to reverse.

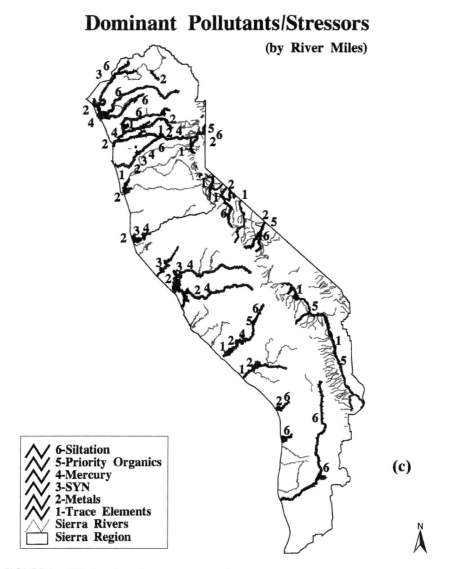

**FIGURE 2c** The location of the streams affected by the six most prevalent pollutants/stressors are depicted for the Sierra Nevada Bioregion. The rivers and streams affected by these pollutants are highlighted with thick lines. The rivers and streams are also numbered to indicate the specific pollutant(s) causing impairment. For example, a river with a 2 and a 6 is affected by heavy metals and siltation.

### PREDOMINANT SOURCES OF WATER QUALITY POLLUTION

In each instance of rating a stream or river, the Regional Water Quality Control Board staff also attempt to identify the sources of the water quality impairment. These sources are determined from the best available information and include the natural and disturbance factors that lead to water quality pollution. In the Sierra

Nevada, the most frequently cited pollutant sources were resource extraction, natural sources, urban runoff, hydromodification (dams and diversions), agriculture, and range land (Figure 3a). When we calculate sources by river miles, the relative importance of these categories shifts: agriculture and hydromodification dominate (Figure 3b). This suggests that these sources are associated with the larger rivers. Generally, the identified sources are consistent with the pollutants identified in the previous analysis. Resource extraction causes heavy metals to be the predominant source of water quality pollution, while agriculture is the most likely cause of pesticide problems. Hydromodification is often responsible for reduced flows and intensifies all types of pollutants/stressors including trace elements from natural sources. Perhaps more surprising is that urban runoff and development are now cited as dominant contributors to water quality problems in the Sierra Nevada. This was not apparent in the Water Quality Assessment Report published in 1992 (California State Water Resources Control Board, 1992).

## ARE RESTORATION EFFORTS LINKED TO WATER QUALITY IMPAIRMENT?

The California Watershed Projects Inventory (CWPI) (Pawley and Quinn, 1994–1997) and its successor, the Watershed Projects Inventory, is to date the most comprehensive list of restoration and conservation efforts in California. Projects included in the inventory include water quality nonpoint source protection programs, habitat conservation efforts, and riparian enhancement and fisheries restoration projects. Though CWPI projects address numerous types of environmental problems, most focus on resource issues that either directly or indirectly affect water quality. Therefore, we chose to compare these independent databases (the CWQA and CWPI) to determine if the most impaired waterbodies and surrounding watersheds were subjects of restoration and to evaluate whether the problems and resource issues addressed by projects identified by the CWPI were consistent with the water quality pollutant sources identified by the CWQA. Our preliminary results indicate that nearly half of the 41 restoration projects identified by the CWPI in the Sierra Nevada Bioregion were located in Sierra Nevada watersheds with rivers listed as having impaired water quality (Figure 1).

When we evaluated the resource issues that CWPI projects address, we noted discrepancies in the problems identified by the CWQA and those addressed by CWPI projects. The resource issues addressed by CWPI projects in the Sierra Nevada region are listed in Figures 4a and 4b. The majority of the projects addressed riparian degradation, soil erosion control, recreation issues, fisheries, and water quality impairment. Sediment and nutrients were overwhelmingly the major issues addressed by those projects that cited water quality issues as a concern. These results are not consistent with those of the CWQA where resource extraction, natural sources, urban runoff, hydromodification (dams and diversions), and agriculture were cited as the dominant sources of water quality impairment. This discrepancy may reflect the type and scope of environmental problems that restoration project leaders are willing to address. Project planners and managers appear to choose problems that are more readily solved through on-site remediation such as soil erosion control measures, bank stabilization, vegetation enhancement, and grazing control. Larger-scale problems such as mining, urban growth, and hydromodification may seem insurmountable with the short time

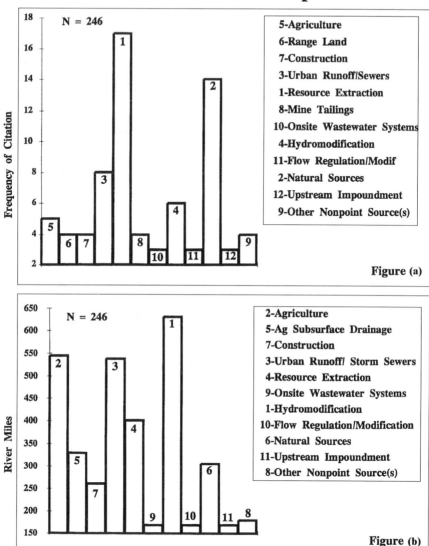

**FIGURE 3** (a) This bar chart summarizes the number of citations recorded for streams and rivers within the Sierra Nevada Bioregion affected by specific sources of water quality impairment according to the 1994 California Water Quality Assessment. (b) This chart depicts the number of river miles impacted by each of the dominant sources of water quality impairment. (c) The location of the streams affected by the six dominant sources of water quality impairment are depicted for the Sierra Nevada Bioregion. The rivers and streams affected by these pollutants are highlighted with thick lines. The rivers and streams are also numbered to indicate the specific sources of pollution. For example, a river with a 2 and a 4 is subject to impairment due to agriculture and resource extraction (mining).

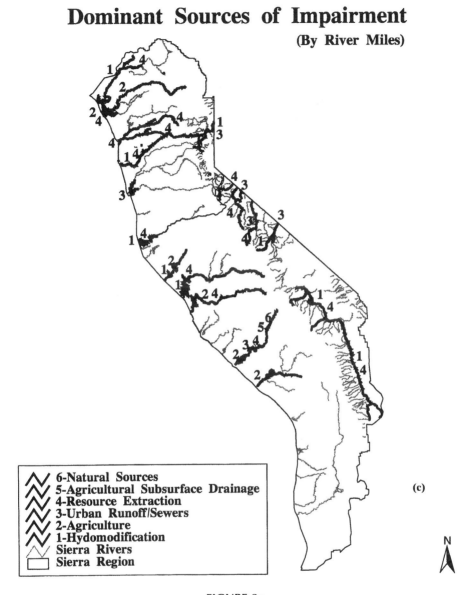

**FIGURE 3c**

scales and limited budgets most projects experience. Problems such as heavy metals may require extremely costly large-scale technological solutions. In addition, solutions to problems such as urban growth, dams, and diversions may require policy changes and funding mechanisms that extend beyond the jurisdiction of a single watershed project. Though these issues are complex and require more detailed analysis, this comparison sheds light on inconsistencies in the problems identified by the CWQA and the solutions proposed by current restoration programs.

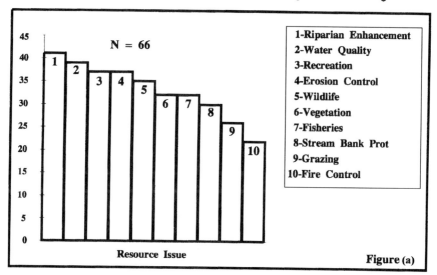

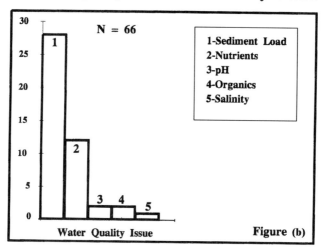

**FIGURE 4** (a) This bar chart summarizes the dominant resource issues addressed by projects located in the Sierra Nevada Bioregion as indicated by the CWPI. (b) This bar chart summarizes the dominant water quality issues addressed by projects located in the Sierra Nevada Bioregion as indicated by the CWPI.

## FUTURE DIRECTIONS

These results demonstrate the utility of the 1994 CWQA for identifying problems in a spatial context (Figures 1, 2c, and 3c) and facilitate comparisons with other spatial datasets. Because the CWQA is an attempt to summarize water quality problems and track general trends, the spatial coarseness of 1994 CWQA data limits its value as a means for identifying proximate causes, particularly for large rivers. The lack of spatial resolution also creates problems for regional data interpretation because many of the river ratings extended across regional boundaries. If river ratings referred to portions of the river outside of the Sierra Nevada Bioregion, the current form of waterbody identification does not allow easy identification of these instances. In addition, the CWQA is far from complete, because many Sierra Nevada rivers, streams, and lakes were not evaluated.

It is because of these inadequacies that this first phase of CWQA GIS database development has suggested many possible future refinements and additional analyses. For example, to develop methodologies to improve the assessment process, we will work with the North Coast Regional Water Quality Control Board to provide a more spatially refined CWQA for California's North Coast and to improve our techniques for data entry and extraction. It is believed that this process will enable resource managers to improve the accuracy of the assessment and provide a basis for the TMDL (Total Maximum Daily Load) calculations required for impaired waterbodies (303(d) list, Figure 1). In addition, we will be providing a system for Internet access to encourage the public to view a portion of the data and provide input on its accuracy. The California Water Quality web-site will provide information on further elaboration and details on this work and will be on-line at the Information Center for the Environment (http://ice.ucdavis.edu) in the near future.

## ACKNOWLEDGMENTS

We wish to thank the individuals who have been instrumental in guiding the creation of the California Water Quality Assessment spatial database. Agency support and guidance has come from the following individuals: Janet Hashimoto and Terry Fleming (Environmental Protection Agency), Nancy Richard (State Water Resources Control Board), and Paul Veisze (California Department of Fish and Game). Karen Beardsley and Eric Lehmer (Information Center for the Environment, U.C. Davis) provided advice in the development of the GIS database. Mike McCoy provided administrative support. Our special thanks goes to Andrea Thode (U.C. Davis) who assisted with poster development and figures. The project, during its development, has received financial assistance from the U.S. Environmental Protection Agency (Region IX) and the Center for Ecological Health Research, U.C. Davis.

For questions or further information regarding the work described in this chapter, please contact:

**Anita Pawley**
709 N Street
Davis, CA 95616
e-mail: alpawley@dcn.davis.ca.us

# REFERENCES

1. CALFED Bay-Delta Program. 1997. Executive Summary and Tables Working Draft. CALFED Bay-Delta Program.
2. California Environmental Resource Evaluation System. 1996. California Bioregions. http://www.ceres.ca.gov/.
3. California State Water Resources Control Board, California Environmental Protection Agency. 1992. Draft Water Quality Assessment. California State Water Resources Control Board, Sacramento, CA.
4. California State Water Resources Control Board, California Environmental Protection Agency. 1994. Draft Water Quality Assessment. California State Water Resources Control Board, Sacramento, CA.
5. Clifford, J., W. Wheaton and R. Curry. 1996. EPA's Reach Indexing Project — Using GIS to Improve Water Quality Assessment. URL=http://www.epa.gov/.
6. DeWald, T.G., S.A. Hanson and W. Wheaton. 1997. Managing Watershed Data with the USEPA Reach File. http://www.epa.gov/.
7. Environmental Protection Agency, Office of Water. 1995. Guidelines for the Preparation of the 1996 State Water Quality Assessments (305b) Reports.
8. Environmental Protection Agency, Office of Water. 1998. Clean Water Initiative. http://www.epa.gov/cleanwater/qa.html.
9. Horn, R. C. and W. Grayman. 1993. Water Quality Modeling with the EPA Reach File System. *Journal of Water Resources Planning and Management.* 119 (2):262–275.
10. Kattelman, R. 1996. Hydrology and Water Resources. In: *Sierra Nevada Ecosystem Project: Final Report to Congress, vol. II, Assessments and scientific basis for management options.* Davis: University of California, Centers for Water and Wildland Resources. pp. 855–908.
11. Moyle, P.B. and P.J. Randall. 1996. Biotic Integrity of Watersheds. In: *Sierra Nevada Ecosystem Project: Final Report to Congress, vol. II, Assessments and scientific basis for management options.* Davis: University of California, Centers for Water and Wildland Resources. pp. 975–982.
12. Pawley, A. and J. Quinn. 1994–1997. The California Watershed Projects Inventory Website. URL=http://ice.ucdavis.edu.
13. Sierra Nevada Ecosystem Project, Final Report to Congress. 1996. *Status of the Sierra Nevada. Volume 1: Assessment Summaries and Management Strategies.* Wildland Resources Center Report 36.

# Index

## A

*Abies concolor,* see White fir
Abiotic injury, to pine needles, 241-243, 244-246
ACE (alternating conditional expectation) modeling, 296
Air pollutants, 95-96
    case study results, 103-111
    criteria for risk assessment of ambient, 307-320, see also Ozone; Ozone injury
    deposition models for, 111-112, see also Dry deposition
    critical doses/loads concept, 112-114
    and effects on conifer forests, 103, 176, 233-237, see also Ozone injury index
        biochemical effects, 178-187
        growth and morphologic effects, 176
        at landscape level, 233-234
        physiologic effects, 176-178
        in San Bernardino mountains, 17-24, 94-95, see also San Bernardino mountains
        in Sierra Nevada, 94-95, see also Sierra Nevada
    future research needs, 114
    monitoring methods for, 97-103
    monitoring networks, 103-111
    nitrogenous
        dry deposition of
            in Barton Flats, 19-23
            in coastal sage scrub ecosystems, 111
            on plant surfaces, 17-24, 101-103
            in San Bernardino Mountains, 17-24, 108-109
            in San Gorgonio Wilderness Area, 109-111
            in Sierra Nevada, 108
        gaseous and particulate, in Sierra Nevada, 107-108
        monitoring, 98
    ozone, see also Ozone
        critical exposure levels of, 112
        effects of, on conifer forests, 103, 176
            biochemical effects, 178-187
            growth and morphologic effects, 176
            physiologic effects, 176-178
        particulate, 193-203
    transport of, 96-97
AIRS database, 312
Ambient exposure characterization, 309-314
Ammonium nitrate, see Nitrogenous air pollutants
Amplitude time series, 295
Annular denuder samplers, 98
Appalachian mountains, risk assessment for ambient ozone, 315-316
ARDRA (amplified ribosomal DNA restriction analysis), for microbial DNA identification, 199-202
*Aulocoseiria distans,* 133
Average squared residual (ASR), 301-303

## B

Barton Flats, 17, 18-19, 108
    model for nitrogenous deposition in, 20-21
        estimates and empirical data, 21-23
    ozone critical exposure levels at, 112
Benthic invertebrates, in Clear Lake aquatic ecosystem, 252
    mercury in, 257-259, 259-265
Big Leaf multiple resistance model, 111
BIOLOG, 224
Biomarkers, see also DNA fingerprinting
    of aquatic ecosystem health, 207-209, 215-219
        conceptual approach to, 210
        enzyme protein extraction of, 214-215
        examples of, 211
        protein electrophoresis of, 212, 214
        sampling process for, 209, 211-212
Bioremediation, intrinsic, see Intrinsic bioremediation
Boise National Forest
    fire hazard/risk assessment model for, 273-279
        design goals, 279
        fire ignition submodel, 275
        fisheries condition submodel, 277-278
        forested vegetation outside HRV submodel, 273-275
        practical goals, 280-283
        watershed hazard submodel, 277
        wildlife habitat persistence submodel, 275-277

ponderosa pine forests of, 271-273
  fire history, 272-273
  potential for uncharacteristic wildfire, 280
  watersheds at risk by wildfire, 281
Branch rinse, and throughfall data, 20
Branch wash deposition, 20
Brundtland Report, 6-7
Buell patterns, 291

## C

CADMP (California Acid Deposition Monitoring Program), 99
California black oak, 18-19
  dry deposition predictor for, 17-24, 112
Canadian Environmental Assessment Act, 150
Canadian Vegetation Objective Working Group (VOWG), 310-311
Cantara pesticide spill, 71-92
CASTNet sites, 311-312
*Catostomus commersoni,* see Trout perch
Central Valley, DNA fingerprinting of soil-derived dust, 193-203
Chlorotic mottle, 235-236, 242
Clear Lake aquatic ecosystem, 119, 250-251
  benthic invertebrates in, 252
  calcium carbonate and alkalinity of, 141
  ecological stresses on, recent, 141-142
  farming/livestock influences on, 139-140
  geology of, 120-121
  human history of, 121-123
  human impact on, 120
  mercury levels in, 259-265
    analytical laboratory procedures for, 252
    benthic invertebrate, 257-259
    plankton, 253-257
    sampling for, 251-252
    statistical analysis of, 252-253
  mercury loading in, 137-138
  plankton in, 251-252
  sediment cores of, 123-124
    analysis of, 124-127
      carbon and nitrogen, 124
      diatom enumeration, 126
      lead radionuclide dating, 124, 125
      mercury, 125
      phosphorus, 124-125
      pollen counts, 125-126
    obtaining, 124
    results on, analytical, 127-137
      chemical changes, 129-133
      diatom profile, 133, 137
      lead radionuclide dating, 127-129
      physical changes, 129-133

pollen profile, 133, 134, 135
sediment deposition in, 138, 139-140
sulfur loading of, 140-141
vegetation changes in, 139
Coastal sage scrub ecosystems, nitrogenous deposition in, 111
Conifer forests
  as biomonitors of ozone effects, 235, see also Ozone injury index
  ozone and effects on, 103, 176, 233-237, see also Ozone injury index
    biochemical effects, 178-187
    growth and morphologic effects, 176
    at landscape level, 233-234
    physiologic effects, 176-178
  in San Bernardino mountains
    air pollutants and effects on, 17-24, 94-95
    nitrogenous pollutants and effects on, 17-24, 108-109
    ozone and effects on, 103, 233-246
  in Sierra Nevada
    air pollutants and effects on, 94-95
    ozone and effects on, 103, 233-246
Conifer needles
  abiotic injury to, 241-246, see also Ozone injury index
  biotic and abiotic injury determinations, 239
  ozone effects on
    dendrobiological analysis of, 183-187
    historical scale analysis of, 182-187
    *in situ* biochemical analysis of, 180-182
  symptoms of ozone injury, 235-236
Constanza's homeostatic concept, 208-209
Contributor contacts
  Arbaugh, Michael J, 23
  Brown, Larry R, 59
  Bytnerowicz, Andrzej, 114
  Cairns, John, 11
  Fan, Teresa, 188
  Hooper, Scott, 70
  Jassby, Alan, 304
  Lefohn, Allen, 317
  McGinnis, Michael Vincent, 331
  Miller, Paul R, 246
  Munkittrick, Kelly R, 170
  Ogunseitan, Oladele A, 219
  Orlab, Gerald T, 91
  Pauley, Anita, 347
  Richerson, Peter, 135
  Scow, Kate M, 203
  Suchanek, Thomas H, 266
  Sudarshana, Padma, 230
  Weiland, Cydney A, 283
Crown condition, symptoms of ozone injury, 236

Crown position classes, and ozone injury index distribution, 241
Cumulative effects assessment, 149-152
    cause and effect relationships in, 170-171
    choice of indicators for, 152-155
    issues of, 150
    latitudinal habitat survey results in, 163-165
    longitudinal comparison of developed sites in, 166-170
    longitudinal comparison of undeveloped sites in, 165-166
    model development for, 170
    objectives of, 155
    performance criteria in, 152-155
        baseline conditions for, 162-163
CWPI (California Watershed Projects Inventory), 336, 343
CWQA (California water quality assessment) spatial database, 333-337
    GIS contributions to, 336
    pollutant information provided by, 339-347
        geographical overview, 340-342
        identification of pollutants, 339-342
        sources of pollution, 342-343
            dominant, 343-346
    practical and future use of, 346-347
    Regional Water Quality Control Board contributions to, 337-338
    River Reach File contributions to, 336-337
    River Reach File structural constraints and, 338-339
*Cyclostephanos costalimbus*, 133

# D

DataDesk®, 253
DCA analysis (detrended correspondence analysis), 34
Delta Graph®, 253
Dendrobiochemical analysis, of ozone effects on pines, 183-187
Denuder samplers, 98
Deposition models, 111-114, see also Dry deposition; Wet deposition
    geographic information systems (GIS), 113-114
    nitrogen critical loads, 112-113
    ozone critical exposure levels, 112
DGGE/TGGE (denaturing/thermal gradient gel electrophoresis), for microbial DNA identification, 199-202
Diatom profile, of sediment cores, 133, 137
Diazinon study, of San Joaquin river drainage area, 37-41, 52-53

DNA fingerprinting, see also Biomarkers
    of agricultural soil microbiota, 223-231
        DNA extraction and purification, 225
        DNA isolation, 226-227
        materials and methods, 224-225
        microbial fingerprint comparisons, 227-230
        polymerase chain reaction amplification, 225-226
    of soil-derived dust microbiota, 193-205
        DNA extraction and purification, 195
        DNA yields, 196-198
        polymerase chain reaction, 195-196
        reliability of, 198
        thermal gradient gel electrophoresis, 196
Dose/exposure-response assessment, 314-315
Dose-response interactions, 209
Doses/loads concept, 112-114
DRIS (Diagnosis and Recommendation Integrated System), 113
Dry deposition, 18
    of nitrogenous pollutants
        on plant surfaces, 17-24, 101-103
        in San Bernardino Mountains, 17-24, 108-109
        in San Gorgonio Wilderness Area, 109-111
        in Sierra Nevada, 108
Dust, see Dry deposition; Soil microbiota, particulate

# E

Ecocentrism, 323
*Ecological Applications,* 3
*Ecological Economics,* 3
*Ecological Engineering,* 3
Ecosystem services, 8-9
Ecotoxicology, 1-13
    current journals in, 3
    development of field of, 2
    ecosystem health and, 9-10
    ecosystem services and, 8-9
    interdisciplinary nature of, 2-3
    methods used in, 3-6
    scale and, 5-6
    sustainability and, 6-8
    techniques for every scale of, 5
*Ecotoxicology,* 3
Eigenvalue, 288
Eigenvectors, 288
Electrophoresis protein profiles, of microbial biomarkers, 212, 214

Empirical orthogonal function (EOF), 288
Environmental sinks, 64
Enzyme protein extraction, of microbial biomarkers, 214-215
Enzymes, inducible, 211

## F

Fire hazard/risk assessment model, for Boise National Forest, 273-279
  design goals of, 279
  fire ignition submodel, 275
  fisheries condition submodel, 277-278
  forested vegetation outside HRV submodel, 273-275
  practical goals of, 280-283
  watershed hazard submodel, 277
  wildlife habitat persistence submodel, 275-277
Fish, performance indicators of adult, 152-155
  baseline conditions for, 162-163
Fish populations
  of Moose River drainage system
    performance indicators of, 152-155, 159, 162-163
    species in, 160-161
    studies of, 159
    stressors of, 159
  of San Joaquin river drainage area, 41-45, 53-55
    species in, 40-41
Fog collectors, 99-100
FOREST (Forest Ozone Response Study), 103, 234-247
  biotic and abiotic injury determinations, 238-239
  changes in ozone injury index, by location of plot, 239-241
  current results of, 243-246
  distribution of ozone injury index, in relation to crown position classes, 241
  methods used in, 237-239
  plot locations of Jeffrey and ponderosa pine in, 238
Forest Pest Management (FPM) score, 236, 237

## G

Generalized additive modeling (GAM), 296-299
  loess smoothers, 297
  smoothing splines, 297

GIS (Geographic Information Systems) based analysis
  in fire hazard risk assessment in Boise National Forest, 272
  model development, 273-279
  model practicality, 280-283
  of ozone injury in western U.S. pine forests, 113-114, 235
  in water quality assessment, 336

## H

Historical scale analysis, of ozone effects on pines, 182-187
Honeycomb denuder samplers, 98
HRV (historical range of variability), 272, see also Fire hazard/risk assessment model

## I

IMPROVE particulate samplers, 195
Indicators, of contaminant removal, 66
Inducible enzymes, 211
*In situ* biochemical analysis, of plant health, 175-192
*In situ* methods, of demonstrating metabolism, 68-70
Interannual variability modeling, 285-286, 303-304
  causal factor identification in, 295-299
    amplitude time series, 295
    general additive modeling (GAM), 296-299
    local regression modeling, 295
    tree-based modeling, 295-296
  estimating prediction error in, 301-303
  linear expression and testing of, 299-301
  principal components analysis in
    aggregation of data within bins, 293-294
    Buell patterns, 291
    orthogonal and oblique rotations, 292-293
    problems associated with, 294-295
    Rule N, 290-291
    scree test, 290
    Seasonal and Trend Decomposition using Loess (STL), 287-289
    simple structure, 291
Intrinsic bioremediation
  criteria for, 65-70
    biodegradation potential, 67-68
    decreasing concentration, 66-67
    developing, 64-65

Index 353

proof of biodegradation, 68-70
subsurface contamination and, 64-65

**J**

Jeffrey pine, 18-19, 95
   ozone injury to, 233-234, 236-237, see also FOREST
JMP® data visualization program, 252
*Journal of Ecosystem Health,* 3

**L**

Lake Tahoe Basin, ozone injury to forests in, 237
*Landscape Toxicology,* 3
Lead radionuclide dating, of sediment cores, 127-129
Leaf area estimation, 20
Leaf needle, see Pine needles
Local regression modeling, 295
Loess smoothers, 297

**M**

Mercury analysis, of plankton and benthic invertebrates, 259-265
   methods of, 251-253
   results of, 253-259
Metabolism, *in situ,* methods of demonstrating, 68-70
Metam sodium, 72-73
   decomposition in water, 80-81
   transport modeling, 74-75
      riverine spill simulations, 85-89
   water modeling, 78-79
Methyl isothiocyanate (MITC), 72
   volatilization of, 81-83
Microbial DNA fingerprinting
   of agricultural soil, 223-224
      RAPD analysis, 223-224
         DNA extraction and purification, 225
         DNA isolation, 226-227
         materials and methods, 224-225
         microbial fingerprint comparisons, 227-230
         polymerase chain reaction amplification, 225-226
   comparisons by RAPD analysis, 227-230
   of soil-derived dust, 193-194, 198-203
      DNA extraction and purification, 195

      DNA yields, 196-198
      polymerase chain reaction amplification, 195-196, 199
      RAPD analysis, 199-202
      reliability, 198
      sampling, 194-195
      thermal gradient gel electrophoresis, 196
Microorganisms, subsurface, in biodegradation, 67-68
MITC (methyl isothiocyanate), 72
   volatilization of, 81-83
Modeling
   ACE (alternating conditional expectation), 296
   AVAS (additivity and variance stabilization), 296
   Big Leaf multiple resistance, 111
   cumulative effects assessment, 149-152
      cause and effect relationships in, 170-171
      choice of indicators for, 152-155
      issues of, 150
      latitudinal habitat survey results in, 163-165
      longitudinal comparison of developed sites in, 166-170
      longitudinal comparison of undeveloped sites in, 165-166
      model development for, 170
      objectives of, 155
      performance criteria in, 152-155
         baseline conditions for, 162-163
   deposition of air pollutants, 19-20, 111-114, see also Dry deposition; Wet deposition
      geographic information systems in, 113-114
      nitrogen critical loads, 112-113
      nitrogenous, 20-21
      ozone critical exposure levels, 112
   fire hazard/risk assessment, 273-279
      design goals, 279
      fire ignition submodel, 275
      fisheries condition submodel, 277-278
      forested vegetation outside HRV submodel, 273-275
      GIS analysis in, 273-279, 280-283
      practical goals, 280-283
      watershed hazard submodel, 277
      wildlife habitat persistence submodel, 275-277
   GAM (generalized additive modeling), 296-299
      loess smoothers, 297
      smoothing splines, 297

interannual variability, 285-286, 303-304
   causal factor identification in, 295-299
      amplitude time series, 295
      general additive modeling (GAM), 296-299
      local regression modeling, 295
      tree-based modeling, 295-296
   estimating prediction error in, 301-303
   linear expression and testing of, 299-301
   principal components analysis in
      aggregation of data within bins, 293-294
      Buell patterns, 291
      orthogonal and oblique rotations, 292-293
      problems associated with, 294-295
      Rule N, 290-291
      scree test, 290
      Seasonal and Trend Decomposition using Loess (STL), 287-289
      simple structure, 291
local regression, 295
prediction error (PE) in, 301-303
riverine transport, 74-89
   constituent relations, 79
   hydrodynamic boundary conditions, 77-78
   hydrodynamic modeling, 76
   model calibration, 83-84
   results and spill simulations, 84-91
   river geometry, 75-76
   run-riffle-pool modification, 76-77
   spill simulations, 85-89
   water modeling, 78-79
   water quality boundary conditions, 83-84
   water quality modeling, 78-79
tree-based, 295-296
water, 78-79
water quality, 78-79
Moose River drainage system, 150-152
   cumulative effects assessment of, 149-152
      cause and effect relationships in, 170-171
      choice of indicators for, 152-155
      issues of, 150
      latitudinal habitat survey results in, 163-165
      longitudinal comparison of developed sites in, 166-170
      longitudinal comparison of undeveloped sites in, 165-166
      model development for, 170
      objectives of, 155
      performance criteria in, 152-155
         baseline conditions for, 162-163

      fish populations of
         performance indicators of, 152-155, 159, 162-163
         species in, 160-161
         studies of, 159
         stressors of, 159
      geography of, 155-159
      industrial sites in, 151

## N

NADP (National Atmospheric Deposition Program), 99
National Water Quality Assessment (NAWQA) Program, 27-28
   sample collection procedures in, 32-33
   study design of, 28-32
Natural Heritage Program, 336
Natural Step USA Program, 7
NCLAN program, 312, 313, 315
Nitrogen critical loads, analysis of, 112-113
Nitrogenous air pollutants
   deposition in coastal sage scrub ecosystems, 111
   dry deposition of, in San Bernardino Mountains, 17-24, 108-109
      model of, 19-20
      model parameters for Barton Flats, 20-21
         empirical data and, 21-23
   gaseous and particulate, in Sierra Nevada, 107-108
   monitoring, 98
N pollutants, see Nitrogenous air pollutants
Nutrient flow, stable C and N isotope evaluation of, 159, 164-165

## O

Optimism estimation, 301
*Our Common Future*, 6-7
Ozone
   critical exposure levels of, 112
   effects of, on photosynthesis, 177-178
      antioxidant response, 179
      ascorbate and glutathione, 178-179
      dendrobiochemical analysis, 183-187
      enzymes in phenolic pathway, 179
      historical scale analysis, 182-187
      Rubisco, 178
      *in situ* analysis, 180-182
   exposure to ambient, see also Ozone injury index
      criteria for risk assessment of, 307-308

Index 355

dose/exposure-response assessment, 314-316
exposure characterization, 309-314
future research, 316
pollutant identification and concentration, 308-309
and effects on conifer forests, 103, 176, 233-237
biochemical effects, 178-187
growth and morphologic effects, 176
physiologic effects, 176-178
passive samplers of, 98
Ozone injury index (OII), 233-246
biotic and abiotic injury determinations, 239
changes in, by location of plot, 239-241
distribution of, in relation to crown position classes, 241
injury qualifications for, 239
scale of, 238

## P

Particulate matter, see Air pollutants; Dry deposition; Soil microbiota, particulate
PCA, see Principal components analysis
PCR (polymerase chain reaction)
amplification of soil microbial proteins by, 195-196, 225-226
DNA analysis and, 199-202, 226-230
*Percopsis omiscomaycus,* see White sucker
Pesticides
ecological effects of, 58
studies of San Joaquin River drainage, 35-37, 50-52
diazinon, 52-53
transport of, in riverine environment, 71-92
modeling, 74-84
modeling results and spill simulations, 84-91
pesticide spill, 72-74
Photosynthesis,
ozone and biochemical effects on, 178-180
dendrobiochemical analysis of, 182-187
*in situ* analysis of, 180-182
Pine needles
abiotic injury to, 241-246, see also Ozone injury index
biotic and abiotic injury determinations, 239
ozone effects on
dendrobiological analysis of, 183-187
historical scale analysis of, 182-187
*in situ* biochemical analysis of, 180-182

symptoms of ozone injury, 235-236
*Pinus albicaulis,* 108
*Pinus contorta,* 108
*Pinus jeffrey,* see Jeffrey pine
*Pinus ponderosa,* see Ponderosa pine
Plankton, in Clear Lake aquatic ecosystem, 251-252
mercury in, 253-257, 259-265
Plant defense mechanisms, ozone effects on, 178-180
Pluto syndrome, 3
PM10, 194
Pollen profile, of sediment cores, 133, 134, 135
Pollutant fate, see Air pollutants; Water pollutants
Pollutant identification, 308-309, see also specific pollutant
Pollution, early assessment studies of, 3
Ponderosa pine, 18-19, 95
of Boise National Forest, 271-273
dry deposition predictor for, 17-24, 112
ozone injury to, 233-234, 235-237, see also FOREST
Prediction error (PE), in models, 301-303
Principal component loading, 288
Principal components analysis (PCA), 34, 287-295
aggregation of data within bins, 293-294
Buell patterns, 291
orthogonal and oblique rotations, 292-293
problems associated with, 294-295
Rule N, 290-291
scree test, 290
Seasonal and Trend Decomposition using Loess (STL), 287-289
simple structure, 291
Promax method, of oblique rotation, 292
Protein(s)
microbial
electrophoresis of, 212, 214
extraction of, 214-215
profiles of, 212, 214
stress, 211
Pyrolysis/gas chromatography-mass spectrometry (Py/GC-MS), 185-187

## Q

Q survey methodology, in restoration of ecosystems, 323-326, 330-331
communication concourse, 324
conclusions drawn from, 330-331
factors (groups) resulting from, 327-330
matrix design in, 324-325

participants in, 325
statement sorts analysis in, 325-326
*Quercus kelloggii,* see California black oak

# R

Rain collector, 99
RAPD (randomly amplified polymorphic DNA) analysis, of soil microbiota, 199-202, 223-224
  DNA extraction and purification, 225
  DNA isolation, 226-227
  materials and methods, 224-225
  microbial fingerprint comparisons by, 227-230
  polymerase chain reaction amplification, 225-226
Reach Files, see River Reach Files
Regional Water Quality Control Boards, 337-338
REP-PCR (repetitive extragenic palindrome-PCR) analysis, for microbial DNA identification, 199-202
Response patterns, 153-154
Restoration, of ecosystems, 321-322
  Q survey methodology for, 323-326, 330-331
  factors (groups) resulting from, 327-330
  survey methodologies for, 323
Ribosomal Database Project (RDP), 201
RISA (ribosomal intergenic spacer analysis), for microbial DNA identification, 199-202
Risk assessment, see Modeling; Ozone; Ozone injury index
Riverine spill transport modeling, 74-75
  constituent relations, 79
  hydrodynamic boundary conditions, 77-78
  hydrodynamic modeling, 76
  model calibration, 83-84
  results and spill simulations, 84-89
  river geometry, 75-76
  run-riffle-pool modification, 76-77
  spill simulations, 85-89
  water quality boundary conditions, 83-84
  water quality modeling, 78-79
Riverine water quality, CWQA spatial database information on, 339-347, see also Moose River drainage system; Riverine spill transport modeling; San Joaquin river drainage; Sierra Nevada
River Reach Files, 336-337
  structural constraints of, and CWQA spatial database, 338-339
RMA-GEN program, 76

R survey technique, 323
Rubisco
  in photosynthesis, 178
  *in situ* biochemical assessment in pine needles, 180-182
Rule N, 290-291

# S

San Bernardino mountains
  air pollutants in, 17-24, 94-95
  nitrogenous pollutants in conifer forests of, 17-24, 108-109
  ozone injury to conifer forests of, 103, 233-246
San Gorgonio Wilderness Area, wet and dry deposition in, 109-111
San Joaquin river drainage, 26-28
  fish populations of, 41-45, 53-55
  species in, 40-41
  water quality, habitat and fish data from, 28-62
    analysis of, 33-35
    analytical methods used, 33
    diazinon study results, 37-41, 52-53
    fish assemblages, 41-45, 53-55
    integration of, 45-50, 55-58
    pesticide studies, 35-37, 50-52
    species collected in, 40-41
Scores, 288
Scree test, 290
SDS-polyacrylamide gel electrophoresis (SDS-PAGE), 180-181
Seasonal and Trend Decomposition using Loess (STL), 287-289
Sediment cores, 123-124
  analysis of, 124-127
    carbon and nitrogen, 124
    diatom enumeration, 126
    lead radionuclide dating, 124, 125
    mercury, 125
    phosphorus, 124-125
    pollen counts, 125-126
  analytical results on, 127-137
    chemical changes, 129-133
    diatom profile, 133, 137
    lead radionuclide dating, 127-129
    physical changes, 129-133
    pollen profile, 133, 134, 135
  obtaining, 124
Sediment deposition, in Clear Lake watershed, 138, 139-140, see also Clear Lake aquatic ecosystem

# Index

Sequoia Kings Canyon, ozone injury surveys of, 236
Sierra Nevada
　air pollutants in, 94-95
　　dry deposition of nitrogenous, 107-108
　　wet deposition of nitrogenous and sulfurous, 103-107
　　ozone injury to conifer forests of, 103, 233-246
　riverine water quality of
　　CWQA pollutant information, 339-347
　　geographical overview, 340-342
　　identification of pollutants, 339-342
　　sources of pollution, 342-346
Simple structure, 291
Sinks, environmental, 64
Smoothing splines, 297
SNEP (Sierra Nevada Ecosystem Project), 334-335, 336
Soil-derived dust, see Soil microbiota, particulate
Soil microbiota
　agricultural
　　DNA fingerprinting of, 223-231
　　　DNA extraction and purification, 225
　　　DNA isolation, 226-227
　　　materials and methods, 224-225
　　　microbial fingerprint comparisons, 227-230
　　　polymerase chain reaction amplification, 225-226
　particulate, 194-195
　　DNA fingerprinting of, 193-194, 198-203
　　　DNA extraction and purification, 195
　　　DNA yields, 196-198
　　　polymerase chain reaction, 195-196
　　　polymerase chain reaction amplification, 199
　　　reliability, 198
　　　thermal gradient gel electrophoresis, 196
　　sampling, 194-195
Spearman rank correlation, 34, 35
Spill transport modeling, riverine, 74-75
　constituent relations, 79
　hydrodynamic boundary conditions, 77-78
　hydrodynamic modeling, 76
　model calibration, 83-84
　results and spill simulations, 84-89
　river geometry, 75-76
　run-riffle-pool modification, 76-77
　spill simulations, 85-89
　water quality boundary conditions, 83-84

　water quality modeling, 78-79
Stable carbon isotope analysis, 169-170
Stable C/N isotope ratios, in analysis of nutrient relationships, 164-165
*Stephanodiscus vestibulis,* 133
Stress proteins, 211
Stress response, microbial biomarkers of, 207-209, 215-219
　conceptual approach to, 210
　examples of, 211
　sampling process for, 209, 211-212
Subsurface contamination, fate of, 63-70
　intrinsic bioremediation and, 64-65
　criteria for, 65-70
Sulfurous air pollutants, monitoring, 98
Survey methods, for restoration management, 323-331
Sustainability, 150
　and ecotoxicology, 6-8
Sustainable Agriculture Farming Systems (SAFS) Project, 224-225

# T

Teale Data Center, 337
Technocentrism, 323
TGGE (thermal gradient gel electrophoresis), for microbial DNA identification, 196
Time series, ecological, see Interannual variability modeling
Toxic Contaminant Hydrology (TCH) Program, 31-32
Tracer molecules, 66-67
Tree-based modeling, 295-296
Trout perch, in Moose River drainage system, 159
　physiologic data for, 167, 169
　preliminary stressor data for, 170
TWINSPAN analysis, of fish assemblages, 34

# U

Upper Newport Bay Ecological Reserve, 213
　ecosystem health of, 209-222
U.S. National Ambient Air Quality Standard, 309-310

# V

Values, environmental, 323
Variability modeling, see Interannual variability modeling
Varimax method, of orthogonal rotation, 292

## W

Water pollutants, see also CWQA (California water quality assessment) spatial database
   metam sodium, 72
      decomposition in water, 80-81
      water modeling, 78-79
   National Water Quality Assessment (NAWQA) Program, 27-28
      sample collection procedures, 32-33
      study design, 28-32
   riverine spill transport modeling, 74-89
   in San Joaquin river drainage area, 25-28
      water quality, habitat and fish data, 28-62
   in sediment deposition, 123-124
      analysis of, 124-127
      analytical results of, 127-137
      Clear Lake watershed and, 138, 139-140
   and subsurface contamination, 63-70
Water quality data, analysis of, 58-59
Wet and dry deposition collectors, 100-101
Wet deposition
   of nitrogen and sulfur in Sierra Nevada, 103-107
   of nitrogen in San Gorgonio Wilderness Area, 109-111
Wet precipitation collectors, 99-100
White fir, 18-19
   dry deposition predictor for, 17-24, 112
White sucker, in Moose River drainage system, 159, 163-164
   physiologic data for, 168, 169
   preliminary stressor data for, 170
Winter fleck, 239, 241-243, 244
WQA (water quality assessment) database, 333-334, 335-336

## Y

Yosemite National Parks, ozone injury surveys of, 236